THE BOOK OF EVIDENCE

OXFORD STUDIES IN PHILOSOPHY OF SCIENCE

General Editor
   Paul Humphreys, University of Virginia

Advisory Board
   Jeremy Butterfield
   Peter Galison
   Ian Hacking
   Philip Kitcher
   Richard Miller
   James Woodward

The Book of Evidence
Peter Achinstein

THE BOOK OF EVIDENCE

Peter Achinstein

OXFORD
UNIVERSITY PRESS

2001

OXFORD
UNIVERSITY PRESS

Oxford   New York
Athens   Auckland   Bangkok   Bogotá   Buenos Aires   Cape Town
Chennai   Dar es Salaam   Delhi   Florence   Hong Kong   Istanbul   Karachi
Kolkata   Kuala Lumpur   Madrid   Melbourne   Mexico City   Mumbai
Nairobi   Paris   São Paulo   Shanghai   Singapore   Taipei   Tokyo   Toronto   Warsaw

and associated companies in
Berlin   Ibadan

Copyright © 2001 by Peter Achinstein

Published by Oxford University Press, Inc.
198 Madison Avenue, New York, New York 10016

Oxford is a registered trademark of Oxford University Press

All rights reserved. No part of this publication may be reproduced,
stored in a retrieval system, or transmitted, in any form or by any means,
electronic, mechanical, photocopying, recording or otherwise,
without the prior permission of Oxford University Press.

Library of Congress Cataloging-in-Publication Data
Achinstein, Peter.
  The book of evidence / by Peter Achinstein.
    p. cm. —(Oxford studies in philosophy of science)
  Includes bibliographical references.
  ISBN 0-19-514389-2
  1. Evidence.   I. Title.   II. Series.
BC173 .A34 2001
121'.65—dc21   2001032143

9 8 7 6 5 4 3 2 1

Printed in the United States of America
on acid-free paper

FOR MY DAUGHTER, BETTY

Contents

1. *The Dean's Challenge* 3
2. *Concepts of Evidence, or How the Electron Got Its Charge* 13
3. *Two Major Probabilistic Theories of Evidence* 44
4. *What's Wrong with These Probabilistic Theories of Evidence?* 69
5. *Objective Epistemic Probability* 95
6. *Evidence, High Probability, and Belief* 114
7. *The Explanatory Connection* 145
8. *Final Definitions and Realism* 168
9. *Two Paradoxes of Evidence: Ravens and Grue* 185
10. *Explanation versus Prediction: Which Carries More Evidential Weight?* 210
11. *Old-Age and New-Age Holism* 231
12. *Evidence for Molecules: Jean Perrin and Molecular Reality* 243
13. *Who Really Discovered the Electron?* 266

    Index 287

THE BOOK OF EVIDENCE

1

THE DEAN'S CHALLENGE

Once there was a dean at my university who was a scientist with high intelligence but low boiling point. One day at a faculty meeting, after I said something that displeased him, he replied, "Peter, you have never made a contribution of interest to scientists." Naturally, my first thought was to take offense. But trying to maintain a generous spirit, and believing that a highly intelligent dean offers personal insults only in private, I decided what he really meant was not the singular "you" but the plural one. "You philosophers of science," he meant, "have nothing to offer us scientists."

This interpretation at least took some of the sting out of his remarks and enabled me to think about them more clearly. Perhaps the dean is right, I now speculated. Although philosophers of science have carefully worked out views about a range of general concepts scientists employ—such as evidence, explanation, and law, to name just three of many—scientists seem to take little heed of them.

This book is devoted to the first of these concepts, and to the question of what it means to say that some fact is evidence that a certain hypothesis is true. In this introductory chapter I will say why I believe the dean was right: standard philosophical theories about evidence are (and ought to be) ignored by scientists. They ought to be ignored because they propose concepts of evidence that are based on assumptions incompatible with ones scientists make when they speak of, and offer, evidence for hypotheses. The ideas briefly developed here will be expanded and defended much more fully in later chapters when I take up the dean's challenge to provide a concept of more interest to scientists. This chapter serves as an introduction to, and an anticipation of, claims I will develop. I realize that a danger of anticipating later claims is that an unsympathetic reader may go no further. But a danger of not anticipating later claims is that the uninquisitive or unmotivated reader may also go no further. I'll have to take my chances.

1. Disagreements about Evidence

Scientists frequently disagree with one another about whether some fact is evidence that a certain hypothesis is true, or, if it is, about how strong that evidence is. I have in mind cases in which they agree that some fact has been observed or established, or some experimental result obtained. They agree on a description of that fact or result. They also agree on the meaning of the hypothesis in question. Their disagreement lies in whether, or the extent to which, what has been observed, or the experimental result, supports, or provides evidence for, the hypothesis. Moreover, they seem to treat this disagreement as an objective matter—one for which there is a right answer, and not one for which different people can have different right answers.

Let me mention two cases. The first (which will be discussed in detail in chapter 2) involves the discovery of the electrical nature of cathode rays. In 1883 Heinrich Hertz conducted a series of carefully designed experiments with cathode rays to determine whether they are electrically charged. In one experiment cathode rays were made to enter an electrometer, the deflection of which would determine the presence of electricity. The needle of the electrometer remained at rest when cathode rays were produced. In a second experiment Hertz introduced oppositely electrified plates into the cathode tube. If cathode rays were electrically charged they should be deflected by these plates, as indicated in a changed position of the phosphorescence produced by the rays. But no such change occurred. Hertz claimed that these two experimental results are evidence, indeed decisive evidence, that cathode rays are not electrically charged.

Fourteen years later J. J. Thomson repeated the second of Hertz's experiments and got the same results as Hertz, no deflection of the cathode rays. Yet he refused to take this to be evidence—certainly not decisive evidence—that cathode rays are electrically neutral. Thomson hypothesized that if cathode rays are charged particles, then when they pass through the gas in the cathode tube they ionize the gas molecules producing positive and negative charges that will neutralize the charge on the metal plates between which the cathode rays travel. So if the gas in the cathode tube has not been sufficiently evacuated, there will be no deflection of the cathode rays. Indeed, in 1897 Thomson was able to remove a sufficient amount of gas from the tube and demonstrate the electrical deflection of the cathode rays.

Thomson did not dispute that Hertz obtained no electrical deflection of the cathode rays. Indeed, Thomson obtained the same results in initial experiments. What he challenged was the claim that these results were evidence that cathode rays are electrically neutral.

My second example involves an archaeological hypothesis about the earliest campfires used for cooking, for light, and as a protection against animals. For 60 years it had been hypothesized that the first campfires were built by Peking Man in caves in Zhoukoudian, China between 200,000 and 500,000 years ago. The evidence for this hypothesis was the existence of burned animal bones in the same layer of soil as stone tools and the sediment there that looks like wood ash. On July 10, 1998 a group of scientists from Israel, the United States, and China rejected the claim that the existence of the burned animal bones and the existence of sediment together provided strong evidence that campfires existed there.[1] This claim was based on the discovery that the sediment in question is not wood ash but fine minerals and clay deposited by water. These scientists claimed that the burned animal bones by themselves do not constitute very good evidence that a fire was started by humans.

In both of these cases there is an agreement between disputants over what has or has not been observed—at least at some important level of description. J. J. Thomson agreed that in the sorts of experiments conducted by Hertz, and initially by himself, there was no electrical deflection of the cathode rays. The disputants in the campfire case agree that burned animal bones and sediment that looks like wood ash were found in caves in China. The disagreement arises over whether, or to what extent, these observed facts constitute evidence that the hypotheses in question are true.

I mention cases of this sort because philosophers of science have developed theories or definitions of evidence that are designed to do at least two things for scientists: first, to clarify what it means to say that some fact is evidence that a hypothesis is true; second, and relatedly, to help scientists determine whether (and to what extent) putative evidence supports a hypothesis. These goals are championed by a range of philosophers who have developed theories of evidence. A corollary—one emphasized by Carnap[2]—is to develop a theory of evidence that will enable scientists to settle disputes, such as the ones I mentioned, over whether, or to what extent, putative evidence supports a hypothesis.

By and large, however, philosophical theories of evidence are ignored by scientists. You don't find scientists with disagreements of the sort in question turning to philosophers for help. Why not? Is this just a matter of people in very different fields ignoring one another's work? That may be part of the answer, but I don't think that is the main problem or the most interesting one. I think the problem is deeper, and stems from two very basic, but questionable, assumptions philosophers usually make about evidence.

The first assumption is that evidence is a very weak notion. You don't need very much to have evidence that something is the case. The second is that the evidential relation is a priori not empirical. It is a logical, or semantical, or mathematical relation that can be established by "calculation." The philosophers who make both assumptions are concerned with an objective, not a subjective, notion of evidence. On their view, whether some fact e is evidence that hypothesis h is true, and how strong that evidence is, does not depend on what anyone believes about e, h, or their relationship. Not all philosophers who talk about evidence recognize, or are interested in, objective evidence. Subjective Bayesians, for example, reject such a notion. But a range of philosophers accept the idea, including objective Bayesians such as Carnap, hypothetico-deductivists, satisfaction theorists such as Hempel,[3] bootstrappers such as Glymour,[4] and others as well.

1. Steve Weiner, Q. Xu, P. Goldberg, J. Liu, and O. Bar-Yossef, "Evidence of the Use of Fire at Jhoukoudian, China," *Science* (1998), vol. 281, pp. 251–253.
2. Rudolf Carnap, *Logical Foundations of Probability* (Chicago: University of Chicago Press, 2nd ed., 1962).
3. Carl G. Hempel, *Aspects of Scientific Explanation* (New York: Free Press, 1965).
4. Clark Glymour, *Theory and Evidence* (Princeton: Princeton University Press, 1980).

In this book I will defend the claim that although there are occasions on which scientists employ a subjective concept of evidence, the important concept for them is objective. Indeed, in chapter 2 I will argue that there are several different objective concepts in use. I will also argue that the subjective concept is most usefully defined by reference to an objective one.

2. The Weakness Assumption

I will illustrate this assumption by reference to three standard theories of evidence. The first theory is a Bayesian one: for a fact e to be evidence that a hypothesis h is true, it is both necessary and sufficient that e increase h's probability over its prior probability. So, for example, since my buying 1 ticket in a million-ticket lottery increases the probability that I will win, this fact is evidence that I will. To be sure, it is not a lot of evidence; it is certainly not decisive; but it is *some*. According to *The New York Times*, there is 1 elevator accident per 6 million rides. Using this as a basis for a probability judgment, since my riding this elevator today raises the probability that I will be involved in an elevator accident today, it is evidence that I will. Not a lot, but some, perhaps a tiny bit of evidence. Such a notion of evidence, I think the dean would say, is too weak to be taken seriously. To be sure, the bigger the probabilistic boost e gives to h, the stronger the evidence. But the fact remains that for e to be evidence that h, on this view, all that is required is that e raise h's probability.

A second standard theory of objective evidence is hypothetico-deductive (h-d). For e to be evidence that h it suffices for e to be derivable deductively from h. So, for example, since the fact that light travels in straight lines is derivable from the classical wave theory of light, it is evidence that light is a classical wave motion. This is a very weak notion of evidence, because it allows the same fact to be evidence for a range of conflicting theories. (The same is true of the previous Bayesian account.) For example, since the rectilinear propagation of light is also derivable from the classical particle theory, it is evidence that light is composed of classical particles.

A third approach to objective evidence is a "satisfaction" theory of the sort proposed by Hempel. The basic idea is that an observation report is confirming evidence for a hypothesis if the hypothesis is satisfied by the class of individuals mentioned in that report. To use Hempel's famous example, an observation report that a particular raven observed is black is evidence that all ravens are black. So is the fact that a particular nonblack thing observed is a nonraven. Glymour devises a more complex bootstrap approach that takes Hempel's idea of satisfaction as basic.

Let me say why I believe such notions of evidence are too weak for scientists to take an interest in. Why do scientists want objective evidence for their hypotheses? What does evidence give them? My answer (to be developed in chapter 2) is that in the case of the two most important types of objective evidence (which I call potential and veridical evidence) it gives them a good reason to believe their hypotheses. Not necessarily a conclusive one, or the best possible one, but a good one nonetheless. If the results of the biopsy constitute evidence that the patient's tumor is malignant then there is a good reason to believe the patient has cancer. By contrast, if you visit your doctor complaining of a stomach ache persisting for the last few days I don't believe the doctor would or should count this fact by itself as evidence that you have cancer, even if the probability that you do is raised slightly by this symptom. By itself it is not a good reason to believe this hypothesis. Similarly, although the fact that I am entering an elevator increases my chances of being in an elevator accident, it is not evidence that this will be so, even a little bit of evidence, since by itself it fails to provide any reason to believe this hypothesis.

I will argue in chapter 3 that evidence is related to probability, but that it is a "threshold" concept with respect to probability. In order for e to be evidence that h there must be a certain threshold of probability that e gives to h, not just any amount greater than zero. What is the threshold? Returning to the idea that evidence provides a good reason to believe, a basic principle that I will defend in chapter 4 is that if e is a good reason to believe h, then it cannot also be a good reason to believe *not-h* or some proposition incompatible with h. (It might of course be the case that e is an equally good reason to believe h as to believe *not-h*. But that does not make it a good reason to believe both or either.) So, for example, the fact that I am tossing this fair coin is not a good reason to believe that it will land heads, because it is an equally good reason to believe it will land tails; that is, it is not a good reason to believe either hypothesis. If this is right, then h's probability on e must be greater than $\frac{1}{2}$. If it were less than or equal to $\frac{1}{2}$, then, as in the coin tossing case, e could be a good reason to believe both h and *not-h*.

Does this mean that it is impossible to have evidence for conflicting theories? Yes and no. Yes, it is impossible for the same fact to be evidence for conflicting theories. The fact that I am about to toss this fair coin is not evidence that it will land heads and evidence that it will land tails. It is not evidence—not even a little bit of evidence—that either "theory" is true.

Suppose, however, that we consider a different coin, whose fairness we don't yet know. We give the coin to two tossers. The first conducts an experiment making 100 tosses with the coin resulting in 80 heads. The second conducts an experiment also making 100 tosses but obtaining 80 tails, where the conditions of tossing are approximately the same in both cases. Now the results of the first experiment, we might conclude, constitute evidence that the coin is biased in favor of heads, while the results of the second constitute evidence for the conflicting theory that the coin is biased in favor of tails. This conforms in the following way with my claim that evidence provides a good reason for belief: the results of the first experiment, *if considered by themselves*, would provide a good reason to believe the coin is heads-biased, while the results of the second experiment, *if considered by themselves*, would provide a good reason to believe that the coin is tails-biased. I suggest that this is what is meant by having evidence for conflicting theories, viz. having certain information which, if considered by itself, would be evidence for one theory and having other information which, if considered by itself, would be evidence for a conflicting theory.

When the two bodies of evidence are combined, however, the combination may be evidence for neither theory. In our coin tossing example, if we consider the two experiments to have equal probative value and we combine their outcomes in the simplest manner, the result would be 200 tosses with this coin, yielding a total of 100 heads and 100 tails. This combined information by itself would not be evidence that the coin is heads-biased and evidence that it is tails-biased. The combined results in this case do not provide a good reason for believing either or both of the bias theories.

Some philosophers who are objective Bayesians about evidence suggest that there is a concept of evidence according to which e is evidence that h if and only if h's probability on e is sufficiently high, say greater than $\frac{1}{2}$. This notion is much stronger than the weak increase-in-probability account. And it has the advantage of ruling out the unwanted lottery, elevator, and stomach ache cases. However, even if probability greater than $\frac{1}{2}$ is a necessary condition, it is not sufficient. High probability by itself is too weak for evidence, since h's probability may be high with or without e. It may have nothing to do with e. Let e be the fact that Michael Jordan eats Wheaties. (He used to promote Wheaties on TV.) Let h be the hypothesis that Michael Jordan will not become pregnant. Now h's probability with or without e is close to 1. Yet surely the fact that Michael Jordan eats Wheaties is not evidence, or a good reason to believe, that he will avoid pregnancy. For e to be evidence that h, for e to be a good reason to believe h, not only must h's probability on e be sufficiently high, but there must be some other connection between e and h, or the probability of such a connection: e must have "something to do" with h. What that amounts to is a question I will attempt to answer in chapter 7.

All of the views of evidence I have mentioned (increase-in-probability, high probability, h-d, and "satisfaction") are much too weak because they fail to provide a good reason to believe. Although my entering an elevator increases my chances of being in an elevator accident (thereby satisfying the increase-in-probability view), it is not a good reason to believe it will happen. Although (in accordance with the high-probability account) it is highly probable that Michael Jordan will not become pregnant, given that he eats Wheaties, the latter is not a good reason to believe the former. The fact that the classical wave theory of light entails rectilinear propagation, which is observed to be the case (thereby satisfying the h-d view), is not enough to provide a good reason to believe that theory. Or put it this way: it provides an equally good reason—and hence not a good one at all—for believing a range of conflicting theories, including particle theories. Finally, shifting to Hempel's satisfaction view, the fact that the hypothesis that all ravens are black is satisfied by the one black raven I have observed is not by itself a good reason to believe that hypothesis. Surely I need a bigger sample. Even more importantly, it depends crucially on how I selected the raven for observation. If, for example, I purposely selected it from a cage marked "black birds" then the result does not provide a good reason at all for believing that all ravens are black.

On all of these views of evidence it is too easy to get evidence. To be sure, each of these theories could be strengthened. For example, instead of demanding just any increase in probability, that view could be altered to require a very significant increase. In addition to high probability, that view could impose some further condition(s). The h-d and satisfaction views, respectively, could require the hypothesis to entail, or to be satisfied by, not just some single instance but a range of them. In later chapters I will show that "significant" increases in probability do not suffice for evidence. Nor will "many instances" entailed or satisfied by the hypothesis. This is due to the fact that what you get in such cases does not necessarily give you a good reason to believe a hypothesis. High probability is necessary, but we need to consider a lot more than this for the putative evidence to provide a good reason, or indeed any reason, to believe a hypothesis.

This, then, is the first reason scientists do not and should not take such philosophical accounts of evidence seriously: they are too weak to be taken seriously. They do not give scientists what they want, or enough of what they want, when they want evidence.

3. The A Priori Assumption

The second assumption made by many philosophers who try to provide objective accounts of evidence is that the evidential relation is a priori: whether e, if true, is evidence that h, and how strong that evidence is, is a matter to be determined completely by a priori calculation, not by empirical investigation.

I will illustrate this idea with brief references to some philosophical theories, the first one being Carnap's. Carnap embraces both an increase-in-probability and a high-probability concept of evidence. But for Carnap the probability relation is entirely a priori. What h's probability is on e is determined a priori, by reference to the rules of the "linguistic framework" (as Carnap calls it). The h-d view of evidence also makes the evidential relation a priori, since whether h entails e is a priori. (Even more complex and sophisticated h-d views, which appeal in addition to ideas about simplicity or coherence, are a priori, since whether these additional criteria are satisfied is supposed to be settleable without empirical investigation.) Finally, Hempel's satisfaction and Glymour's bootstrapping criteria again yield concepts in accordance with which one calculates a priori whether e is evidence that h.

What's wrong with this a priori assumption? Let's return to the case of Thomson versus Hertz concerning the electrical character of cathode rays. On the basis of his 1883 experiments in which no deflection of the cathode rays was produced, Hertz concluded that his results were evidence, indeed conclusive evidence, that cathode rays are electrically neutral. Thomson in 1897 rejected this claim, not on a priori grounds, but on empirical ones. He assumed, on empirical grounds, that if cathode rays are electrically charged particles, then, since there is gas in the cathode tube, when these charged particles pass through the tube they will ionize the gas molecules producing positive and negative charges that will neutralize the charge on the metal plates between which the cathode rays travel.

Similarly, recent scientists offered an empirical reason for rejecting the claim that the burned animal bones in the same layer as stone tools and sediment that looks like wood ash is evidence that the first culinary campfires were built by Peking Man in caves in China between 200,000 and 500,000 years ago. The empirical reason was the new discovery that although the sediment looks like wood ash, it is in fact not this but fine minerals and clay deposited by water.

I am not claiming that all evidential statements are empirical. There are cases, to be noted in subsequent chapters, where enough information is packed into the e-statement to make the claim that e is evidence that h a priori. But these cases are the exception, not the rule. Nor am I denying that it is possible to transform an empirical evidential claim into an a priori one by incorporating a sufficient amount of additional information of a sort that might be used in defending the empirical evidential claim. But even if this is possible, that will not suffice to alter the empirical character of the original evidential claim, or demonstrate that the original claim is incomplete until this transformation occurs. An empirical claim, whether evidential or not, is not necessarily incomplete or lacking in truth-value if its defense is not provided; nor does the original claim lose its empirical character if it is replaced by an a priori claim. Furthermore, an empirical evidential claim may be true, and very useful, even if a scientist who makes the claim does not have a sufficient defense of it that could transform the claim into an a priori one.

If evidential claims, or many of them, are empirical, not a priori, then it is scientists, not philosophers, who are in the best position to judge whether e, if true, is evidence that h, and how strong that evidence is. If evidential claims are, by and large, empirical, this is and ought to be an important reason why scientists do not consult philosophical theories of evidence when they try to settle disagreements over evidential claims. Philosophical theories would make such disagreements settleable on a priori grounds. Accordingly, scientists may find philosophical theories of evidence wanting because they give a very mistaken idea of how evidential disputes are usually settled.

4. Conclusion and Preview

I have noted two reasons why scientists do and should ignore typical philosophical theories of objective evidence. These theories furnish concepts that are much too weak to give scientists what they want from evidence, viz. a good reason to believe. And they furnish concepts that, I believe, mistakenly make all evidential claims a priori. These problems are related. Frequently scientists try to discover whether evidential claims are true not solely by a priori reasoning but by empirical investigation. This, I suggest, follows from the idea of providing a good reason to believe. Some fact can be a good reason to believe a hypothesis even if this cannot be demonstrated by a priori calculation.

I return to the dean's challenge to philosophers of science. In the case of evidence I take this challenge to be to propose and defend a concept of evidence that is at once empirical and robust. It is empirical because, in general, it renders the question of whether e is evidence that h an empirical question, which scientists can attempt to answer using that concept. It is robust in two ways. It is a strong, not a weak, concept of evidence, and it is one that yields a good reason to believe something. If it is empirical and robust in these ways, then, I think, it should be of interest to scientists. Finally, I extend the dean's challenge by requiring a concept of evidence that is philosophically robust as well. Not only should it meet standards of philosophical clarity, but it should also help resolve paradoxes and other issues about evidence raised by philosophers.

In the next chapter, by means of a historical example involving cathode ray experiments, I introduce four concepts of evidence employed in the sciences. I call them, respectively, potential, veridical, ES (epistemic-situation), and subjective evidence. Each is illustrated by reference to the cathode ray experiments and then given a detailed, albeit preliminary, characterization. More formal and precise definitions for these concepts will not be given until chapter 8. The reason for the delay is that the most basic concept, potential evidence, will be defined by reference to two ideas: probability and explanation. In the intervening chapters I will introduce more standard probabilistic theories of evidence and the concepts of probability they employ (chapter 3); show why these probabilistic theories of evidence need to be abandoned (chapter 4); introduce a concept of objective epistemic probability that will be used for potential and veridical evidence (chapter 5); argue that potential and veridical evidence require high probability, in the objective epistemic sense (chapter 6); show that high probability, although necessary, is not sufficient, and that an "explanation condition" is also necessary (chapter 7). At this point (chapter 8), the formal definition of potential evidence is presented, which is then used to define the other three concepts of evidence.

Chapters 9–13 apply the theory to some well-known philosophical puzzles about evidence and to two scientific cases. The puzzles include the ravens and grue paradoxes (chapter 9); the question of whether explanations or predictions provide stronger evidence, and Glymour's puzzle of "old evidence" (chapter 10); and the Duhem-Quine doctrine of holism (chapter 11), which, if true, might be taken to negate the entire project of this book. The scientific cases involve Jean Perrin's evidence for molecules (chapter 12) and J. J. Thomson's evidence for electrons (chapter 13). Both of these illustrate the use of the concepts of evidence I develop. But each also introduces related philosophical and historical issues.

In previous work I have examined various issues pertaining to evidence. This book goes much deeper. It presents a foundation for my theory of evidence, justifies and unifies claims I have made, and applies the theory in ways not previously done. Whether the concepts of evidence I develop meet the dean's challenge I invite the reader to decide.

5. Acknowledgments

Many individuals gave me sleepless nights by offering powerful criticisms. They must accept some responsibility for the length of the book. These include very smart present and former graduate students Stacie Bumgarner, Steven Gimbel (who made extensive, insightful comments on large chunks of the book), Adam Goldstein, Frederick Kronz, Gregory Morgan, Laura J. Snyder, and Kent Staley. I am also indebted to Paul Humphreys, editor of this series, for his encouragement and wise and very helpful suggestions, as well as to a band of sharp, stern, and sobering referees that he and Peter Ohlin, philosophy editor at Oxford University Press, hired to bring me down to earth. Other individuals who made comments on specific points are acknowledged in footnotes. I thank Kimberlea Brinson Burke, Stacie Bumgarner, and Mary Berk for valuable assistance with the proofs. Finally, I owe a very great deal to Katharine Rowland, Robert Rynasiewicz, Richard Richards, Rita Snyder, Richard Rosenberg, and Donna Rosenberg.

2

CONCEPTS OF EVIDENCE, OR HOW THE ELECTRON GOT ITS CHARGE

When the claim is made that something is evidence that a hypothesis is true, what exactly is being claimed? Is there some unique concept of evidence by reference to which we can understand what is being said? I will argue that there is not one concept of evidence but several in use in the sciences. To do so I begin with a historical example.

1. Heinrich Hertz and Cathode Rays

Cathode rays were discovered in 1859 by the German physicist Julius Plücker. The discovery involved the use of an air-filled glass tube containing positive and negative electrodes. When the air pressure in the tube is reduced to approximately .001 mm of mercury and a source of high electrical potential is connected to the positive electrode (the anode), the glass near the negative electrode (the cathode) glows with a greenish phosphorescence. The position of the glow changes if a magnetic field is introduced. Plücker concluded that something was being emitted from the cathode that was distinct from the ordinary electrical discharge observed in such tubes. In 1869 Plücker's student Johann Wilhelm Hittorf discovered that a solid body placed between the cathode and walls of the tube causes a shadow to be cast on the walls. His conclusion was that there are rays ("cathode rays") emitted from the cathode that travel in straight lines. Later experiments by other physicists showed that these rays are perpendicular to the cathode's surface, that they produce certain chemical reactions, and that they generate the same effects whatever metal the cathode is made of.

Two theories concerning the nature of cathode rays soon developed. According to one, suggested by the British physicists William Crookes and Arthur Schuster, cathode rays are atoms or molecules of the gas in the tube that have become negatively charged. According to the other, supported by the German physicists Eugen Goldstein, Gustav Heinrich Wiedemann, and Heinrich Hertz, cathode rays are not atoms or molecules or particles of any sort, but some type of wave in the ether, similar to light in some respects but not others.

Against the particle theorists the wave theorists argued that if cathode rays are negatively charged particles, then they should exhibit electrical as well as magnetic effects. In 1883 Hertz conducted a series of experiments designed to determine whether electrical effects could be demonstrated. In performing the experiments Hertz wrote that he was trying to answer two questions:

> First: Do the cathode rays give rise to electrostatic forces in their neighborhood? Secondly: In their course are they affected by external electrostatic forces?[1]

To answer the first question, Hertz employed an apparatus illustrated in fig. 1. There is a glass tube, 250 mm long and 25 mm wide, that contains a cathode and anode. The anode consists of three parts: (i) a brass tube that almost completely surrounds the cathode but has opposite it a circular opening 10 mm in diameter through which the cathode rays pass; (ii) a wire gauze about 1 square mm in mesh through which the cathode rays also pass; (iii) a protective metallic case which surrounds most of the cathode tube that screens off that portion of the tube beyond the wire gauze from any electrostatic forces the cathode might produce. This metallic case and a metallic mantle surrounding the entire apparatus are connected to an electrometer, the deflection of which determines the presence of electricity.

When the cathode is connected to a source of electricity, cathode rays are produced that travel from the cathode to the end of the tube and cause a green phosphorescence to appear on the glass. But there is also the ordinary electric current that flows from the cathode to the anode, which Hertz attempted to separate from the cathode rays by means of the wire gauze. The latter captures the current, allowing what Hertz took to be pure cathode rays, unmixed with current, to enter the electrometer. If these pure cathode rays carry a charge, this should be registered by the electrometer.

In this experiment when cathode rays were produced and the electrometer was connected to the apparatus, the needle of the electrometer remained at rest. Hertz concluded:

> As far as the accuracy of the experiment allows, we can conclude with certainty that no electrostatic effect due to the cathode rays can be perceived. (p. 251)

Another experiment was designed to determine an answer to the second question, viz. whether cathode rays are affected by external electrostatic forces. In this experiment Hertz placed a cathode tube similar to the one in the first ex-

---

1. Heinrich Hertz, *Miscellaneous Papers* (London: Macmillan, 1896), p. 249. For an illuminating discussion of Hertz's experiments, see Jed Z. Buchwald, *The Creation of Scientific Effects* (Chicago: University of Chicago Press, 1994), pp. 150–174.

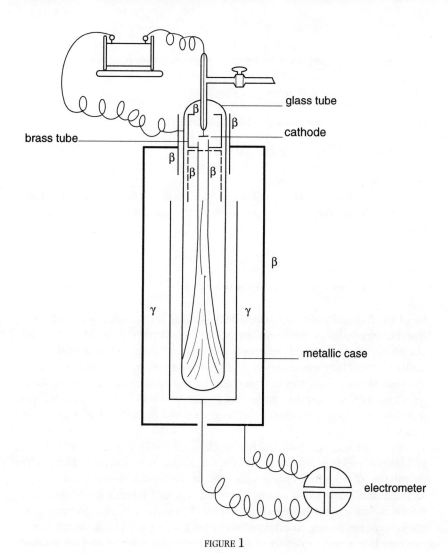

FIGURE 1

periment between oppositely electrified plates. If cathode rays were electrically charged they should be deflected by these plates, as indicated in a changed position of the phosphorescence. Hertz reported that "no effect could be observed in the phosphorescent image" (p. 252). More generally,

> under the conditions of the experiment the cathode rays were not deflected by an electromotive force existing in the space traversed by them. (p. 253)

From both his experiments Hertz concluded:

> These cathode rays are electrically indifferent, and amongst known agents the phenomenon most nearly allied to them is light. (p. 254)

Let

> $e =$ In Hertz's experiments, which were designed to show electrical effects of cathode rays, no such effects were produced.
> $h =$ Cathode rays are not electrically charged.

Hertz was making the following evidential claim:

> *Hertz's claim*: e is evidence that h is true.

What does such a claim mean?

To begin to deal with this question, let us ask a different one: Is Hertz's claim true? To help answer that question we need to continue the story of cathode rays.

2. J. J. Thomson and the Electron

In 1895 (during the year following Hertz's death), the French physical chemist Jean Perrin conducted new experiments on cathode rays, from which he concluded that they are indeed electrically charged. In Perrin's experimental set-up, unlike Hertz's, there was no device, such as the wire gauze, designed to separate the cathode rays from the ordinary electric current. So Hertz, had he lived, might have replied that the charge that Perrin detected in the electrometer was not a charge on the cathode rays but one produced by the regular electric current in the tube.

Two years later, in 1897, the British physicist J. J. Thomson conducted new cathode ray experiments that settled the issue definitively and led to Thomson's postulation of negatively charged particles as the constituents of cathode rays. In the first of these experiments, Thomson repeated Perrin's experiment in a form that would avoid the objection that the cathode rays and the ordinary electric current were being mixed together. By using a magnet he deflected the cathode rays into a set of cylinders connected to an electrometer and determined that the electrometer registered a charge when and only when the pure cathode rays were deflected toward it. In his discussion of the second experiment, Thomson wrote:

> An objection very generally urged against the view that the cathode rays are negatively electrified particles, is that hitherto no deflexion by the rays has been observed under a small electrostatic force. . . . Hertz made the rays travel between two parallel plates of metal placed inside the discharge tube, but found that they were not deflected when the plates were connected with a battery of storage-cells; on repeating the experiment I at first got the same result, but subsequently experiments showed that the absence of deflexion is due to the conductivity conferred on the rarefied gas by the cathode rays. On measuring this conductivity it was found that it diminished very rapidly as the exhaustion increased; it seemed then that on trying Hertz's experiment at very

high exhaustion there might be a chance of detecting the deflexion of the cathode rays by an electrostatic force.[2]

Thomson in 1897 was technically able to produce a much higher vacuum in the tube than was Hertz in 1883. If cathode rays are charged particles, Thomson reasoned, then when they pass through the gas in the tube they ionize the gas molecules, thus producing both positive and negative charge that will neutralize the charge on the metal plates between which the cathode rays travel. So, if the tube is not sufficiently evacuated, there will be no deflection of the cathode rays. Thomson's experiment showed that when the gas in the tube was much more completely evacuated than was done in Hertz's experiment, electrical deflection of the cathode rays was produced.

From these two experiments Thomson concluded:

> As the cathode rays carry a charge of negative electricity, are deflected by an electrostatic force as if they were negatively electrified, and are acted on by a magnetic force in just the way in which this force would act on a negatively electrified body moving along the path of these rays, I can see no escape from the conclusion that they are charges of negative electricity carried by particles of matter.[3]

Next Thomson raised the question of whether these particles are molecules, atoms, or something smaller. For this purpose he performed experiments designed to determine the ratio of the mass of the particles to the charge they carry (m/e). From these experiments he arrived at a value for m/e whose order of magnitude is 10^{-7}. At that time the smallest known ratio of mass to charge was 10^{-4} for hydrogen ions. Thomson concluded that the charged particles constituting cathode rays are much smaller than the smallest atoms. He also speculated that these "corpuscles" (as Thomson called them), or "electrons" (as they were later called by Thomson's student Rutherford), are constituents of all atoms. As a result of his experiments with cathode rays Thomson received a Nobel Prize in 1906. He is credited with the discovery of the electron. (Chapter 13 contains a philosophical account of the concept of discovery as well as a historical discussion of whether Thomson really did discover the electron.)

3. Were Hertz's Results Evidence That Cathode Rays Are Not Electrically Charged?

Let us back up from 1897, the year Thomson discovered that cathode rays carry a negative charge, to the year 1883, when Hertz performed his experiments and concluded that cathode rays carry no electric charge. Again let us ask whether the results of Hertz's experiments, in which no electrical effects of cathode rays

---

2. J. J. Thomson, "Cathode Rays," *Philosophical Magazine* 44 (October, 1897).
3. Ibid, p. 302.

were observed, constitute evidence (indeed, as Hertz claimed, strong evidence) that cathode rays are not electrically charged.

I will suggest three different answers, each of which, I believe, is plausible, and each of which (as I will demonstrate in chapters 3 and 4) can be related to a different philosophical theory of evidence. Here are the three answers, followed by a brief defense of each:

1. The results of Hertz's experiments are (strong) evidence that cathode rays are not charged.
2. From 1883 to 1897 the results of Hertz's experiments were (strong) evidence that cathode rays are not charged; after 1897 they were not so.
3. The results of Hertz's experiments are not and never were (strong) evidence that cathode rays are not charged.

The first answer might be defended, as follows, by appeal to what Hertz was justified in believing:

> Given what was known by Hertz in 1883—including the particular experimental set-up, the results, the techniques available for removing gas from cathode tubes, and other assumptions—Hertz was completely justified, on the basis of his experimental results, in believing that cathode rays carry no electrical charge. Anyone in the type of epistemic situation Hertz was in would be justified in drawing the conclusion he did. That is why we can say that such results are evidence for this conclusion.

The historian of physics Jed Buchwald defends Hertz's reasoning in a way that might tempt one to draw conclusion 1 about evidence (although Buchwald himself does not explicitly address the question of how one should speak of evidence in this case):

> Hertz's arguments and experimental work were tightly closed, carefully wrought to preclude damaging criticism. Even in retrospect it is not possible to destabilize Hertz's experiments by pointing out that something might have been going on that he did not take into account.[4]

The second answer, viz. that from 1883 to 1897, but not thereafter, Hertz's experiments provided (strong) evidence that cathode rays are not charged, can be defended, as follows, by appeal to how Hertz and other physicists viewed his experimental results:

> Hertz's 1883 experiments provided the physics community with the only, or at least the best, information available until 1897 about whether the cathode rays were charged. Hertz and other physicists generally regarded the results of his experiments as evidence of neutral cathode rays. After Thomson's new experiments in 1897 physicists no longer regarded Hertz's experiments as evidence for the electric neutrality of cathode rays.

Finally, can anything be said in favor of the third answer, according to which Hertz's experiments never provided evidence for the neutrality of cathode rays? One might defend this idea as follows:

---

4. Jed Z. Buchwald, *The Creation of Scientific Effects*, p. 168.

Let's face it, Hertz's experiments were based on a mistake, even if Hertz had no means of recognizing this. The gas in Hertz's cathode tubes was not sufficiently evacuated, allowing it to serve as an electrical conductor and to screen off the electrical force from the cathode rays. Moreover, as it turns out, Hertz's hypothesis itself is false. When the gas is more completely removed, as in Thomson's experiments, the screening off effect is sufficiently reduced to permit the cathode rays to be electrically deflected. So, even if from 1883 to 1897 Hertz and other physicists were fully justified, on the basis of Hertz's experimental results, in believing that cathode rays are neutral, Hertz's experiments, as Thomson demonstrated, do not really support this conclusion. They do not in fact provide a good reason to believe the conclusion. They do not, and never did, provide "true," or "genuine," or "veridical" evidence that Hertz's hypothesis is true.

Three different answers have been suggested to the question of whether Hertz's 1883 results constitute evidence that cathode rays are not electrically charged. Which, if any, of them is correct? I believe that all of them are. If this is so, it suggests that there are at least three different concepts of evidence or ways of viewing the concept of evidence in science. (In fact, I will argue, there are at least four, since the third answer can be justified in two different ways.) Before settling on this pluralistic view, however, let us try to say more about what is central to each of these concepts, as reflected in the defense given for each answer. In the four sections that follow I propose to give a preliminary, intuitive characterization of each concept (what Carnap calls a "clarification of the explicandum"[5]). In chapters 3 and 4 these will be related to some standard accounts of evidence proposed by philosophers of science, in order to determine whether deeper, more precise analyses are possible. Such analyses will be provided in chapters 5–8.

4. ES-Evidence

Suppose we give the first answer—that Hertz's 1883 experiments yielded results that constitute evidence that cathode rays are not charged—and offer a defense of the sort given earlier. What concept of evidence would this reflect?

It would be based on the idea of providing an epistemic justification for belief. The results of his experiments justified Hertz's belief that cathode rays are electrically neutral.[6] They did so in an epistemic rather than a pragmatic sense. (A pragmatic justification for belief is concerned solely with the beneficial consequences of believing something for the believer that have nothing to do with establishing the truth or probability of what is believed. The fact, assuming it is

---

5. Rudolf Carnap, *Logical Foundations of Probability* (Chicago: University of Chicago Press, 2nd ed., 1962), p. 3.
6. Here and in what follows I adopt a realist stance toward evidence and speak of believing hypotheses (that is, believing them true). Whether avid (or should I say rabid) antirealists can understand this to mean believing hypotheses to be instrumentally correct, or to save the phenomena, or however an antirealist wants to describe his beliefs, is a question I will explore in chapter 8 after my final definitions of evidence are offered.

one, that I will be rewarded if I believe in God might be held to constitute a pragmatic justification for my believing that God exists, but it does not provide an epistemic justification.) In the case of Hertz the justification in question is relativized to Hertz's epistemic situation in 1883: He was justified in believing what he did *given what he knew and believed in 1883 and given what he did not know or believe and was not in a position to know or believe in 1883.*

Hertz's particular epistemic situation in 1883 involved many things, including knowing the results of his experiments as described in *e* or something similar; knowing how to reason from the fact that no perceptible deflection of cathode rays occurred in Hertz's experiment to the hypothesis that cathode rays are electrically neutral (which requires knowing, for example, that electric forces cause deflection in electrified bodies); and *not* knowing, and not being in a position to know, facts that undermine *e*'s support for *h*. The latter include the fact that the cathode tubes used were not sufficiently evacuated to allow electrical deflection, and, of course, the results of Thomson's 1897 experiments.[7] Hertz in 1883, but not Thomson in 1897, did not know and was not in a position to know such undermining facts. Although Hertz was justified in believing his hypothesis in 1883, Thomson would not have been justified in 1897.

Hertz's particular epistemic situation includes many other facts about what Hertz did not know or believe, and was not in a position to know or believe, in 1883. It might also include facts about the strength of his beliefs. Many of these facts will be completely irrelevant to the question of whether Hertz was justified in believing the hypothesis that cathode rays are neutral. In relativizing Hertz's justification to an epistemic situation we could include both relevant and irrelevant parts of Hertz's epistemic situation in 1883. Or we could abstract those parts, including ones noted above, that are relevant to the question of whether Hertz was justified in believing the hypothesis of interest, and relativize the justification to an epistemic situation containing just those parts. This is the more manageable strategy. Even if no one but Hertz was in his particular epistemic situation (involving his full set of beliefs), other physicists may have been in an epistemic situation containing the relevant parts of Hertz's. In any case, the justification is not necessarily restricted to Hertz. Anyone in an epistemic situation with the (relevant) knowledge and beliefs of the sort Hertz possessed in 1883 would similarly be justified in believing that cathode rays are neutral.

Accordingly, an epistemic situation is different from what philosophers usually call "background information." The latter consists solely of propositions (assumed to be true). The former is an abstract type of situation in which, among other things, one knows or believes that certain propositions are true, one is not in a position to know or believe that others are, and one knows (or does not know) how to reason from the former to the hypothesis of interest, even if such a situation does not in fact obtain for any person.

7. Note that I have put this condition as "not knowing and not being in a position to know" rather than simply "not knowing." An English-speaking physicist in 1900 who knew only Hertz's 1883 experimental results but not those of Thomson in 1897 would not have been justified in believing Hertz's conclusion. Such a physicist was in a position to know of Thomson's 1897 experimental results that undermine the claim that Hertz's results support the neutrality hypothesis.

A person in an epistemic situation may not know that some propositions $P_1, \ldots, P_n$ believed in that situation are true and may not even be justified in believing those propositions. But if the person is to be justified in believing a given hypothesis h on the basis of $P_1, \ldots, P_n$, that person must be justified in believing $P_1, \ldots, P_n$. Although the latter may be *known* to be true, we need not make this a necessary condition.[8] One can be justified in believing a hypothesis on the basis of one's other beliefs, so long as these are justified, even if they are not true. Since justification is relativized to an epistemic situation, so is the concept of evidence based on it. Hertz's experimental results are evidence for the truth of his hypothesis, relative to an epistemic situation of the sort attributable to him in 1883, but not relative to one attributable to Thomson in 1897. More generally, we might say that e is evidence that h relative to (epistemic situation) ES_1 but not ES_2.

The concept of evidence based on the notion of justification in question is *objective* in this sense. Whether or not e is evidence that h does not depend upon whether anyone in fact believes or knows anything about e, h, or their relationship, or upon whether anyone is in fact in the epistemic situation to which the evidence is relativized. To be sure, Hertz knew that the statements reporting his experimental results were true. He believed the truth of the hypothesis that cathode rays are electrically neutral, and he believed that his results were evidence for his hypothesis. But he need not have believed any of these things in order for these results to justify a belief in that hypothesis. Anyone in the epistemic situation we are attributing to Hertz—whether or not anyone in fact was—would have been justified in believing Hertz's hypothesis.

The concept of evidence based on the present notion of justification allows the hypothesis to be false. It is what some epistemologists call a "fallibilist" concept by contrast to an "infallibilist" one.[9] In accordance with this concept, one can be justified in believing a hypothesis, given everything one knows, even if the hypothesis is false. This was the situation with Hertz. If Hertz's 1883 results are to constitute evidence for his hypothesis, and the concept of evidence in question is to be tied to justified belief, then the latter requires a fallibilist interpretation.

Although providing a justification for belief is necessary for evidence of the sort in question, it is not sufficient. This point will be defended after the formal definition of this and other types of evidence is developed in chapter 8.

If we say that

e: In Hertz's experiments, which were designed to show electrical effects, no such effects were produced

is evidence that

h: Cathode rays are not electrically charged,

---

8. I am indebted here to Gregory Morgan.
9. See, for example, Scott Sturgeon, "Knowledge," in A. C. Grayling, ed. *Philosophy* (Oxford: Oxford University Press, 1995), p. 14; Trenton Merricks, "Warrant Entails Truth," *Philosophy and Phenomenological Research* 55 (1995), 841–855.

must *e* be true? This suggests a more general question: What sort of thing is it that is evidence for a hypothesis? We have been understanding *e* above as a proposition. This proposition is true, and it is the fact that it is true, rather than the proposition itself, that is evidence that cathode rays are not electrically charged.[10] However, to simplify matters, and to conform to standard philosophical practice, I shall continue to write sentences of the form "*e* is evidence that *h*." Sentences of this form should be understood to mean "the fact that *e* is true is evidence that *h* is true," which presupposes, of course, that *e* is true. If the truth of *e* is not being presupposed, we can write "*e*, if true, is evidence that *h*," which should be understood to mean "if *e* is true, then the fact that it is true is evidence that *h* is true."

The propositions known or believed by someone in a given epistemic situation may include evidential ones themselves. For example, we are supposing that among Hertz's beliefs in 1883 was the belief that

(1) The absence of electrical effects in Hertz's experiments is evidence that cathode rays are neutral, relative to Hertz's epistemic situation in 1883.

This does not trivialize a claim about Hertz's justification for believing that cathode rays are neutral. It is not being claimed that his belief that (1) is true justifies his belief that cathode rays are neutral. The claim is that it is the absence of electrical effects in his experiments that justifies his belief that cathode rays are neutral, given his epistemic situation (which includes his beliefs about what is evidence for what). Nor does the fact that Hertz's epistemic situation includes the belief that (1) is true render (1) circular. The evidential claim in (1) is not being relativized to the claim that the absence of electrical effects in Hertz's experiments is evidence that cathode rays are neutral.

I have described various features of a concept of evidence that would allow us to say that Hertz's experimental results are evidence that cathode rays are not charged, and would permit a defense of this claim of the sort indicated in section 3. I will call the concept in question *ES-evidence*.

5. Subjective Evidence

I turn now to the second answer to the question of whether Hertz's experimental results constitute evidence that cathode rays are not charged, and to the defense of that answer. The answer was that from 1883 to 1897 they were indeed evidence, but after 1897 they were no longer evidence. The defense appealed to how Hertz and others regarded the results of Hertz's experiments: From 1883 to

---

10. Some writers on evidence take propositions (or sentences) themselves to be evidence that other propositions are true. This, I think, is a very odd way of speaking. These writers want to be able to say that Hertz's report *e* is evidence that *h* is true whether or not *e* is true. We can accommodate this desire, however, without having to accept the idea that propositions are evidence. Where *e* is a true proposition we can say that the fact that *e* is true is evidence that *h* is true. Where *e* is a false proposition, or where it is unknown whether *e* is true or false, we can say that *if e* is (or were) true, then that fact is (or would be) evidence that *h* is true.

1897 they were considered evidence for the electric neutrality of cathode rays; after Thomson's 1897 experiments they were not.

This answer and its defense suggest a concept of evidence that is relativized to a specific person or group: e is evidence that h *for such and such a person or group;* moreover, it may be evidence for that person or group *at one time but not another.* How can this be understood?

First, we might understand this as implying that the person or group at the time in question *believes* that e is evidence that h. Hertz clearly believed this about his experimental results in 1883. Thomson, following his own more conclusive experiments in 1897, did not believe that Hertz's results were evidence that cathode rays are electrically neutral.

Second, it may be understood as implying that the person or group in fact believes that the hypothesis h is true or at least probable. Following his 1883 experiments Hertz believed that cathode rays are electrically neutral. In 1897, following his own experiments, Thomson believed that cathode rays are negatively charged. Thomson in 1897 knew about Hertz's results and he was aware of Hertz's hypothesis that cathode rays are neutral. Yet the results of Hertz's experiments were not *Thomson's* evidence that cathode rays are neutral, since Thomson did not believe that Hertz's hypothesis was true or probable.

Third, it may be understood as implying that the person or group has as a reason for believing that h is true or probable that e is true. Hertz's reason for believing that cathode rays are electrically neutral is that no electrical effects were produced in his experiments. If this had not been Hertz's reason for believing this hypothesis, then the experimental results in question would not have been *his* evidence for that hypothesis. Exactly analogous claims can be made for Thomson's evidence for the hypothesis that cathode rays are negatively charged.

Let me speak of evidence in such cases as *X's evidence at time t*, where X is a person or group. We might say that e is X's evidence that h at time t if and only if at time t

1. X believes that e is evidence that h;
2. X believes that h is true or probable; and
3. X's reason for believing that h is true or probable is that e is true.

The term "evidence" in condition 1 is not tied to any person or time. Whether this is to be construed as ES-evidence or something else is a question I leave for section 9.

I call the concept of evidence satisfying the three conditions above *subjective*. (The question of how this concept is related to one endorsed by subjective probability theorists will be examined in chapter 3.) Something can be evidence, in this sense, only if it is someone's evidence at some time, which, as the three conditions indicate, requires that someone actually believes certain things about e and h, including that e is evidence that h. This is unlike ES-evidence: e can be ES-evidence that h, even if no one in fact believes that e or h is true or that e is ES-evidence that h. ES-evidence is relativized with respect to a *type* of epistemic situation. It is not required that anyone be in that epistemic situation. For subjective evidence someone (or some group) must be in a certain epistemic situation with regard to that evidence.

There is another difference as well. Subjective evidence, as reflected in the three conditions above, does not require that e be true, only that X believe that it is. Suppose that Hertz's evidence report e had been false, so that there were electrical effects produced in the experiment which he failed to observe. We might still say that Hertz's evidence that cathode rays are neutral was that no electrical effects were produced in his experiments. This would be true in virtue of the fact that Hertz believed that none were.

Finally, subjective evidence, unlike ES-evidence, does not require that the person or group whose evidence it is be justified in believing h on the basis of e. Even if Hertz's experimental results had not warranted his belief that cathode rays are neutral, those results could have been his evidence for this hypothesis. The crucial point for subjective evidence is how that evidence was viewed by the person or group in question. In section 3 above, the second (subjective) answer to the question of whether Hertz's results are evidence for his hypothesis was that from 1883 to 1897 they were so. The defense of this answer consisted in saying how these results were viewed, which leaves open the question of whether those who viewed them this way were justified in doing so.

6. Veridical Evidence

The third answer to the question of whether Hertz's experimental results constitute evidence that cathode rays are not charged was that these results never really were evidence for this hypothesis and still are not. Two concepts of evidence, one stronger, one weaker, can be distinguished here, both of which yield this answer. In the present section I will describe the stronger one ("veridical" evidence), and in section 7 the weaker ("potential" evidence).

In the earlier defense of this answer we noted two facts: first, that Hertz's experiments were based on the mistaken assumption that the cathode tubes were sufficiently evacuated to permit the rays to be electrically deflected should they be charged; second, that, as Thomson's later experiments showed, Hertz's conclusion is false. We said that even though Hertz, given his knowledge, was justified in believing what he did, his experimental results do not in fact provide a good reason to believe his conclusion. Accordingly, Hertz's results never were true, or genuine, or veridical evidence for his conclusion. What concept of evidence is suggested by this defense?

It is one for which I have invoked the expression "providing a good reason to believe." If e is evidence that h (in the present sense), then e provides a good reason to believe h.[11] (This is necessary but not sufficient.) As in the case of ES-evidence, the idea required is epistemic rather than pragmatic. Yet it is importantly different from the idea of justification used to characterize ES-evidence. In 1883 Hertz was justified in believing that cathode rays are neutral on the basis of his experimental results. But these results, as Thomson demonstrated, do not provide a

---

11. In chapter 6, sec. 5, I consider (and reject) the idea that veridical evidence (and potential evidence, to be discussed next) need not supply a good reason to believe, but only some reason.

good reason to believe this hypothesis.[12] In this respect the present concept of "good reason to believe" functions like "sign" or "symptom." A rash may be a sign or symptom of a certain disease, or a good reason to believe the disease is present, even if the medical experts are completely unaware of the connection and so are not justified in believing that this disease is present, given what they know.

Experiments and tests sometimes have flaws serious enough to impugn conclusions that are drawn. (See the appendix for a discussion of evidential flaws.) Let me distinguish two types of flaws: those that could have been expected given all the information available, and those that could not have been expected on the basis of this information. As Thomson showed in 1897 there was a flaw in Hertz's 1883 experiments: Hertz's cathode tubes were not sufficiently evacuated to produce electrical deflection. This flaw, we are supposing, could not have been expected by Hertz in 1883. Nevertheless, since his results were based on this flaw, however unexpected, they do not provide a good reason for believing Hertz's conclusion.

The present concept of "good reason to believe," like the earlier idea of "justifying belief," is objective: whether e is a good reason to believe h does not depend on whether in fact anyone believes e or h or that e is a good reason for believing h. However, unlike the case with the earlier concept of justification, no epistemic situation is invoked or implicit. Someone who knows Hertz's results is justified in believing his hypothesis only if that person's epistemic situation meets certain conditions (such as, that person does not know, and is not in a position to know, Thomson's later results). By contrast, Hertz's experimental results either are, or are not, a good reason to believe that cathode rays are electrically neutral, no matter what knowledge is imagined available. In this respect, again, "good reason to believe" functions like "sign" and "symptom." If this rash is a sign or symptom of that disease, it is so whatever epistemic situation is being imagined.[13]

Accordingly, unlike the earlier notion of justification, a claim that e is a good reason to believe h can be vitiated by discovering information not available to a person or community. The claim that Hertz was justified in believing his conclusion on the basis of his experimental results is not refuted by Thomson's later ex-

---

12. In what follows I could have continued to use the term "justification," but distinguished two types: that which justifies belief for someone in a certain epistemic situation, and that which does so without regard to epistemic situations. I choose a new expression for the latter, in order to emphasize the difference between the two cases.
13. There is also a use of "sign" (or "symptom") that is relativized to a person or epistemic situation or both (sign for such and such a person). Similarly, there is a use of the expression "good reason to believe" that is to be understood as relativized to an epistemic situation and/or to a person in such a situation. (In chapter 5 the latter will be spelled out for the related concept of "being reasonable to believe.") In what follows, however, when speaking of veridical and potential evidence, the unrelativized use is meant. For a perceptive discussion of this unrelativized concept, see Laura J. Snyder, "Is Evidence Historical?," in Peter Achinstein and Laura J. Snyder, eds., *Scientific Methods: Conceptual and Historical Problems* (Malabar, Florida: Krieger Publishing Co., 1994). I am indebted to her for very helpful suggestions in this chapter.

perimental results, since that claim is to be understood as applicable to someone in Hertz's epistemic situation. On the other hand, the claim that Hertz's reason for believing his conclusion is a good reason for doing so *is* refuted by Thomson's later experimental results. This means that "good reason for believing" (and the concept of evidence based on it) is potentially *empirically incomplete* in this sense. Whether e is a good reason to believe h can depend on empirical facts in addition to those reported in e. If those facts do, or do not, obtain, then the claim that e is a good reason to believe h will be true or false, as the case may be. The empirical fact that Hertz's cathode tubes were not sufficiently evacuated to produce electrical deflection falsifies the claim that the absence of deflection in Hertz's experiments is a good reason to believe that cathode rays are electrically neutral.

What about the truth or falsity of h itself? In the case of ES-evidence, h can turn out to be false. As with Hertz, one may be justified in believing a false hypothesis on the basis of e. Is this the case with "good reason to believe" as the latter should be understood for the present notion of evidence? Thomson's 1897 experiments showed that Hertz's 1883 experiments contained a flaw, however unexpected. But Thomson's experiments showed something even more important, viz. that Hertz's hypothesis that cathode rays are electrically neutral is false. And, we might say, the fact that it is false shows that although Hertz's 1883 experimental results seemed to provide a good reason to believe the hypothesis is true, they do not really do so.

There is a corresponding sense of "sign" and "symptom." If the rash in this patient is of a sort that is almost always caused by a certain disease, but in this case the patient does not have that disease, we might conclude that the rash was not in fact a sign or symptom of the disease in that patient although it seemed to be. In accordance with this usage, if x is a sign or symptom of y, then y is present.[14] Similarly, we might say, if e is a good reason to believe h, then h is true. (Weaker senses of "sign," "symptom," and "good reason to believe" will be given in the next section.) Because it requires the truth of the hypothesis h, I call a concept of evidence that is based on this idea of a good reason to believe "veridical" evidence.[15]

As characterized so far, then, if e is veridical evidence that h, then e is a good reason to believe h in a sense requiring (i) objectivity (if e is veridical evidence that h, it is so whether or not anyone knows of e or h or believes that e is evidence that h); (ii) nonrelativization (veridical evidence is not relativized either to persons, as is subjective evidence, or to epistemic situations, as is ES-evidence); (iii) possible empirical incompleteness (whether e is veridical evidence that h can de-

---

14. Suppose both the disease and the rash are present but the rash was caused not by the disease but by something else. Is the rash a sign of the disease? Is it evidence that the disease is present? This complication will be discussed in chapter 8 when I offer the more precise definition of this type of evidence.
15. A number of writers in epistemology have made an analogous claim for the concept of "warrant" or "epistemic justification." See, for example, Trenton Merricks, "Warrant Entails Truth," *Philosophy and Phenomenological Research*, LV (1995), 841–855; John Pollock, *Contemporary Theories of Knowledge* (Totowa, NJ: Rowman and Littlefield, 1986), 183–190.

pend on empirical facts in addition to those reported in *e* and *h*); and (iv) the truth of *h*.

Finally, we need to distinguish *veridical* from *conclusive* evidence. If *e* is conclusive evidence that *h*, then not only is *h* true, but *e* establishes *h* with certainty. Invoking probabilities, we might say that if *e* is conclusive evidence that *h*, then the probability of *h* given *e* is 1 (the maximum value). If *e* is veridical evidence that *h*, then *h* is true, but *e* may or may not establish *h* with certainty. (The probability of *h* given *e* need not be 1.) Suppose John owned 999 of the 1000 tickets sold in a lottery, one of which was drawn at random. If John won the lottery, then the fact that he owned 999 of the 1000 tickets is veridical evidence that he won. However, it is not conclusive evidence that he did. By analogy, Sam's spots may be (genuine) signs or symptoms of the measles virus without being conclusive signs or symptoms. He may have the measles virus, which is causing those spots, without its being the case that those spots conclusively establish that he has the measles virus.

7. Potential Evidence

In the previous section I introduced a notion of a good reason to believe *h* that requires the truth of *h*. However, this is not the only possible use of "good reason to believe." We might compare the situation with a case in which something other than a reason to believe (or to do *x*) is being evaluated. Suppose that some medical treatment is being administered to a patient with a certain disease; it is the only treatment for that disease; and it is effective 95% of the time. However, the treatment turns out not to work for the patient in question; and the only way of determining this, let us suppose, is to administer the treatment. *Was that a good treatment to use?* There are two different ways to answer this question: (1) Yes, because that sort of treatment, which is the only effective one, almost always works; (2) No, because it did not work on that patient. Answer (1) does not carry the implication that the treatment will be successful, only that it probably will be, it is usually successful. A good treatment to use for this patient is one with a high probability of success for that patient. Answer (2) carries the implication of success in this case. A good treatment to use for this patient is one that will work.

Earlier I claimed that there is a sense of "sign" and "symptom" in accordance with which, if *x* is a sign or symptom of *y*, then *y* must obtain. There is also a sense of these terms which does not require that *y* obtains, only that it probably does. Suppose that Sam has the spots typically associated with the measles virus but does not have the measles virus. We might say (*a*) that he nevertheless has the signs or symptoms of the measles virus, or (*b*) that whatever those spots are they are not signs or symptoms *of the measles virus* since no such virus is present. In an exactly analogous fashion we might say (*a*) that the fact that Sam has the spots typically associated with the measles virus is a good reason to believe (and evidence) that he has the measles virus. Or we might say (*b*) that this fact is not a good reason to believe (or evidence) that this is so, because, as it turns out, he does not have the measles virus. Response (*b*) employs veridical evidence. What concept is involved in response (*a*)?

First, as just noted, this concept of evidence is related to a notion of "good reason to believe" that is weaker than that required for veridical evidence. It is one that does not presuppose the truth of the hypothesis. Some fact *e* can be evidence that *h* (in the present sense), and hence a good reason to believe *h* (in the present sense), even if *h* is false. This concept of evidence, like ES-evidence, is a fallibilist one.

Second, unlike ES-evidence, it is not relativized to any epistemic situation. If the fact that Sam has those spots is evidence (in the present sense) that he has the measles virus, it is so independently of the type of epistemic situation envisaged. Just as a sign or symptom need not be thought of as a sign or symptom *for* someone in a particular type of epistemic situation, an analogous claim holds for evidence.

Third, the concept is objective in the same sense that ES-evidence is: whether or not *e* is evidence that *h* does not depend upon whether anyone believes or knows anything about, *e*, *h*, or their relationship.

Fourth, as with ES-evidence, the present concept requires *e* to be true. (Where *e*'s truth-value is not known we can write that *e*, if true, would be evidence that *h*.)

Fifth, like veridical evidence, the present concept, and that of a good reason for belief underlying it, are potentially empirically incomplete: whether *e* is evidence that *h*, and hence a good reason to believe *h* (in a sense that does not require its truth), can depend on empirical facts in addition to *e*. Whether my hygrometer's registering 45% is evidence, and hence a good reason for believing, that the relative humidity is 45% depends on whether my hygrometer is working properly. If it is not, if the pointer is stuck on 45%, then the fact that it registers what it does is not a good reason for believing the hypothesis in question, independent of whether the hypothesis is true.

Accordingly, as with veridical evidence, an important distinction emerges between the present concept and ES-evidence in the case of an unexpected flaw. Whether some experimental, observational, or test results *e* constitute a good reason to believe *h* (even in a sense of "good reason" that does not require *h* to be true) depends on whether there is some flaw in the design or execution of the experiment or test. Suppose such a flaw exists but is unexpected given the epistemic situation ES. Then someone in ES may be justified in believing a hypothesis *h* on the basis of result *e*, even though *e* is not a good reason to believe *h*. These results may *seem* to be a good reason, and someone (in a given epistemic situation) may be justified in believing they are a good reason, but in fact they are not. Such was the case with Hertz's experimental results in 1883. Flawed experimental, observational, or test results do not constitute a good reason to believe a hypothesis, even in a sense of "good reason" that does not require the hypothesis to be true.

I shall call the present concept *potential evidence*. Such evidence, although different from ES-evidence and veridical evidence, is more akin to the latter than to the former. It is potentially veridical evidence.

8. Thomson's Evidence

When scientists seek evidence for hypotheses, which concept, if any, best reflects what they are seeking? Are all these concepts in fact used? How are they related? These questions will be addressed in the present section and the two that follow.

To answer them let us invoke a case in which it seems plausible to employ all four concepts: Thomson's 1897 experimental results regarding the presence of electric charge. As noted in section 2, Thomson, unlike Hertz in 1883, was able to achieve a much higher evacuation of gas in the cathode tubes. In his major 1897 paper Thomson includes a diagram of his experimental apparatus given in fig. 2.

In the diagram, C is the cathode from which the cathode rays emanate. A is an anode with a slit in the center through which the cathode rays pass. B is a metal plug with a slit. D and E are parallel aluminum plates connected to a series of batteries. There are markings on the right end of the tube to note where the cathode rays land and produce a narrow phosphorescent patch. At the high exhaustions Thomson was able to achieve, when the aluminum plate D was connected to the negative terminal of the batteries and the plate E to the positive terminals, Thomson reported that the cathode rays were deflected toward E, as indicated by the appearance of the phosphorescent patch. When D was connected to the positive terminal and E to the negative, the cathode rays were deflected toward D. From the fact that such deflection was produced by an electrical force, Thomson concluded that cathode rays are electrically charged. And from the fact that the cathode rays were deflected away from the negatively charged plate and toward the positively charged one, Thomson concluded that they are negatively charged (since negative charges repel).

Thomson's hypothesis is

h = Cathode rays are charged with negative electricity.

Thomson's experimental results might be formulated as

e = In Thomson's experiment, as described above, cathode rays were deflected by an electric force in the direction of the positively charged plate.

The fact that e is true is evidence that h is. What sort of evidence?

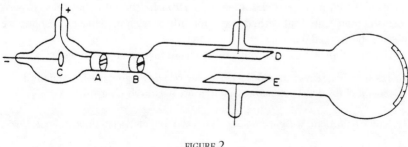

FIGURE 2

It is at least *subjective* evidence. It is, or was, *Thomson's* evidence in 1897. The three conditions for subjective evidence cited in section 5 are satisfied: in 1897 Thomson believed that *e* is evidence that *h*; he believed that *h* is true; and his reason for believing that *h* is true is that *e* is true. Indeed, Thomson took a generalization of *e*, viz. the fact that cathode rays are deflected electrically in the direction of the positive plate, together with the earlier established fact that they are deflected by a magnetic field, to be evidence that cathode rays are negatively charged particles.

Is *e* ES-evidence that *h*, relative to Thomson's epistemic situation in 1897? The justificatory features of ES-evidence noted earlier are present here. Given the results of his experiment, and all the other information available to him in 1897, Thomson was justified in believing that cathode rays are negatively charged. Even more strongly, this justification would hold for anyone with Thomson's knowledge who was aware of the results reported in *e*. Accordingly, Thomson's evidence that *h*, in the subjective sense, was also ES-evidence that *h*, relative to an epistemic situation of the sort in question.

Is *e* potential evidence that *h*? Is it veridical? The crucial test here is whether the results reported in *e* provide a good reason to believe *h*, and whether they do so in a sense that requires *h* to be true. At least this much can be said. Physicists from Thomson to the present believe that cathode rays are negatively charged. Moreover, they regard the fact that cathode rays are deflected by electric and magnetic fields as establishing that cathode rays are electrically charged, and the fact that they are electrically deflected toward the positive anode as establishing that they are negatively charged.

In 1905 Ernest Rutherford (Thomson's student) writes:

> The nature of these [cathode] rays was successfully demonstrated by J. J. Thomson in 1897. If the rays consisted of negatively electrified particles, they should be deflected in their passage through an electric as well as through a magnetic field. Such an experiment had been tried by Hertz, but with negative results. J. J. Thomson, however, found that the rays were deflected by an electric field in the direction to be expected for a negatively charged particle, and showed that the failure by Hertz to detect the same was due to the masking of the electric field by the strong ionizations produced in the gas by the cathode stream. This effect was got rid of by reducing the pressure of the gas in the tube.[16]

Rutherford (pp. 74–75) goes on to describe the types of experiments Thomson used to demonstrate electrical deflection and to determine the ratio of charge-to-mass of the cathode particles. Clearly in 1905 he thought that these experiments demonstrate that cathode rays not only carry a negative charge but are negatively charged particles.

Here is a typical passage from a more recent text on atomic physics:

> These cathode rays can be deflected by electric and magnetic fields, and the direction of the deflection shows that they are negatively charged.[17]

---

16. Ernest Rutherford, *Radio-Activity*, 2nd ed. (Cambridge: Cambridge University Press, 1905), p. 73.
17. Henry Semat, *Introduction to Atomic and Nuclear Physics*, 4th ed. (New York: Holt, Rinehart, and Winston, 1962), p. 67.

The text goes on to say that "J. J. Thomson (1897) first successfully determined the nature of cathode rays," and it uses a cathode ray diagram very similar to the one Thomson himself used.

In short, physicists believe that Thomson's experimental results provide a *conclusive* reason for believing that cathode rays are negatively charged. (They claim that these results "showed," or "established," or "demonstrated" this.) As noted earlier, conclusive evidence is one type of veridical evidence, viz. that in which not only is e a good reason to believe h, but in which e establishes h with certainty. To be sure, it is logically possible that physicists could be wrong about these things. Some new facts could conceivably be discovered that would show that cathode rays are not negatively charged and that their deflection is being caused by something else. In that case e would turn out not to be veridical evidence that h.

If Thomson's experimental results do provide a conclusive reason to believe his hypothesis h, then there can be no flaw in the experimental procedure. If there is none, then Thomson's results also provide a good reason to believe h even in a sense of "good reason" that does not require h to be true. His experimental results are potential evidence that cathode rays are negatively charged. By analogy, if these spots are symptoms of the measles virus, in a sense of "symptom" that requires the presence of the measles virus, then they are also symptoms in a sense that does not require the presence of the measles virus; they are the sort of thing generally associated with the measles virus, even if they were not in this case.

9. How Are the Four Concepts of Evidence Related?

My answer is given in the following claims.

1. If e is veridical evidence that h, then e is also potential evidence that h, but not necessarily the reverse.
2. If e is veridical evidence that h, then e may or may not be ES-evidence that h with respect to some particular epistemic situation; but if e is veridical evidence that h, then there will be some epistemic situation with respect to which e is ES-evidence.
3. If e is veridical, or potential, or ES-evidence that h, then e may or may not be any person X's (subjective) evidence that h. And if e is X's (subjective) evidence that h, then e may or may not be potential, or veridical, or ES-evidence that h.
4. If e is person X's (subjective) evidence that h, then X believes that e is (probably) veridical evidence that h, and not simply potential, or ES-evidence, that h.
5. If e is ES-evidence that h, relative to an epistemic situation ES, then anyone in ES is justified in believing that e is (probably) veridical evidence that h.

Claim 1 should be obvious. The first part says that whatever is veridical evidence is also potential evidence. This is based on the idea, noted at the end of section 8, that if e provides a good reason to believe h in a sense that requires h's truth, then it also provides a good reason in a sense that does not require h's truth. As for the second part of claim 1, since potential evidence does not require the truth of the hypothesis, but veridical evidence does, something could be potential evidence for a hypothesis without being veridical evidence for it.

The first part of claim 2 should also be obvious. Veridical evidence need not be ES-evidence for some specific ES. Thomson's 1897 experimental results are veridical evidence that cathode rays are negatively charged, but they are not ES-evidence for someone whose epistemic situation does not include the rudiments of physics. The second part of claim 2 says that veridical evidence is ES-evidence for some epistemic situation or other. If this claim were false, then there would be some e that provides a good reason to believe some true h, without providing any justification for believing h for anyone in any type of epistemic situation, which seems absurd.

Turning to claim 3, if e is veridical, or potential, or ES-evidence that h, then e may or may not be any person X's evidence that h. And if e is X's evidence that h, e may or may not be veridical, or potential, or ES-evidence that h. The first part of this claim holds in virtue of the objectivity of veridical, potential, and ES-evidence. The fact that e is the case can be evidence, of any of these types, that h is true even if no one knows or believes anything about e, h, or their relationship. (This holds for ES-evidence, since no one may be in the epistemic situation ES). Conversely, if e is X's evidence that h, nothing follows about whether e is veridical, potential, or ES-evidence that h. My evidence that I will eventually win at poker tonight (h) may be that I have been losing so far (e). This is my evidence in virtue of the fact that I believe that e is evidence that h, that I believe that h is true, and that my reason for believing that h is true is that e is true. But I have committed the gambler's fallacy. The fact that e is true does not provide a good reason for believing the hypothesis in question (in either sense of "good reason"), or any justification for doing so, even for someone in my epistemic situation.

Claim 4 may be regarded as the most interesting because of its assertion that if e is someone's evidence that h, then that person believes that e is (probably) veridical, and not simply potential or ES-evidence that h, and hence that h is (probably) true.[18] Thomson's (subjective) evidence that cathode rays are negatively charged consisted in the fact that in his experiments these rays were deflected by an electrostatic force. Thomson clearly believed his hypothesis to be true ("I can see no escape from the conclusion that they are charges of negative electricity carried by particles of matter.") He explicitly stated that his reason for believing this is that he obtained the experimental results he did. And since Thomson believed that his experiments demonstrated this conclusion, he regarded his results as providing a good reason to believe his hypothesis in a sense requiring its truth.

Could something be X's evidence for a hypothesis if X believes it is ES- or potential evidence but not veridical evidence for that hypothesis? Recall that in his discussion Thomson admits that on repeating Hertz's experiments he at first got the same results as Hertz did, viz. no electrical deflection. It seems very plausible

---

18. Claims 4 and 5 involve a so-called *de re* sense of belief, according to which if X believes that a has property P, and if $a = b$, then X believes that b has P.

to suppose that Thomson believed that these negative results constituted ES-evidence, for a Hertzian ES, that cathode rays are not electrically charged: Anyone, such as Hertz, who obtained these results and was in the sort of epistemic situation Hertz was, would have been justified in believing the "electrically neutral" hypothesis. However, Thomson does not claim, nor would we, that his initial results on repeating Hertz's experiment constituted *his* (Thomson's) evidence that cathode rays are electrically neutral. Thomson appears never to have believed this hypothesis, even before his later experiments. And even if he had, after the results of these later experiments we would not say that in 1897 Thomson's evidence that cathode rays are electrically neutral is that prior to this he, like Hertz, obtained no electrical deflection. Yet it seems reasonable to suppose that even in 1897 after his new experiments Thomson believed (or could have believed) that his initial results were ES-evidence for the "electrically neutral" hypothesis (with respect to a Hertzian epistemic situation): that if such initial results together with information of the sort Hertz had possessed were what was known, and no undermining facts such as Thomson's later results were known, a belief in the neutrality hypothesis would be justified.

To be *Thomson's* evidence in 1897 for the neutrality hypothesis, however, Thomson in 1897 would have had to believe much more than this. He would have to have believed that Hertz's 1883 results and his own preliminary ones provide a good reason to believe that cathode rays are neutral. But in 1897, at least after his own experiments, Thomson clearly believed just the opposite. He believed that Hertz's results were based on flawed experiments, and hence that they did not in fact provide a good reason for believing the neutrality hypothesis (in either sense of "good reason"). In short, in 1897 he did not believe that Hertz's results constituted potential or veridical evidence for this hypothesis.

According to claim 4, if e is X's evidence that h, then X believes that e is veridical evidence that h, or at least X believes that this is probably so. In Thomson's case the former condition is satisfied. Thomson believed that his experiments demonstrated the truth of his electrical hypothesis. In other cases what may be believed is only a probability. For example, let

$e = $ John owned 850 of the 1000 lottery tickets sold in this fair lottery, one of which was drawn at random.
$h = $ John won.

The fact that e is true may be *my* evidence that h is true. If it is, then I believe that e is probably veridical evidence that h, and hence that h is probably true. The numbers here are not sufficiently compelling for me simply to believe that h is true.

Suppose that someone believes that e is potential evidence that h but does not believe that e is veridical evidence. Could e be that person's evidence that h? (Since Thomson did not believe the former about Hertz's evidence we need a different example.) Suppose that on Monday the doctor believes that the fact that Sam has those spots is both potential and veridical evidence that he has the measles virus. On Friday, after tests, he believes that Sam does not have the measles virus but some other rare virus which produces identical spots in 1% of the cases. On Friday, let us suppose, the doctor still believes that the fact that Sam has those spots is potential, but not veridical, evidence of the measles virus. He believes that those spots are of the sort almost always caused by the measles virus, but, as it happens, not in this case. We can say that on Monday the fact that Sam has those spots was the doctor's evidence that Sam has the measles virus. On Friday, however, it would not, I think, be correct to speak of this as the doctor's evidence for this hypothesis, even though on Friday he still believes it is potential evidence. If it is to be the doctor's evidence, he needs to believe not only that it is potential evidence but that it is veridical as well.

When subjective evidence was introduced in section 5, one of the conditions noted for e's being X's evidence that h was that X believes that e is evidence that h. What concept of evidence is involved in such a belief? If e is X's evidence that h, then X believes at least that e is potential evidence that h and hence that e is ES-evidence for some epistemic situation. But if claim 4 is right then X must believe something even stronger, viz. that e is (probably) veridical evidence that h. Even though Thomson's early experiments showed no electrical deflection, the results of these experiments were not Thomson's evidence that cathode rays are neutral. To be so Thomson would have had to believe they were veridical evidence for this hypothesis, which he never did.

Finally, claim 5 seems obvious. Hertz's experimental results were ES-evidence, relative to Hertz's epistemic situation, that cathode rays are neutral. Hertz was justified in believing what he did on the basis of his experiments, given what he knew. If he was, then he was justified in believing his hypothesis to be true (or probable), and he was justified in believing that his experimental results probably provided a good reason (in the strong sense) for believing his hypothesis.

10. What Type of Evidence Is Most Important for Scientists?

Scientists seek veridical evidence, not just potential or ES-evidence. They want their hypotheses to be true.[19] And they want to provide a good reason for believing them in a sense of "good reason" that requires truth. They are not satisfied with providing only a justification of belief for those in certain epistemic situations, even their own, since the belief might turn out to be false. Nor are they satisfied with providing a good reason for belief in the weaker sense. Thomson could have provided a justification of belief in the electrical neutrality hypothesis simply by citing the results of Hertz's experiments and his own preliminary results. That would have sufficed to justify a belief in the neutrality hypothesis

---

19. As indicated in footnote 6, I am adopting a realist stance. Antirealists would say that scientists want their hypotheses to save the phenomena or to be instrumentally correct. Accordingly, the evidence sentences of interest to them are ones of the form "e is evidence that h saves the phenomena." In chapter 8 I will discuss the question of whether it is possible to understand sentences of this form without making realist assumptions.

on the part of those physicists (up to 1897) in roughly Hertz's epistemic situation. Thomson wanted to do something more powerful, viz. to provide a good reason to believe that cathode rays are negatively charged—a reason not tied to any actual or hypothetical epistemic situation. To be sure, not everyone would be capable of realizing that it is a good reason, or would be justified in believing it for that reason. So what Thomson may need to do, and what a good physics text does, is to set the stage by including sufficient information so that others can be in an appropriate epistemic situation, so that others can become justified in believing the hypothesis for that reason. But whether or not this is done, Thomson is claiming that the results of his experiments do in fact provide a good reason for believing his conclusion.

Moreover, Thomson claimed that such reasoning was conclusive, and hence that he was providing veridical evidence. His aim was not simply to provide potential evidence. It was rather to provide potential evidence *that is also veridical*. Again, recall the measles case. The test on Friday shows that the measles virus is not causing the spots. Still the spots are potential evidence of the virus. Yet on Friday the doctor no longer cites the spots as his evidence that the patient has measles. Its being potential evidence is not good enough. He seeks veridical evidence.

If veridical and not simply ES- or potential evidence is what scientists desire, is it attainable? One might think this is too high to aim, since veridical evidence requires the truth of the hypothesis, and not simply probability. Let us be clear regarding what is and is not required to obtain veridical evidence. It is not required that one obtain conclusive evidence: it is not required that the evidence prove the hypothesis (that the probability of the hypothesis on the evidence be 1). Nor is it even required that one knows that one has obtained veridical evidence. To show that veridical evidence is not attainable one would need to show either that truth, or that good reasons for belief, or both, are not attainable. Even if one is a philosophical sceptic who denies that knowledge of the truth of any hypothesis is possible, that would not commit one to the view that veridical evidence is unattainable, but only to the view that it is impossible to know that one has it. Scepticism does not preclude veridical evidence. Scientists, who tend to reject scepticism, certainly believe that they have obtained veridical evidence for many of their hypotheses. These scientists believe that such evidence is both desirable and attainable.

11. Are All Four Concepts Employed? When?

My answer to the first question is yes, at least in this sense. Statements of the following types, each of which corresponds to one of the concepts distinguished, make good sense:

(i) Anyone who knows or justifiably believes such and such would be justified in believing that e is evidence that h. For such a person, e should count as evidence that h.
(ii) e is X's evidence that h; e is what X proposes as, takes to be, evidence that h.

(iii) Potentially, *e* is evidence that *h*; *e* is the sort of fact that is usually evidence that a hypothesis such as *h* is true.
(iv) *h* is true, since *e* is genuine evidence that it is.

More important, however, is the issue of whether sentences of the form "*e* is evidence that *h*" are ever in fact used to express the ideas conveyed by (i)–(iv). Let's return to Hertz. In section 3 the question was raised concerning whether the results of Hertz's experiments were evidence that cathode rays are not charged. Three responses were proposed, each of which seems reasonable. If, as suggested by response 1, we claim that the results of Hertz's experiments are (strong) evidence that cathode rays are not charged, and if we defend that claim the way we did, then we are understanding a sentence of the form "*e* is evidence that *h*" in the manner of (i) above (ES-evidence). In this case, we are claiming that anyone in Hertz's epistemic situation would be justified in believing what Hertz did. Suppose, as suggested by response 2, we want to claim that from 1883 to 1897 the results of Hertz's experiments were (strong) evidence that cathode rays are not charged, but after 1897 they were not so; and suppose we defend this in the way we did. Then we are understanding the evidence claim in question in the manner of (ii)—as subjective evidence. During these years, but not later, the results of Hertz's experiments were what Hertz and other physicists took to be evidence.

What about response 3, that the results of Hertz's experiments are not and never were evidence that cathode rays are not charged? Depending on the defense, this response can be understood by reference to (iii) or (iv)—as a denial of either or both. Since Hertz's experiments, which yielded the results they did, were flawed, those results do not provide a good reason to believe Hertz's hypothesis, even in the weaker sense of "good reason." Despite what they seemed from 1883 to 1897, they are not potential evidence for Hertz's hypothesis. This fact, together with the fact that Hertz's hypothesis is false, makes it also false to claim that the results provide veridical evidence for Hertz's hypothesis. Response 3, then, does not allow us to choose between a veridical and a potential sense of evidence, since both are applicable.

We need a case where there is no experimental flaw and where veridical and potential evidence yield different responses. The previous measles example is such a case. If Sam has the spots almost always caused by the measles virus but his spots were caused by something else, then shall we say that Sam's spots are, or are not, evidence that the measles virus is present? Either answer seems reasonable. If so, then the concepts of veridical and potential evidence are both employed.

Under what conditions is each concept employed? That is, why are four concepts needed? At the simplest level, there are occasions when all one wants to know are the reasons some person has or had for believing something, whether or not these are good reasons, whether or not the person was justified in believing a hypothesis for those reasons. This is so in historical contexts when an investigator simply wants to learn what some scientist of the past proposed as evidence. Such a concept is also employed by scientists concerned with a hypothesis *h* who want to know what evidence some other scientist has proposed for *h*. In such situations one may ask for the scientist's evidence without implying or presupposing that it is ES-, potential, or veridical evidence. When one speaks of evidence in these cases no evaluation of that evidence is being made.

The other three concepts involve evaluations of different types. With ES-evidence one's concern is with justification of belief for someone in a certain epistemic situation. The interest is usually at least partly historical, as in the case of Hertz. A historian may want to know not only Hertz's reasons for believing cathode rays neutral, but also, given what Hertz knew, whether he was justified in his belief. Recall the previous quotation from the historian Jed Buchwald, who seeks not only to give Hertz's reasons, but to evaluate them from this perspective as well.

Whether e is (was) ES-evidence for a particular scientist can be a difficult historical question to answer, since one needs to know what epistemic situation the scientist was actually in with respect to the hypothesis, as well as whether one in that epistemic situation is justified in believing the hypothesis. For this reason scientists themselves, in contrast to historians, do not have very much interest in this question. Scientists do not really care very much about what Hertz knew in 1883 or about whether anyone who knew what he did would be justified in believing in the neutrality of cathode rays. Even though Thomson in 1897 cites Hertz's experiments of 1883, Thomson's concern was not with the question of whether Hertz, given what he knew, was justified in believing the neutrality hypothesis. Thomson's concern was with the question of whether Hertz's experiments really do provide a good reason to believe that hypothesis.

The usual interest of scientists is not historical, nor is it restricted to particular or types of epistemic situation, not even their own. To be sure, when a scientist claims that e is evidence that h, he believes and hopes that, given his knowledge, he is justified in believing h on the basis of e. But he believes and hopes for something much more. That is why for scientists potential and (particularly) veridical evidence are crucial.

A scientist wants to know whether some experimental results reported in e provide a good reason for believing a hypothesis h—not a good reason for someone in some particular epistemic situation, and not just a good reason for him, but a good reason period, independent of epistemic situations. And he wants to know whether e is a good reason in the strong sense. His goal is to obtain veridical evidence, since he seeks true hypotheses. He is not satisifed with potential evidence that he knows or believes is not veridical. And whether it is veridical evidence has nothing to do with what he or anyone else knows or believes. It is not veridical evidence for one type of epistemic situation but not another.

What role remains for potential evidence? The main one is to serve as a vehicle for obtaining veridical evidence and for demonstrating that some experimental result is not veridical evidence. A necessary condition for e's being veridical evidence is that it is potential evidence. A necessary condition for e's being a good reason in the strong sense is that it is a good reason in the weak sense. Accordingly, in searching for veridical evidence one must search for something that is at least potential evidence. One can refute someone's claim that e is veridical evidence by showing that it is not even potential evidence. This is one of the things Thomson did with respect to Hertz's claim. Thomson showed that Hertz's (subjective) evidence that cathode rays are neutral is not even potential evidence, since there was a serious flaw in the design of the experiment.

12. The Empirical Character of Evidence Statements

My final question is this: Are evidence statements of the form

(a) e is evidence that h,

as these are understood in terms of the four concepts of evidence, empirical statements?[20]

My initial answer is very simple: All evidence statements of all four types are empirical. Their truth is not establishable by a priori calculation. The reason can be stated briefly. Potential, veridical, and ES-evidence all require e to be true. Whether e is true (for instance, whether no electrical deflection was produced in Hertz's experiments) is an empirical question, not settleable a priori. Subjective evidence does not require e to be true. But it does require that the person or community whose evidence it is hold certain beliefs about e and h (including the belief that e is evidence that h, and the belief that h is true or probable). Whether such beliefs are in fact held is an empirical matter, not an a priori one.

A priorists about evidence will not be satisfied with this simple answer. They may agree that when an evidence statement of form (a) is uttered usually the truth of e is presupposed. So, they may claim, their concern is with evidence statements of this sort:

(b) e, if true, is evidence that h,

which do not presuppose e's truth. My claim is that even evidence statements of type (b) are, generally speaking, empirical, if construed in terms of our four concepts. Let me say why this is so in each case, starting with the most obvious.

Potential and veridical evidence. Evidence of these sorts, we said, is potentially empirically incomplete: whether e, if true, is potential or veridical evidence that h can depend on empirical facts in addition to e and h. Again, let

e = In Hertz's experiments, which were designed to show electrical effects of cathode rays, no such effects were produced.
h = Cathode rays are not electrically charged.

Assuming that Hertz's experiments were performed, whether e, if true, is potential or veridical evidence that h depends on whether Hertz's cathode tubes were sufficiently evacuated of gas to allow electrical deflection to occur. Whether they were is an empirical not an a priori matter.

This is not to say that all potential or veridical evidence claims of type (b) are empirical. In some cases it is possible to formulate e in such a way that it is empirically complete, that is, so that whether e (if true) is potential evidence that h

---

20. This question will become particularly important in the next chapter, when we discuss various theories of evidence in the literature, some of which maintain that evidence statements are never empirical but establishable solely by logical or mathematical calculations.

does not depend on any empirical facts (other than *e*), and whether *e* (if true) is veridical evidence that *h* (if true) does not depend on any empirical facts (other than *h* and *e*).[21]

Subjective evidence. To construct a subjective evidence statement analogous to (*b*), we might write:

(c) If X believed *e* to be true, *e* would be X's evidence that *h*.

But the truth of (*c*), like that of the subjective version of (*a*), depends on empirical facts about X, including whether if X believed *e* to be true X would believe that *e* is evidence that *h*, and X would believe that *h* is true or probable.

ES-evidence. ES-evidence statements are to be understood as relativized to an epistemic situation. Expanding (*b*) to reflect this we can write:

(d) *e*, if true, is evidence that *h* for anyone in epistemic situation ES.

My claim here is the same as in the case of potential and veridical evidence. Some statements of these types are empirical, some are a priori. It all depends on how ES is described. Here is an empirical example. As we did previously, suppose we let ES be described in this way: Hertz's epistemic situation in 1883. Then, where *e* and *h* are given as above, the following evidence claim of form (*d*) is empirical, not a priori:

(e) *e*, if true, is evidence that *h* for anyone in Hertz's epistemic situation in 1883.

This claim is empirical, since its truth can be determined only by empirically determining what Hertz's epistemic situation was in 1883.

Finally, an a priorist about evidence may agree with me that evidential statements as customarily formulated are frequently empirical. But he may say that they can and should all be rendered a priori by packing enough information into them to make them empirically complete. My response is that even if it could be done, why do so? No doubt rendering a claim a priori is one way of establishing its truth (or falsity). But it is not the only way or necessarily the best. Thomson discovered the falsity of Hertz's (potential) evidence claim not by turning it into a (false) a priori claim, but by conducting the experiments he did.

13. Appendix: Evidential Flaws

In this chapter I have spoken of flaws in evidence gathering that affect whether the result obtained counts as evidence for a hypothesis. In this appendix I will make this idea more precise.

---

21. A previous lottery example will illustrate how. Let

> *e* = John owned 850 of the 1000 lottery tickets sold in this fair lottery, one of which was drawn at random.
> *h* = John won.
>
> Whether in this case *e* (if true) is potential or veridical evidence that *h* does not depend on any empirical facts (other than *e* and *h*).

A scientist may conduct some test, or experiment, or set of observations, that results in some fact e being true. For example, Hertz performed an experiment that resulted in the fact that no electrical deflection of cathode rays was produced. By contrast, there are things scientists and others do to establish that e is the case but do not result in e's being the case. (For example, I might establish that Hertz's experimental result e obtained by reading Hertz's paper; my activity, in contrast to Hertz's, did not result in e's being true.) In what follows I am concerned only with the former sort of case.

When a scientist conducts some test, or experiment, or observations he may be employing what I will call a *selection procedure*. This is simply a rule or set of instructions he is following. For example, in testing the efficacy of a new drug D designed to relieve headaches, the investigator may be employing a selection procedure (SP) such as

SP_1: Test drug D by administering it to 1000 headache sufferers of both sexes, various ages, in varying dosages, who have headaches of varying degrees of severity. Administer a placebo, not drug D, to members of a control group otherwise like the first group.

Another possible selection procedure is this

SP_2: Test drug D by administering it to 1000 headache sufferers, all of whom are young girls, age 5, who have mild symptoms. Don't employ a control group.

Consider the hypothesis

h: Drug D is 95% effective in relieving headaches.

Suppose there are two investigators, one following SP_1, the other SP_2. The former produces a study with the following result:

e_1: Of the 1000 headache sufferers tested with drug D in test 1, 950 had relief of their headache.

The tester following SP_2 obtains the following result:

e_2: Of the 1000 headache sufferers tested with drug D in test 2, 950 had relief of their headache.

SP_2 is a flawed selection procedure for e_2 with respect to hypothesis h. By contrast, SP_1 is not a flawed selection procedure for e_1 with respect to h. SP_2, by contrast with SP_1, restricts those tested to 5-year-old girls with mild symptoms, and has no control group to see what would happen without administering drug D.

Now let us consider probabilities.[22] Suppose that a selection procedure SP used to generate result e is completely flawed. If so this should cancel any effect e has on the probability of h. That is,

22. In chapters 3 and 5 various interpretations of probability will be introduced. One that will be employed for the definitions of potential and veridical evidence is what I call objective epistemic probability, a concept developed in chapter 5. It is an objective measure of how reasonable it is to believe something. Probabilities invoked in this appendix can be understood in this sense.

If SP, which is used to generate e, is completely flawed, then $p(h/e$ & SP is used$) = p(h)$.

Flaw is a matter of degree. So we can say that the more that SP is flawed with respect to e and h the more this fact diminishes the effect e has on the probability of h. That is,

If SP, which is used to generate e, is flawed to some extent, then the greater the flaw the closer is $p(h/e$ & SP is used$)$ to $p(h)$.

Suppose there is no information concerning any selection procedure used to generate e; in particular, suppose there is no information concerning whether, or to what extent, such a selection procedure is flawed. If it is completely flawed, then $p(h/e) = p(h)$. If there is no flaw whatever in the selection procedure, then $p(h/e) = p(h/e$ & the SP used is unflawed$)$. So, if there is no information concerning whether, or to what extent, the selection procedure used to generate e is flawed, then $p(h/e)$ will be between $p(h)$ and $p(h/e$ & the SP used is unflawed$)$.

Accordingly, there are two possible cases.

Case 1: $p(h/e$ & the SP used is unflawed$) \geq p(h)$. Since $p(h/e)$ lies between these two values, $p(h/e$ & the SP used is unflawed$) \geq p(h/e)$.

Case 2: $p(h/e$ & the SP used is unflawed$) \leq p(h)$. Since $p(h/e)$ lies between these two values, $p(h/e$ & the SP used is unflawed$) \leq p(h/e)$.

Therefore, if an unflawed selection procedure used to generate e does not decrease, and possibly increases, h's probability over its prior probability, then it does not decrease, and possibly increases, h's probability over its probability on e alone. And if an unflawed selection procedure used to generate e does not increase, and possibly decreases, h's probability over its prior probability, then it does not increase, and possibly decreases, h's probability over its probability on e alone.

Here is how this works for the drug case above. The fact that the flawed selection procedure SP_2 was used to generate e_2 cancels any effect that e_2 has on h's probability, so that

$p(h/e_2$ & SP_2 is used$) = p(h)$.

Now consider the unflawed SP_1 used to generate e_1. In this case the result e_1 together with the fact that the unflawed SP_1 was used to generate this result increases h's probability over its prior probability. That is,

$p(h/e_1$ & SP_1 was used$) > p(h)$.

Since $p(h/e_1)$ lies between these values,

$p(h/e_1$ & SP_1 is used$) \geq p(h/e_1)$.

Therefore, the fact that the unflawed SP_1 is used to generate result e_1 does not decrease, and possibly increases, h's probability over its probability given result e_1 alone. This fits case 1 above.

Let us consider a result that fits case 2. Suppose that there is a third test whose selection procedure is SP_1. Suppose this test yields the following result:

e_3: Of the 1000 headache sufferers tested with drug D in test 3, only 10 had relief of their headache.

In this case, let us suppose, e_3, together with the fact that the unflawed SP_1 was used to generate e_3, lowers h's probability over its prior probability. That is,

$p(h/e_3 \,\&\, SP_1 \text{ is used}) < p(h)$.

Since $p(h/e_3)$ lies between these values,

$p(h/e_3 \,\&\, SP_1 \text{ is used}) \leq p(h/e_3)$.

So, in this case, by contrast to the previous one, the fact that an unflawed selection procedure is used to generate a result does not increase, and possibly decreases, h's probability on that result alone.

A thesis propounded in the present chapter is that if e is flawed evidence—that is, if some selection procedure that results in e's being true is flawed with respect to hypothesis h—then e is not a good reason for believing h. Can this thesis be related to probabilities? Two proposals for relating "good reasons for belief" to probability are prominent in the literature. (These will be critically explored in chapters 3 and 4.) The first is this.

A. *Increase in Probability*: Fact f is a good reason to believe hypothesis h only if f increases h's probability, that is, only if $p(h/f) > p(h)$.

The second is this.

B. *High Probability*: Fact f is a good reason to believe hypothesis h only if h has high probability on f (say, greater than $\frac{1}{2}$), that is, only if $p(h/f) > \frac{1}{2}$.

Now suppose that some selection procedure used to generate e is completely flawed with respect to h. Then, we said,

$p(h/e \,\&\, SP \text{ is used}) = p(h)$.

So by the increase-in-probability idea A, the fact that result e was obtained using the flawed SP is not a good reason for believing h. Suppose, further, that the probability of h independently of result e is low, say less than $\frac{1}{2}$. Then, since SP is completely flawed,

$p(h/e \,\&\, SP \text{ is used}) = p(h) < \frac{1}{2}$.

So, by the high-probability idea B, the fact that result e was obtained using the flawed SP is (again) not a good reason for believing h.

These ideas can be stated in a way that allows us to speak of differences in the degree of flaw. On the basis of A, one might say that the more (less) that f increases h's probability the stronger (weaker) is f as a reason to believe h. Similarly, on the basis of B, one might say that the higher (lower) the probability of h on f the stronger (weaker) is f as a reason to believe h. Now, we said that the greater the flaw in SP the closer is $p(h/e \,\&\, SP \text{ is used})$ to $p(h)$. So, on the basis of A, the greater the flaw in SP the less good is the fact that e was obtained using SP as a reason for believing h. And if $p(h)$ is less than $\frac{1}{2}$, the same can be said on the basis of B.

Returning, finally, to experiments with cathode rays, Hertz's experimental results, by contrast with Thomson's, were flawed. Hertz's selection procedure (call it *SPH*) that yielded his experimental results e (no electrical deflection) was flawed with respect to the hypothesis h that cathode rays are electrically neutral. So,

$p(h/e\&\ SPH$ is used) is close to $p(h)$.

Using idea *A*, depending on how close these probabilities are, we can say that the fact that Hertz obtained the results he did, given his selection procedure, is not a (particularly) good reason for believing Hertz's hypothesis. We can say the same thing, using *B*, if we can suppose that the probability of Hertz's hypothesis independently of his results is not high, say, not greater than $\frac{1}{2}$.

3

Two Major Probabilistic Theories of Evidence

In the previous chapter I distinguished four concepts of evidence, each of which was given a preliminary characterization. Now we need to dig deeper. Is one of these four concepts more basic than the others, allowing each of the others to be defined in terms of it? Can we define this basic concept in a way that will be more precise and illuminating than the characterization offered in chapter 2?

One of the main aims of this book is to provide an affirmative answer to each of these questions. The basic concept of evidence I will define is potential evidence; the others will then be defined by reference to it. This does not make potential evidence the type that scientists seek primarily. As I claimed in chapter 2 that honor is reserved for veridical evidence. It is only more basic from a definitional perspective.

The concept of potential evidence that I will advocate is defined, in part, by reference to probability. In order to arrive at this definition I will need to critically examine two of the most prominent and widely accepted probabilistic definitions of evidence in the philosophical literature: (i) evidence as information that increases the probability of a hypothesis, and (ii) evidence as information on the basis of which a hypothesis has high probability. In order to motivate the theory I advocate we need to see why the first of these definitions provides a condition that is neither necessary nor sufficient for evidence and why the second of these definitions provides a condition that is necessary but is not sufficient for evidence. If the second definition does give a necessary condition we will need to indicate something not offered by advocates of this definition, viz. an answer to the question of how high is "high probability"?

However, before delving into the question of where and why these theories go wrong—which I will do in chapter 4—an important task awaits us. Probabilistic definitions of evidence are incomplete until the concept of probability that is being employed is itself interpreted. This will be done in the present chapter, where four standard interpretations of probability will be presented: logical, frequency, propensity, and subjective. The two major probabilistic definitions of evidence to be examined in this chapter and the next can then be understood more completely by reference to each of these interpretations. Those who advocate one or the other of the two probabilistic definitions of evidence subscribe to one or the other of these interpretations of probability. In chapter 5 an alternative interpretation of probability is developed that will be used for my positive account of potential and veridical evidence in chapters 6 through 8.

1. Two Probabilistic Definitions of Evidence

Rudolf Carnap, a leading proponent of this approach, distinguishes three different sorts of sentences involving evidence (or "confirmation," as he generally calls it).[1] There are sentences of the form

(1) e is evidence that h (or, e confirms h),

which Carnap calls *qualitative*. Second, there are sentences with forms such as

(2) e is stronger evidence for h than for h'
(3) e is stronger evidence for h than is e',

which Carnap calls *comparative*. Finally, there are sentences of the form

(4) the degree of support or confirmation that e confers upon h is r (where r is a number),

which Carnap calls *quantitative*.

In this book I am principally concerned with "qualitative" evidence statements of form (1). Carnap's view is that qualitative evidence statements of form (1) are to be understood by invoking quantitative ones of form (4), and that the latter are to be understood in terms of probability. For Carnap, a sentence of form (4) means the same as this:

(5) The probability of h, given e, is r.

Carnap sees his main task, then, as one of explicating the concept of probability needed for (5). (His explication will be given in section 3.)

With a suitable concept of probability in hand, Carnap and many others proceed to define both "qualitative" and "comparative" evidence. Carnap offers two different probabilistic definitions of the former. The first, by far the most popular with probabilists, states that e is evidence that h if and only if e increases h's probability:

(6) *First Probability Definition of Evidence (increase in probability):* e is evidence that h if and only if $p(h/e) > p(h)$,

where $p(h/e)$ means the probability of h, given e (or on the assumption of e). This

---

1. Rudolf Carnap, *Logical Foundations of Probability*, 2nd ed. (Chicago: University of Chicago Press, 1962).

is called the "positive relevance" definition, where e is relevant to h if e changes h's probability, and it is positively relevant if it increases it.

Without settling yet on any interpretation of probability, consider an example—Thomson's experimental results described in the previous chapter. We let

> e = In Thomson's electrical experiment with cathode rays (described in chapter 2) the rays were deflected by an electric force in the direction of the positively charged plate.
> h = Cathode rays are charged with negative electricity.

Thomson and the physics community generally took the fact that cathode rays are deflected by an electrical force toward the positively charged plate to increase (indeed vastly so) the probability that cathode rays are negatively charged, thus satisfying the condition for evidence in (6).

Probability (and evidential) claims are usually made in the light of additional information—so-called background assumptions. For example, in the Thomson case, the reason that physicists took deflection toward the *positive* plate as indicating that cathode rays are *negatively* charged is that they assumed that opposite charges attract. This is part of their background information. Probabilists may then relativize evidence statements to some set b of background assumptions and say that

> (7) e is evidence that h, given b, if and only if $p(h/e\&b) > p(h/b)$.

If in 1897 Thomson is asserting a sentence of the form "e is evidence that h, given b," b will include those assumptions (in addition to the one noted above) that Thomson made in 1897. On the probabilist version (7), when Thomson claims that his experimental results constitute evidence that cathode rays are negatively charged, he is right to do so if those results increase the probability of his hypothesis over what it is on the background assumptions alone.

If we employ (7) and substitute *not-h* ($-h$) for h, we obtain

e is evidence that $-h$ if and only if $p(-h/e\&b) > p(-h/b)$.

But $p(-h/e\&b) > p(-h/b)$ if and only if $p(h/e\&b) < p(h/b)$. So on this conception of evidence, e is evidence that $-h$, that is, that h is false (or we might also say, e is evidence *against* h) if and only if e decreases h's probability.

Although (6) (or the background-relativized (7)) represents the most widely held probabilistic definition of evidence, Carnap offers a second definition. It states that e is evidence that h if and only if h's probability on e is high. We can write this as follows:

> (8)   *Second Probability Definition of Evidence (high probability)*: e is evidence that h if and only if $p(h/e) > k$, where k represents some threshold of high probability.

Carnap calls k a "fixed number" but does not assign any specific value to it.[2] He seems to have in mind a single number fixed for all contexts of evidence assessment rather than a number that can vary from one context of evaluation to an-

---

2. Carnap, p. xvi.

other. For the sake of simplicity, let us choose the number $\frac{1}{2}$, and let us suppose that this is constant for all contexts of inquiry. (My critical discussion of (8) will not be affected by this choice. In chapter 6, when I defend the claim that definition (8) provides a necessary but not a sufficient condition for evidence, I will argue that $\frac{1}{2}$ is indeed a reasonable choice.) In accordance with this second probability definition, we will say that e is evidence that h if and only if the probability of h on e is greater than $\frac{1}{2}$, or what comes to the same thing, if and only if h is more probable than its negation on e.

The Thomson case appears to satisfy this definition as well. Thomson and other physicists certainly believed that Thomson's hypothesis h is very probable given Thomson's experimental results e, much more probable than is the claim that cathode rays are not negatively charged.

If we employ (8) and substitute $-h$ for h, we obtain

e is evidence that $-h$ if and only if $p(-h/e) > k$.

Choosing $k = \frac{1}{2}$, $p(-h/e) > \frac{1}{2}$ if and only if $p(h/e) < \frac{1}{2}$. So on this conception of evidence, e is evidence that $-h$ if and only if the probability of h on e is less than $\frac{1}{2}$.

Which of the two probability definitions of evidence should we use? As noted, probabilists generally prefer the first. Carnap, however, thinks that both concepts of evidence are in fact used in science. The second, Carnap claims, expresses a sense of evidence associated with the idea of something that makes a hypothesis "firm"; the first with something that makes it "firmer." (These ideas of "firmness" will be examined in chapter 4.)

Before concluding this section, let me briefly address the question of how probabilists understand comparative evidence statements such as

(2) e is stronger evidence for h than for h'
(3) e is stronger evidence for h than is e'.

The simplest probabilistic definitions for these are based on the positive relevance concept (6). If, to be evidence for h, e must increase h's probability, then for e to be stronger evidence for h than for h', e must increase the probability of h more than it increases the probability of h'. That is,

(9) e is stronger evidence for h than for h' if and only if $p(h/e) - p(h) > p(h'/e) - p(h')$.

And for e to be stronger evidence for h than is e', e must increase h's probability more than e' does, that is,

(10) e is stronger evidence for h than is e' if and only if $p(h/e) - p(h) > p(h/e') - p(h)$; that is, if and only if $p(h/e) > p(h/e')$.

As Carnap notes, however, comparative definitions are also possible based on analogies with the second definition of (qualitative) evidence (8). If to be evidence for h, e must simply make h's probability "high," then for e to be stronger evidence for h than for h' what is required is that e make the probability of h higher than it makes the probability of h', that is,

(11) e is stronger evidence for h than for h' if and only if $p(h/e) > p(h'/e)$,

and

(12) e is stronger evidence for h than is e' if and only if $p(h/e) > p(h/e')$.[3]

More complex probabilistic definitions of "stronger evidence" have been offered, but I will not pursue this topic here.[4]

2. The Probability Calculus and Types of Interpretation

Our two probabilistic definitions of evidence remain incomplete until we say how the concept of probability is to be interpreted. Those who offer these interpretations insist that the concept of probability employed must obey the formal rules of the probability calculus, which include these:

Rules of Probability: For any sentences h and e, where e is not a logical contradiction,

1. $0 \leq p(h/e) \leq 1$
2. If h and h' are logically equivalent, and if e and e' are logically equivalent, then $p(h/e) = p(h'/e')$
3. $p(-h/e) = 1 - p(h/e)$
4. $p(h \text{ or } h'/e) = p(h/e) + p(h'/e) - p(h\&h'/e)$
5. $p(h\&h'/e) = p(h/e) \times p(h'/h\&e)$.

Rule 1 states that probabilities assume values between 0 and 1. Rule 2 requires that logically equivalent sentences have the same probability on logically equivalent assumptions. Rule 3 (the law of negation) states that the probability of the negation of h equals 1 minus the probability of h. Rules 4 and 5 impose conditions on the probabilities of disjunctions and conjunctions of sentences. The probability calculus can be axiomatized so that some of the rules above (or others) are treated as fundamental and the rest derived mathematically.

Sentences of the form

(13) $p(h/e) = r$

are to be understood so as to satisfy these rules. But various interpretations of probability can be shown to do this, and if we seek a definition of evidence in terms of probability, more needs to be said about how the latter should be understood. In what follows in this chapter I will show how the two probabilistic definitions of evidence will be understood under four standard interpretations of probability.

These theories of probability can be classified into two types: objective and subjective. According to objective theories, whether a probability sentence of form (13) is true is an objective fact that has nothing to do with what any person knows or believes. If a sentence of this form is true (or false) it is so whether

---

3. Note that in this case, (10) = (12), so where we compare the strength of e and e' with respect to the same hypothesis h, the two probability definitions give the same result.
4. See Branden Fitelson, "The Plurality of Bayesian Measures of Confirmation and the Problem of Measure Sensitivity," *Philosophy of Science*, Supplement to vol. 66, no. 3 (1999), 362–378.

or not anyone believes, or even knows about, *e* or *h* or their relationship, and whether or not anyone believes or knows that (13) is true (or false). By contrast, on subjective theories, the truth of statements of form (13) depends upon, and varies with, beliefs of individuals.

Objective theories come in two varieties: a priori and empirical. According to the former, whether a probability sentence of form (13) is true is to be determined solely by logical or mathematical calculation, independently of experience. Carnap is the most prominent representative of this interpretation. It is to his view that I now turn.

3. Carnap's A Priori Theory

On Carnap's view, a sentence of form (13) expresses a logical relationship between *h* and *e*. Whether a sentence of this form is true or false is to be determined a priori solely by appeal to linguistic rules governing the language in which *h* and *e* are expressed.

We can formulate Carnap's theory by means of an example.[5] The ancient Greek natural philosopher Thales (around 585 B.C.) believed that everything is composed of water. Let us imagine a very simple language that might be used to express this idea. The language contains one predicate *P* (for "is composed of water"). To simplify further, suppose that the language contains two terms, *a* and *b*, that are used to name specific objects in the universe (for example, *a* might designate the earth and *b* the sun). Expressible in this language are four descriptions of possible worlds (Carnap calls these "state-descriptions"):

Possible world	State-description
1	$Pa \& Pb$
2	$-Pa \& Pb$
3	$Pa \& -Pb$
4	$-Pa \& -Pb$

In world 1 both objects *a* and *b* have *P*. In world 2 object *a* does not have *P* but *b* does. And so forth. Any sentence expressible in this language is equivalent to a state-description or to a disjunction of state-descriptions. (For example, "*Pa*" is equivalent to "*Pa&Pb* or *Pa&−Pb*.") Two state-descriptions belong to the same *structure* if and only if one can be transformed into the other by permuting the individual names *a* and *b*. (So, for example, state-descriptions 2 and 3 above belong to the same structure.)

Carnap wants to assign a measure to each structure and each state-description within a structure. This measure represents the probability that this structure or state-description describes the actual world. There are many different ways such a measure can be assigned so that the rules of probability are satisfied. On Carnap's preferred method, each structure will receive the same measure in such a way that the sum of the measures is equal to 1.[6] Within a given structure each state-description is to receive the same measure.

---

5. I choose here Carnap's earliest and most famous version of the theory expounded in *Logical Foundations of Probability*, pp. 562–577.

Here is how this works for our example:

State-description	Structure	Measure of State-description
$Pa \& Pb$	1	$\frac{1}{3}$
$-Pa \& Pb$	2	$\frac{1}{6}$
$Pa \& -Pb$	2	$\frac{1}{6}$
$-Pa \& -Pb$	3	$\frac{1}{3}$

The measure assigned to any sentence of the language that is not a state-description is the sum of the measures of the state-descriptions comprising the disjunction equivalent to the original sentence. For example

$$m(Pa) = m(Pa \& Pb \text{ or } Pa \& -Pb) = \tfrac{1}{3} + \tfrac{1}{6} = \tfrac{1}{2}.$$

Finally, the probability of any sentence h, given any sentence e, is simply defined as the measure of the sentence $h \& e$ divided by the measure of e, that is,

$$p(h/e) = \frac{m(h \& e)}{m(e)}.$$

So, to use our simple example, suppose that Thales wishes to determine the probability that object b (the sun) has property P (is composed of water), given that object a (the earth) has P. That is, he seeks to compute

$$p(Pb/Pa) = \frac{m(Pb \& Pa)}{m(Pa)} = \frac{m(Pa \& Pb)}{m(Pa)}.$$

Using the measures supplied by Carnap for our simple language, $m(Pa \& Pb) = \tfrac{1}{3}$, and $m(Pa) = \tfrac{1}{2}$, as we just saw. So

$$p(Pb/Pa) = \frac{\tfrac{1}{3}}{\tfrac{1}{2}} = \tfrac{2}{3}.$$

Determining what probability is to be assigned to a sentence h, on another sentence e, is an a priori matter settleable by appeal to rules of the language we are using. These rules determine the number of predicates and individual constants, as well as the measures to be assigned to state-descriptions. Once these are fixed, probabilities can be calculated a priori.

4. Carnapian Evidence

Turning now to the concept of evidence, consider

(14) *Pa* is evidence that *Pb*.

---

6. In his later work, *The Continuum of Inductive Methods* (Chicago: University of Chicago Press, 1952), Carnap constructs an infinite class of probability functions. The choice of a particular one is based on nontheoretical grounds. This choice may vary from one individual to another, so that in this later version Carnap's theory becomes closer to a subjective interpretation of probability (outlined later in this chapter). In discussing Carnap's theory in the present chapter I will focus on the earlier nonsubjective account. In chapter 5 a comparison will be offered between the objective epistemic account of probability I will develop there and Carnap's later more subjective theory.

50   *The Book of Evidence*

(The fact that the earth is composed of water is evidence that the sun is also.) On the positive relevance definition of evidence (which Carnap adopts) (14) is true if and only if

$p(Pb/Pa) > p(Pb)$.

For the simple language above, this is true, since $p(Pb/Pa) = \frac{2}{3}$, and $p(Pb) = \frac{1}{2}$. Similarly, on the high-probability definition of evidence (which Carnap also defends), (14) is true if and only if

$p(Pb/Pa) > \frac{1}{2}$.

For the simple language above this is also true.

We are now in a position to characterize the concepts of evidence supplied by the positive relevance and high probability definitions if Carnap's logical theory of probability is employed. The resulting concepts of evidence have these features:

(i) *Objectivity*. Whether e is evidence that h does not depend upon what anyone believes or knows about e, h, or anything else.

(ii) *A priori character.* On Carnap's view, the numerical values of $p(h/e)$ and $p(h)$ are determined entirely by rules of the language in which h and e are expressed. They do not depend on, or vary with, empirical facts discoverable about the world. Accordingly, whether $p(h/e) > p(h)$, whether $p(h/e) > \frac{1}{2}$, and hence whether e is evidence that h, are determined a priori by rules of language and mathematics.

(iii) *Relationship to belief.* On Carnap's theory, probability statements, although defined by reference to state-descriptions, can be given an epistemic interpretation in terms of what one is justified in believing. Suppose that $p(h/e) = r$ is true for some h and e. Then, writes Carnap,

> If e expresses the total knowledge of [a person] X at the time t, that is to say, his total knowledge of the results of his observations, then X is justified at this time to believe h to the degree r.[7]

So for Carnap a true probability statement of the form $p(h/e) = r$ justifies not a belief in h, but a *degree of belief* in h (equal to r); and it does so for any person for whom e represents the total observational knowledge available. Using terminology from chapter 2, for Carnap a true probability statement of the form in question provides a justification for believing h to degree r for anyone in an epistemic situation in which the total knowledge is represented by e. (Carnap restricts e to "observational" knowledge, but we need not do so here.)

What is the relationship, on Carnap's theory, between *evidence* statements of the form "*e* is evidence that *h*" and what one is justified in believing? The two definitions of evidence yield different relationships:

1. *Positive relevance* (this is more usefully expressed in a way that invokes background information b): If e is evidence that h, given b, then anyone in an epistemic situation in which the total knowledge is represented by $e\&b$ is justified in believing h to a higher degree than is anyone in an epistemic situation in which the total knowledge is represented by b without e.

---

7. Carnap, *Logical Foundations of Probability*, p. 211.

2. *High probability:* If *e* is evidence that *h*, then anyone in an epistemic situation in which the total knowledge is represented by *e* is justified in believing *h* to a higher degree than *not-h*.

The "justified in believing" idea in 1 and 2 above is entirely objective. No one need be in the epistemic situations mentioned. And even if someone is, that person may or may not believe anything about *h*, *e*, or their relationship.[8]

(iv) *Truth of h and e.* On the logical (as well as any other) theory of probability, it can be the case that $p(h/e) > p(h)$, and that $p(h/e) > \frac{1}{2}$, whether or not *e* and *h* are true. Accordingly, from definitions (6) and (8), *e* can be evidence that *h* even if both *h* and *e* are false.

5. The Frequency Theory

We have seen what the two standard probability definitions of evidence mean under a logical, a priori interpretation of probability, such as Carnap's. I turn now to what they mean when probability is given an empirical interpretation. The most prominent empirical version of objective theories of probability is the frequency theory. This is much more restricted than Carnap's a priori theory, since it is applicable only to cases involving types of events (such as getting heads with this coin), and not to cases involving specific events (such as getting heads with this coin on the next toss). The theory is expressed by taking the probability arguments not to be sentences but types of events.

Letting *S* be an infinite sequence of events, we introduce the expression $p^S(H)$ for the probability in the sequence *S* of a type-*H* event occurring. We define $p^S(H) = r$ as follows:

(15) $\lim_{n \to \infty} f_{n,s}(H) = r$

This means that in the sequence *S* of events, the limit of the relative frequency *f* of type *H* events is *r*. ($f_{n,s}(H)$ is the relative frequency of *H* at the *n*th member of the sequence *S*. The limit of the relative frequency of *H* is *r* if and only if for any number *x* there is a place in the sequence after which the relative frequency of *H* is within *x* of *r*.) On this theory, probability is undefined if there is no sequence *S* or if there is no limit of the relative frequency.

---

8. I have given what Carnap takes to be the relationship between probability (and evidence) statements and belief, on his own theory of probability. But, I believe, Carnap is mistaken. Suppose that $p(h/e) = r$ is true for some *h* and *e*, in accordance with one of Carnap's probability functions. Suppose that *e* represents my total observational knowledge. Am I justified in believing *h* to the degree *r*, as Carnap claims? Not necessarily. Suppose that neither I, nor anyone I know, can determine the truth of $p(h/e) = r$ (because, say, the computations are too complex, or because we do not know how to compute probabilities, or because we do not understand Carnap's system). Then even if $p(h/e) = r$, and *e* is my total observational knowledge, if I do happen to believe *h* to the degree *r*, I am not justified in doing so. What Carnap needs to say is something like this: If $p(h/e) = r$, and if *e* expresses the total knowledge of person *X* at time *t*, *and if X knows that* $p(h/e) = r$, then *X* is justified at *t* in believing *h* to the degree *r*. A corresponding claim needs to be made for the relationship between evidence (construed as positive relevance) and belief.

Can this theory of probability be applied to the two probabilistic definitions of evidence (6) and (8)? The latter require the arguments h and e in probability statements $p(h/e) = r$ to be *sentences* not event-types. With evidence we want h to be a sentence formulating the hypothesis and e to be a sentence describing the putative evidence.[9] Salmon, following Reichenbach, proposes to extend the frequency theory to cover hypotheses, as follows. We are to place the hypothesis h into some sequence of hypotheses of that type and determine the limit of the relative frequency of true hypotheses in that sequence as the latter is extended indefinitely. The resulting number is to be used as a "weight" to be attached to the specific hypothesis h. It is not a probability on the frequency view (since it is not a frequency), but it does help us to decide what practical steps to take concerning whether and to what extent we should believe h or what odds to use in betting on its truth.

We might write this as follows. Let H be the following type of "event": a hypothesis of the same type as h turning out to be true. We let S be an infinite sequence of hypotheses of the same type as h. Then $p^S(H)$ is, indeed, a probability. It is defined as the limit of the relative frequency of true hypotheses in the sequence S. Suppose that this limit exists and is equal to r. Then we can write $w(h) = r$ to express the claim that the weight to be attached to the specific hypothesis h is r.

Even though $w(h)$ is not to be interpreted as a probability, it is derived from a probability $p^S(H)$, and we might seek to use it in the definitions of evidence (6) and (8). Can this be done? The problem in doing so is that the idea of "weight" introduced by Salmon and Reichenbach is not defined *conditional on evidence e*, which is required by definitions (6) and (8).

To see what this means, consider the following sentences

e: In the first 1000 tosses this coin lands heads 950 times.
h: This coin lands heads on the 1001st toss.

Now consider a sequence S consisting of hypotheses of this form:

h_i: This coin lands heads on the i-th toss ($i = 1, 2, 3, \ldots$).

We suppose that the limit of the relative frequency of *true* hypotheses of this form in the sequence S exists and that it is some number r. We now take r to be the weight to be assigned to the hypothesis h above ($w(h) = r$).

What meaning should be assigned to the conditional $w(h/e)$ needed for definitions (6) and (8)? One possibility is that this conditional is to be given no meaning, in which case the frequency theory is prevented from explicating the evidence definitions (6) and (8). Another possibility is to define $w(h/e)$ simply as

9. Accordingly, it is possible for a frequentist to adopt the following dualist position. When speaking of probabilities of event types, the frequency theory is to be used to interpret the probability. When speaking of the probability of a hypothesis, a different theory is to be employed, e.g., Carnap's logical theory of probability. Since evidence statements involve the probability of hypotheses, the frequency theory of probability is not to be used to understand such statements. (This, indeed, is the view of Carnap, who adopts a logical theory of probability for the probability of hypotheses and a frequency theory for the probability of events.) In what follows I want to consider how a nondualistic frequentist can use the frequency theory to interpret evidence statements and thus provide a more comprehensive and unified theory.

$w(h)$. This seems plausible if the limit of the relative frequency of true hypotheses in the sequence S is $\frac{950}{1000}$, that is, if it matches the observed relative frequency reported in e. But what if it doesn't? Suppose the limit of the relative frequency is $\frac{1}{10}$. Then, we might argue, $w(h/e)$ should still be equal to $w(h)$, since the information about the limit of the relative frequency of true hypotheses in S (and hence the information that when this coin is tossed indefinitely the limit of the relative frequency of heads is $\frac{1}{10}$) should completely swamp e, which is a very limited piece of information about the first 1000 tosses of this coin. If so, this immediately rules out the positive relevance definition of evidence (6), which, in terms of weights, requires that e is evidence that h if and only if $w(h/e) > w(h)$. On the current proposal, no e could be evidence that $h!$ If we are to use weights, then we want to allow $w(h/e)$ to be different from $w(h)$.

Perhaps $w(h/e)$ should simply be equal to the relative frequency reported in e (in our example this is $\frac{950}{1000}$). Or perhaps it should be some weighted mean between $w(h)$, the limit of the relative frequency of true hypotheses in S, and the relative frequency reported in e. In our example, the former is $\frac{1}{10}$. So $w(h/e)$ would lie between $\frac{1}{10}$ and $\frac{950}{1000}$, and thus would be greater than $w(h)$. I will assume that this can be done.

6. Determining a Sequence

The question remains how to determine the sequence S into which the hypothesis is to be placed. This is a generally recognized problem for the frequency theory when it is applied to particular events (rather than types) and to hypotheses. Reichenbach's solution is to place the hypothesis in the narrowest class of hypotheses for which we have the most reliable statistics.[10] Suppose I know that e is true and that the class consisting of hypotheses of the form h_i is the narrowest class containing h for which I have statistics indicating the observed relative frequency of true hypotheses, which in this case is $\frac{950}{1000}$. Suppose this is indeed the limit of the relative frequency. Then for me the weight to be assigned to h is $\frac{950}{1000}$.

Suppose, by contrast, that you know a good deal more than I do. You know that e is true, but you also know this:

e': This coin is tossed 100,000 times, and on each toss after a multiple of 1000 it lands heads only $\frac{1}{10}$th of the time.

Now consider the following sequence of hypotheses:

$h_{i(1000)+1}$: This coin lands heads on toss $i(1000) + 1$
$(i = 1,2,3,\ldots)$

Hypothesis h (this coin lands heads on the 1001st toss) is a member of this sequence of hypotheses. Let us suppose that the limit of the relative frequency of true hypotheses in this sequence is equal to its observed relative frequency $\frac{1}{10}$. And finally, let us suppose that this is the narrowest reference class for hypothe-

---

10. Hans Reichenbach, *The Theory of Probability* (Berkeley: University of California Press, 1949), p. 374.

sis h for which you have reliable statistics. Then for you the weight to be assigned to h is $\frac{1}{10}$.

If we do this, however, then the frequency interpretation of evidence will yield a concept of evidence relativized to the knowledge of particular persons or groups. On the present frequency interpretation, understood in terms of weights, e is evidence that h if and only if either $w(h/e) > w(h)$ (increase-in-weight), or $w(h/e) > \frac{1}{2}$ (high weight). $w(h/e)$ is to be a weighted mean between $w(h)$ and the observed relative frequency of true hypotheses of the same type as h. In either case, whether e is evidence that h will depend on $w(h)$. But, on Reichenbach's idea, the value of $w(h)$ will depend on the narrowest class containing h for which statistics are available, which can vary from one individual or community to another. Given my knowledge (and that of my community), the narrowest class for which statistics are available may be the class of hypotheses h_i. Given your knowledge (and that of your community) it may be $h_{i(1000)+1}$. And the limits of relative frequencies of true hypotheses in these classes may be very different. The differences may be such that on the definition of evidence as increase-in-weight, as well as on the definition of evidence as high weight, e is evidence that h for me, but not for you. Now frequentists, of course, espouse a concept of probability that is not relativized to the knowledge of individuals or groups. For any frequentist who wants to supply a corresponding notion of evidence derived from probabilities construed as frequencies, these concepts of evidence should be unacceptable.

Salmon, unlike Reichenbach, proposes a method for selecting a sequence that is not relativized to anyone's knowledge. It is to place the hypothesis in the broadest homogeneous reference class or sequence of which it is a member. A class or sequence is homogeneous with respect to a property P (P is truth, in the case of hypotheses) if the limit of the relative frequency of P in any arbitrarily chosen subclass of the class or sequence is the same as the limit of the relative frequency of P in the class itself. Whether a class or sequence is homogeneous with respect to a property P does not depend on, or vary with, who knows what. This will also be true, then, of the weight to be assigned to a hypothesis. This has the advantage of rendering the truth or falsity of evidence statements that are based on weights independent of who knows what.

Suppose, then, I want to determine whether the fact that (e) in the first 1000 tosses this coin lands heads 950 times is evidence that (h) it lands heads on the 1001st toss. What do I do, on Salmon's proposal? I must place h into some appropriate reference class of hypotheses. Such a reference class might contain hypotheses of this sort:

$h_{i,j}$: Coin C_j lands heads on the i-th toss,

where C_j refers to coins in a class containing the coin in question. I must determine whether this reference class is homogeneous with respect to the truth of hypotheses of the form $h_{i,j}$. I must determine whether it is the broadest such reference class that contains h. And I must determine the limit of the relative frequency of truth of hypotheses in this class. How do I do all these things?

I proceed by making use of a rule of inference proposed by both Reichenbach and Salmon. This is the so-called straight rule, which requires us to make an inference from some number's being the observed relative frequency of a property in a sequence to its being the limit of the relative frequency as the sequence is extended indefinitely. Let us see how this will work in the present case.

First, what weight do I assign to h *not conditional on e*? (What is $w(h)$?) Consider the set $h_{i,j}$ containing hypotheses not only about the coin mentioned in h but about other coins as well. Let us suppose that the coin mentioned in h has not been tossed, but that others have, and that each of these other coins has landed heads half the time. On this assumption, the observed relative frequency of true hypotheses in the set $h_{i,j}$ is $\frac{1}{2}$. Since the observed relative frequency is the same for each of these coins that has been tossed, it is reasonable to infer that the set $h_{i,j}$ is homogeneous with respect to truth. It may also be reasonable to suppose that this is the broadest homogeneous class containing h. Using the straight rule, under the assumptions in question (where our coin has not yet been tossed), we infer that the limit of the relative frequency of truth in the set $h_{i,j}$ is $\frac{1}{2}$. We are then to use this inferred limit as a weight to be assigned to the hypothesis h concerning the 1001st toss of this coin. We will assign $w(h)$ the number $\frac{1}{2}$.

Second, what weight do I assign to h conditional on e? (What is $W(h/e)$?) Earlier, to avoid the problem for the positive relevance definition of evidence if $w(h/e)$ is taken always to be $w(h)$, we considered taking $w(h/e)$ to be equal to the observed relative frequency reported in e, or to some weighted mean of this observed relative frequency and $w(h)$. In our example, the observed relative frequency reported in e is $\frac{950}{1000}$. So $w(h/e)$ is either $\frac{950}{1000}$ or some number between $\frac{950}{1000}$ and $\frac{1}{2}$ ($w(h)$). In either case, $w(h/e)$ is assigned a number larger than that assigned to $w(h)$. So we can conclude that e is evidence that h on the increase-in-weight definition of evidence. Since $w(h/e)$ is assigned a number greater than $\frac{1}{2}$, we can also conclude that e is evidence that h on the high weight definition.

This has a curious, and, I think, objectionable consequence. Suppose e is in fact true (viz. that this coin is tossed 1000 times, resulting in 950 heads). Suppose that $\frac{950}{1000}$ is not just the inferred but the true limit of the relative frequency of truth of hypotheses h_i (about this coin). Suppose that $\frac{1}{2}$ is not just the inferred but the true limit of the relative frequency of truth of hypotheses $h_{i,j}$ (containing a large set of coins). Then the class $h_{i,j}$ is not homogeneous with respect to truth (since there is a subclass h_i with a different limit of the relative frequency of truth). Suppose, finally, that the class h_i is in fact the broadest homogeneous class containing hypothesis h (about the 1001st toss with this coin). Then, on Salmon's view, since the limit of the relative frequency of truth in h_i is in fact $\frac{950}{1000}$, this is the correct weight to be assigned to the hypothesis h; that is, $w(h) = \frac{950}{1000}$. What is the conditional weight to be assigned to h on e? We are supposing that $w(h/e)$ should be equal to the observed relative frequency reported in e or to some weighted mean between this observed relative frequency and $w(h)$, where the latter is equal to the limit of the relative frequency of true hypotheses in the set h_i. But in this case the limit in question and the observed relative frequency are the same. So $w(h/e) = w(h)$, and e will not be evidence that h on the increase-in-weight definition of evidence. This means that if the limits of the relative frequency inferred from the observed relative frequencies actually obtain, then the fact that (e) this coin lands heads 950 out of 1000 times will not be evidence that

(h) it lands heads on the 1001st toss, on the increase-in-weight definition of evidence. That seems like a lot to swallow!

The same problem arises for Reichenbach. Suppose that h_i is the narrowest reference class for which we have statistics. (We don't have information about other coins, let us assume.) Let the observed relative frequency of heads reported in e be $\frac{950}{1000}$. Let this also be the limit of the relative frequency. Then, for Reichenbach, the correct weight to assign h is $\frac{950}{1000}$. If the conditional weight $w(h/e)$ is simply the relative frequency reported in e or a weighted mean of the latter and $w(h)$, then $w(h/e) = \frac{950}{1000}$. Again, e will not be evidence that h, on the increase-in-weight definition of evidence.

I will not dwell further on these and other problems that might be raised with respect to these particular frequency interpretations of evidence. Instead I will simply characterize the resulting concepts of evidence (focusing mainly on Salmon's version).

7. Frequency Evidence

On a frequency interpretation of probability, the concepts of evidence given at the beginning of this chapter have the following features.

(i) *Relationship to probability.* Evidence is defined only indirectly in terms of probability. On the frequency view, probabilities are not defined for hypotheses. If "weights" (as understood by Reichenbach and Salmon) are to be used for the positive relevance and high-probability definitions of evidence, these are not probabilities in the frequency sense, although they are derived from probabilities.

(ii) *Objectivity.* Whether e is evidence that h depends upon whether the weights are such that $w(h/e) > w(h)$, or $w(h/e) > \frac{1}{2}$. On Salmon's version (but not Reichenbach's), this is an objective matter that does not depend on anyone's knowledge or beliefs. What weight $w(h)$ is to be assigned to h depends on the limit of the relative frequency of truth of hypotheses in the broadest homogeneous class of hypotheses containing h. What weight $w(h/e)$ to assign to h given e depends (in addition) on the relative frequency reported in e. Both considerations are objective.

(iii) *Empirical character.* For the frequency theorist, probability statements are empirical, not a priori, since they make claims about limiting frequencies. This means that statements such as $w(h) = r_1$ and $w(h/e) = r_2$ are empirical as well, since their truth is determined by that of probability statements. The same holds true, then, of statements of the form $w(h/e) > w(h)$, and $w(h/e) > \frac{1}{2}$, which, on the present view, reflect the two definitions of evidence.

(iv) *Relationship to belief.* For the frequency theorist probability statements are not epistemic. "The probability of heads with this coin is $\frac{1}{2}$" is a statement which says that the limit of the relative frequency of heads is $\frac{1}{2}$ (in some sequence S). It does not say or imply anything about what one should believe about any particular toss or set of tosses (or about anything, for that matter). Let us focus, then, on weight statements of the form $w(h/e) = r$, as I have been describing these on Salmon's version. Consider an example used earlier.

e: In the first 1000 tosses this coin lands heads 950 times.
h: This coin lands heads on the 1001st toss.

We said that $w(h/e)$ should be equal to the relative frequency reported in e ($\frac{950}{1000}$), or be a weighted mean between this and $w(h)$. To take the most general case (and the only one that will allow the positive relevance definition (6)), let $w(h/e)$ be a weighted mean of $w(h)$ and the relative frequency reported in e. Suppose that $w(h) = \frac{1}{2}$ and that

(a) $w(h/e) = .8$.

Is there any relationship between statement (a) and belief?

If (a) is true, then, we might say, anyone in an epistemic situation in which what is known is represented by e is justified in believing h to the degree .8. Can a stronger statement be made? If (a) is true can we say that e, if true, constitutes a good reason to believe h to the degree .8? (This is to be understood independently of any particular epistemic situation.) With one important caveat, I believe that this is what Salmon has in mind, especially if he is to accept one or both of the probability definitions of evidence.[11] This sense of "good reason," unlike that employed for veridical evidence in chapter 2, does not carry a commitment to the truth of h. The reason I attribute a "fallibilist" notion of "good reason" to Salmon rather than the "veridical" idea is that the two probability definitions of evidence do not require h to be true, even if e is, in order that e be evidence that h. Nor do these definitions require h's truth if "probability" is replaced by "weight." That is, as weights are being construed, it can be the case that $w(h/e) > w(h)$, and that $w(h/e) > \frac{1}{2}$, even if h is false.

In short, on the present proposal, if (a) is true, then the fact that in the first 1000 tosses this coin lands heads 950 times is a good reason to believe to the degree .8 that it will do so on the 1001st toss. Even if the coin lands tails on the 1001st toss, this fact still was a good reason to believe to the degree .8 that it would land heads on the 1001st toss. What would refute this claim is not the fact that the coin lands tails, but the fact that the broadest homogeneous class containing h has a limit of relative frequency different from $\frac{1}{2}$ (and therefore that $w(h/e)$ is not equal to .8).

Accordingly, Salmon's version of the frequency theory, together with the positive relevance definition of evidence, yields the following thesis concerning the relationship between evidence and belief:

If e is evidence that h, given b, then e together with b constitutes a good (nonveridical) reason to believe h to a higher degree than does b without e.

Using the high probability definition, we get

If e is evidence that h, given b, then e constitutes a good (nonveridical) reason to believe h to a higher degree than *not-h*.

---

11. As will be shown in chapter 7, this claim, which I am attributing to Salmon, is too strong. The most that can be said is that if e is true, then, assuming that (a) is true, *there is* a good reason to believe h to the degree .8. We cannot necessarily assume that e is that reason.

8. The Propensity Theory

The propensity theory of probability, first introduced by Karl Popper,[12] is another empirical, objective account. It construes probability as a disposition or propensity of a system to yield a certain result under specified conditions. Think of a nonprobabilistic disposition, inertia. This is the disposition of a body to continue with uniform speed in a straight line. A probabilistic disposition or propensity is similar, except it need not always yield the same result. For example, we might say that this coin (or perhaps a system consisting of this coin, a tosser, and the conditions of the tosses) has a disposition or propensity to land heads half the time. This is a property of the coin (or the system), and not (as on the frequency theory) a property of the sequence of tosses with the coin. On the present view, it is this propensity that is being referred to by probability. And whether the coin has a propensity to land heads half the time is an empirical, not an a priori, question.

Propensity theorists claim several advantages over frequency theorists: (i) They apply their theory not just to types of events, but to particular events as well. According to them, this coin has not only a certain propensity to land heads (type of event), but also a certain propensity to land heads on the next toss (specific event). These are different propensities, and may indeed be represented by different numbers. (ii) Unlike frequency theorists, propensity theorists can apply their theory even to cases in which no outcomes will in fact occur. (This coin has a certain propensity to land heads even if it is never tossed.)

However, a question arises concerning whether the propensity theory is applicable to all cases in which we may wish to speak of evidence. Is it applicable to cases in which outcomes of a trial or experiment are not envisaged? For example, we may want to say that there is evidence that neutrinos have mass, and that cathode rays are negatively charged. In such cases we may be willing to speak of the probability of these hypotheses without referring to possible outcomes of some trial or experiment.

One writer who *may be* willing to apply (his version) of the propensity theory quite broadly is David Lewis.[13] Lewis writes that "it is only caution, not any definite reason to think otherwise, that stops me from assuming that chance of truth [propensity] applies to any proposition whatever" (p. 270). Lewis's concept of *chance* is a function of three arguments: a proposition, a time, and a world (p. 271). To this triple a real number (that is a probability) is assigned.

Let me write

$$C_{w,t}(h) = r$$

to mean that the chance of the truth of h at time t in world w is equal to r. On Lewis's view, chance is something objective, not subjective. In propensity terms,

---

12. Karl Popper, "The Propensity Interpretation of Probability," *British Journal for the Philosophy of Science*, 10 (1959), 25–42.
13. David Lewis, "A Subjectivist's Guide to Objective Chances," in Richard Jeffrey, ed., *Studies in Inductive Logic and Probability*, vol. 2 (Berkeley: University of California Press, 1981), 263–293.

we might think of it as the propensity of a proposition to be true at a certain time in a certain world.

9. Propensity Evidence

To see how the two probability definitions of evidence fare on the propensity theory, consider a lottery case. Let

    h: John wins the lottery.
    e: John buys 999 of the 1000 tickets at 11 A.M. today, and no more after that time.
    b: This is a fair lottery containing 1000 tickets, one to be drawn at noon today.

We might suppose that at 10 A.M. today (before he buys any tickets) John has very little chance of winning. That is, $C_{our\ world,\ 10AM}(h)$ is close to 0.[14] We might also say that at 10 A.M. today, assuming John would buy 999 of the 1000 tickets at 11 A.M. and no more after that, his objective chance of winning would be $\frac{999}{1000}$. That is, $C_{our\ world,\ 10AM}(h/e) = \frac{999}{1000}$. Since $C_{our\ world,\ 10AM}(h/e) > C_{our\ world,\ 10AM}(h)$, we can say that, given the way the world is at 10 A.M., e if true counts as evidence that h, on the positive relevance definition of evidence. Similarly, since $C_{our\ world,\ 10AM}(h/e) > \frac{1}{2}$, we can say that, given the way the world is at 10 A.M., e if true counts as evidence that h, on the high-probability definition. On Lewis's account, since chances are relativized to a world and a time, presumably the same holds for evidence statements based on such chances. Accordingly, to be precise, we should say that e, if true, counts as evidence that h, relative to the world (we are imagining) and the time of 10 A.M.

This version of the propensity theory has the advantage of being applicable to propositions—indeed, to a wide range of them (perhaps all). But there are consequences that defenders of the two probability definitions may not want. Suppose that John wins the lottery at noon, so that h is true. Then for Lewis,

$$C_{our\ world, noon}(h) = 1,$$

since given the way the world is at noon, John's chance of winning is certainty.[15] If, as we are supposing, chances are probabilities that obey the probability calculus, then it follows that

$$C_{our\ world, noon}(h/e) = 1.$$

Since $C_{our\ world, noon}(h/e) = C_{our\ world, noon}(h)$, e cannot count as evidence that h relative to our world at noon, on the positive relevance definition of evidence. This means that given the way our world is at noon, the fact that John has bought 999 of the 1000 lottery tickets cannot count as evidence that he won the lottery. This claim may seem dubious.

    14. The chance is not 0, since, let us suppose, John has some (very small) chance of buying some, or even many, tickets.
    15. This is what Lewis himself claims about an analogous example. See p. 271.

To see why, recall the three objective concepts of evidence of chapter 1: ES-evidence, potential evidence, and veridical evidence. A fundamental idea underlying the claim that *e* is ES-evidence that *h* is that *e* justifies a belief in *h* for someone in an epistemic situation ES. Suppose that the epistemic situation at noon includes knowing the way the world is at noon, in particular knowing that *h* is true. Assume that I am in such an epistemic situation at noon, having heard John declared the winner. At noon the fact that he owned 999 of the 1000 tickets would still justify my belief that he won, even if at noon I now have even greater justification, having heard what I did, and even if at noon my reason for believing he won is no longer (simply or at all) that he owned 999 of the 1000 tickets. Nevertheless, the fact that he did own so many tickets still justifies my belief, even given the new more conclusive information. So a crucial element of ES-evidence is satisfied here.

Something quite analogous can be said for potential and veridical evidence. The fact that John owned 999 of the 1000 tickets is a good reason to believe he won, even if the fact that at noon he is announced the winner is a better reason. The latter reason does not cancel the former one.

However, a defender of the positive relevance definition of evidence can reply that what is crucial to evidence is not that it provides a justification or reason for believing *h* but that it provides a justification or reason for believing *h more than without that evidence*. So if we consider an epistemic situation at noon that includes knowing the way the world is at noon, and hence knowing that John was declared the winner, then the information *e*—that he owned 999 of the 1000 tickets—does not provide a justification for believing *h* (John won) *more* than without *e*. And that is what is crucial for evidence, not providing a justification for believing *h*. It is this question—the relationship of evidence to belief on the positive relevance definition of evidence—that needs to be examined, and will be in the next chapter.

Lewis's view also leads to a consequence on the high probability definition of evidence that may seem questionable. Let us imagine a situation in which John buys just 1 ticket, not 999. Let

e': John buys 1 ticket at 11 A.M. today, and none later.

And let us suppose that, as luck would have it, John wins with this ticket, so that *h* is again true. Then, on Lewis's conception, $C_{our\ world, noon}(h) = 1$. Again, if chances are probabilities, it follows that $C_{our\ world, noon}(h/e') = 1$. So on the high-probability definition of evidence, e', the fact that John buys just 1 ticket in a 1000 ticket lottery, is evidence that he wins the lottery, given the way the world is at noon. To some this may be too much to concede, a conclusion reinforced by appeal to central ideas underlying the three objective concepts of chapter 2. The fact that John buys just 1 ticket in a 1000 ticket lottery is neither a good reason for, nor does it justify, believing that John won the lottery.[16] However, someone

---

16. Even more problematic is the fact that irrelevant information will count as evidence that *h* in this case. If $C_{our\ world}(h) = 1$, then $C_{our\ world}(h/e) = 1$, for any logically consistent *e*. So if *e* = the earth is round, then *e* counts as evidence that John won the lottery, on the high-probability definition. This sort of "irrelevant information" counterexample will be discussed in chapter 4.

defending the positive relevance definition of evidence who wants to use Lewis's account of probability may reply that the fact that John buys just 1 ticket in a 1000 ticket lottery is transformed into a good reason for believing that he won by the way the world is at noon (which includes the fact that he won). This strikes me as dubious, but I leave a fuller discussion of claims such as this for chapter 4.

Suppose, by contrast, that h is false. Then for Lewis, at noon the chance that John wins is 0, that is,

$$C_{our\ world, noon}(h) = 0.$$

Let us suppose, as we did in our first case, that John bought 999 of the 1000 tickets at noon, so that e is true. If $C_{our\ world, noon}(h) = 0$, then

$$C_{our\ world, noon}(h/e) = 0.$$

So, on the high probability definition of evidence, given the way the world is at noon, the fact that John bought 999 of the 1000 tickets is not evidence that he won. Again, some will want to reject this conclusion. It is acceptable if restricted to *veridical* evidence. (Since h in this case is false, e cannot be veridical evidence that h.) But it is not acceptable for ES- and potential evidence, where the concepts of justification and (nonveridical) reason for belief are central.

On Lewis's version of the propensity theory we get interpretations for the two probability definitions of evidence that are similar to but not exactly the same as those for the frequency theory. First, we need to relativize evidence statements to a world and a time:

e is evidence that h in world w at time t.

Using the positive relevance and high-probability definitions of evidence, evidence statements of this form would have the following features:

(i) *Objectivity.* Whether e is evidence that h in world w at time t depends on whether the chances or propensities are such that $C_{w,t}(h/e) > C_{w,t}(h)$ or $C_{w,t}(h/e) > \frac{1}{2}$. These are objective matters that are independent of anyone's knowledge or beliefs.

(ii) *Empirical character.* On Lewis's propensity theory, statements of the form $C_{w,t}(h/e) = r$ are empirical, not a priori, since they make claims about features of the world that can only be determined empirically. The same holds for evidence statements if these are to be understood in terms of positive relevance or high probability.

(iii) *Relationship to belief.* A probability statement interpreted as a propensity describes something about the physical world, not about the believability of anything. Nevertheless, if $C_{w,t}(h/e) = r$, then, it might be claimed, in world w at time t, e if true would constitute a good reason to believe h to the degree r. This is a sense of "good reason" that does not require h's truth. Indeed, Lewis adopts what he calls "the Principal Principle,"[17] which relates nonepistemic to epistemic probability. We might formulate it as follows:

---

17. "A Subjectivist's Guide to Objective Chance," p. 266.

Principal Principle: Let X be the proposition that $C_{w,t}(h/e) = r$. Let p be some rationally corrected subjective probability function.[18] Then $p(h/X\&e) = r$.

That is, given that in world w at time t the chance that h is true, given e, is r, and given that e is true, it is reasonable at t in world w to believe h to the degree r. The relationship between evidence statements (understood in terms of positive relevance or high probability) to belief can readily be constructed from this.

(iv) *Truth of* h *and* e. Neither is required on Lewis's view.

10. The Subjective Theory

The theories of probability considered above are all objective: if a probability statement of the form $p(h/e) = r$ is true (or false) it is so because of the language in which it is expressed (Carnap's a priori view), or because of certain empirical facts about the world (frequency and propensity views), independently of what anyone believes or knows about h or e or their relationship. By contrast, according to a standard subjective theory, developed initially by Ramsey, de Finetti, and Savage,[19] a probability statement is to be construed as describing how much belief a person has in some proposition. If I claim that the probability that I will get heads with this coin on the next toss is $\frac{1}{2}$, what I mean is that my degree of belief in the proposition that I will get heads on the next toss is half-way between the maximum and minimum possible degrees of belief.

Subjective probability theorists suppose that the degree of belief in a proposition, which is to be measured between the values of 0 and 1, can differ from one person to another. On this theory, a standard method of determining what degree of belief I in fact have in a proposition h is by considering what betting odds I am willing to accept on the truth of h. Suppose that b is the most money I am willing to bet, and hence lose, if h is false, in order to win amount a if h is true. Then $b:a$ represents my betting odds on proposition h, and $b/a+b$ (my possible loss divided by the sum of my possible loss and gain) represents my degree of belief in the proposition h. So, in order to win \$1 if this coin lands heads next time, if I am willing to bet (and hence lose) a maximum of \$1, then my betting odds on the proposition "this coin will land heads next time" are 1:1, and my degree of belief in this proposition is $1/1+1 = \frac{1}{2}$.

A set of such degrees of belief in various propositions is said to be *coherent* if and only if no system of bets based on these degrees of belief is bound to lose no matter what in fact happens. For example, let

h: It will rain today.

Suppose that your degree of belief in $h = \frac{4}{5}$, and that your degree of belief in *not-h* (it won't rain today) $= \frac{2}{5}$. (These degrees of belief are determined by the bet-

---

18. This idea will be explicated in the next section. Briefly, it is a function satisfying the probability calculus that represents some person's set of degrees of belief in any set of propositions.
19. Some of the original papers are reprinted in Henry E. Kyburg and Howard Smokler, eds., *Studies in Subjective Probability* (New York: Wiley, 1964).

ting odds you are willing to accept on these propositions, 4:1 on *h*, 2:3 on *not-h*.) Your set of degrees of belief is not coherent, since, whether it rains or not today, you are bound to lose with these odds. If it rains, then you win $1 on your bet on *h*, but lose $2 on your bet on *not-h*; total loss: $1. If it doesn't rain then you lose $4 on your bet on *h*, while winning $3 on your bet on *not-h*; total loss: $1. So with these degrees of belief and the bets they generate, you lose whether it rains or not.

The purpose of introducing this idea of coherence is to tie it to the *rationality* of a set of degrees of belief. Subjectivists claim that a set of degrees of belief is rational if and only if it is coherent, that is, if and only if no system of bets based on these degrees of belief is bound to lose. An important theorem is now provable which states that a set of degrees of belief is coherent (and hence rational) if and only if the degrees of belief in the set satisfy the rules of the probability calculus (given at the beginning of this chapter). As we have just seen in the rain example above, one may have a set of degrees of belief that is not coherent. If one does, then subjective theorists require that one modify one's degrees of belief so that they become coherent, and hence satisfy the rules of probability. No rules for such modification are offered, and different individuals may rationally alter degrees of belief differently. The only requirement of rationality is that the corrected set of degrees of belief satisfy the probability rules. Then and only then will one avoid a system of bets that is bound to lose.

For the subjectivist a probability statement is to be understood as relativized to a given individual at a given time. Accordingly, we might write $p_{x,t}(h) = r$, which means that at time t person x has a rationally corrected degree of belief r in the proposition h. In the case of conditional probability we write $p_{x,t}(h/e) = r$. This means that at time t, person x has a rationally corrected degree of belief in h, on the assumption of e, equal to r. By "rationally corrected" we mean "corrected in such a way that, at time t, x's set of degrees of belief is coherent." If one's actual set of degrees of belief at time t is coherent, then a probability statement of the form $p_{x,t}(h) = r$ states what degree of belief x in fact has in h at time t. If one's actual set of degrees of belief at time t is not coherent, then we can say that $p_{x,t}(h) = r$ is undefined for x at t, or we can say that if x were to make his set of degrees of belief coherent at t, he would do so in such a way that his actual degree of belief in h at time t would be r.

11. Evidence on the Subjective Theory

If one adopts the subjective theory of probability, then the positive relevance and high probability concepts of evidence will have these features:

(i) *Subjectivity.* Whether e is evidence that h depends upon, and varies with, who believes what, and on the time at which it is believed. Since one's beliefs may change over time, for a subjectivist statements of the form "*e* is evidence that *h*" should be understood as relativized to a person and a time:

e is evidence that *h* for person x at time t.

Using the positive relevance definition of evidence, a sentence of this form would be understood as saying this:

At time t person x's (rationally corrected) degree of belief in h, conditional on e, is greater than x's (rationally corrected) degree of belief in h without the assumption of e.

Using the high-probability definition, we would understand an evidence statement as

At time t, person x's (rationally corrected) degree of belief in h, conditional on e, is greater than $\frac{1}{2}$.

(ii) *Empirical character.* On the subjective theory whether a probability statement of the form $p_{x,t}(h) = r$ is true, whether x has that degree of belief in h, is an empirical, not an a priori, fact about x. It is discoverable by examining x's behavior. Accordingly, whether e is evidence that h for person x at time t, that is, whether $p_{x,t}(h/e) > p_{x,t}(h)$, or whether (to use the high probability definition) $p_{x,t}(h/e) > \frac{1}{2}$, is also an empirical fact about x.

(iii) *Relationship to belief.* For a subjectivist, an evidence statement has a direct connection to belief. It expresses what some person does or would believe. But because of the introduction of the idea of a "rationally corrected" degree of belief it expresses something stronger as well. Let us return to a simple probability statement of the form $p_{x,t}(h) = r$. For a subjectivist this means not only that, at time t, x in fact has (or would have) a degree of belief in h equal to r, but that this is a "rationally corrected" one. The latter condition indicates that x's degree of belief in h is probabilistically consistent with x's degree of belief in all other propositions.

Let me call the set of x's degrees of belief in various propositions his belief-set. According to the subjectivist, x's belief set is rational if and only if it satisfies the rules of probability. Accordingly, $p_{x,t}(h) = r$ should be understood as entailing or presupposing that at time t, x has a rational belief-set. The latter is to be understood as implying that each belief in that set is rational or reasonable for x at time t. So for the subjectivist, if $p_{x,t}(h) = r$ is true, then not only is it the case that x does (or would) in fact believe h to the degree r, but it is rational or reasonable for x to do so. What does this mean for the probability definitions of evidence?

Using the positive relevance definition, if e is evidence that h for x at time t, then at t not only does x have a higher degree of belief in h on e than without e, but it is rational or reasonable for x to have such a higher degree of belief. On the high-probability definition, if e is evidence that h for person x at time t, then at t not only does x have a higher degree of belief in h on e than x has in *not-h* on e, but it is rational or reasonable for x to have such a higher degree of belief.

(iv) *Truth of h and e.* On the subjective theory, it can be the case that $p_{x,t}(h/e) > p_{x,t}(h)$, and that $p_{x,t}(h/e) > \frac{1}{2}$, whether or not e or h is true or false. So, e can be evidence that h even if both e and h are false.

12. Probability Definitions and the Four Concepts of Evidence

To what extent do the two probability definitions of evidence, understood in terms of the various theories of probability, conform to any of the four concepts of evidence in the second chapter? As we will see, there are some similarities and some important differences.

a. *Carnap's logical theory.* This theory of probability, as applied to the positive relevance and the high-probability definitions of evidence, yields concepts of evidence that are most akin to, but not identical to, *ES-evidence*. Carnap's theory furnishes two concepts of evidence, each of which, like ES-evidence,

(i) is objective (in the sense that it does not depend on what anyone in fact believes or knows);
(ii) is related to justification of belief;
(iii) does not require the truth of h.

The differences are these:

1. In ES-evidence statements of the form "e is evidence that h" e is required to be true. Carnap's theory has no such requirement. His theory is concerned with evidence statements which might better be understood to be of the form "e, if true, is evidence that h."

2. ES-evidence statements are relativized to some type of epistemic situation, whereas Carnap's evidence statements are not explicitly relativized in this way. They are, however, implicitly so under Carnap's epistemic interpretation of probability statements, according to which $p(h/e)=r$ means that if e expresses the total knowledge of a person at time t then that person at t is justified in believing h to the degree r. But there is this difference. In Carnap's case all the propositions known in the epistemic situation are explicitly formulated as e (or $e\&b$) in the evidence statement itself. This need not be so (and usually will not be) with ES-evidence statements. This is reflected in the ES-evidence statements we have been using for Hertz's experimental results. We claimed that these results were ES-evidence that cathode rays are electrically neutral, relative to an epistemic situation of the sort attributable to Hertz. The latter statement is true even though it does not indicate the content of Hertz's epistemic situation.

3. Carnap's evidence statements, unlike corresponding ES-evidence statements, are all a priori. ES-evidence statements can be empirical or a priori. Since Carnap's evidence statements contain explicitly all the propositions comprising the content of the epistemic situation, and since (Carnap assumes) whether this explicitly formulated content justifies a certain degree of belief in a hypothesis h is settleable a priori, evidence statements themselves are a priori. By contrast, since ES-evidence statements will usually not express all the propositions known in ES, what the latter are is only empirically discoverable. Hence, such ES-evidence statements will be empirical.

4. Although the concepts of evidence generated by Carnap's theory are like ES-evidence in being related to what one is justified in believing, there are important differences. If e is ES-evidence that h, given b, then, given b, e (if true) justifies a belief in h for anyone in ES. However, using the positive relevance definition of evidence together with Carnap's theory of probability yields a different relationship between evidence and belief: If e is evidence that h, given b, then anyone whose epistemic situation is one in which both e and b are known (and who knows that $p(h/e\&b) > p(h/b)$) is justified in believing h to a higher degree than is anyone whose epistemic situation is one in which b but not e is known.

Using the high probability definition of evidence together with Carnap's theory of probability, we obtain this relationship: If e is evidence that h, given b, then anyone whose epistemic situation is one in which e and b are known (and who knows that $p(h/e\&b) > \frac{1}{2}$) is justified in believing h to a higher degree than not-h. These differences concerning the relationship of evidence to the justification of belief lie at the heart of a controversy over the adequacy of the probability definitions of evidence to be examined in the next chapter.

b. *Salmon's frequency theory and Lewis's propensity theory.* Salmon's theory (more specifically the theory of weights I attribute to him), as well as Lewis's propensity theory, when applied to the positive relevance and high-probability definitions of evidence, yield concepts of evidence most akin to *potential evidence*. Both theories generate two concepts of evidence each of which, as with potential evidence,

(i) is objective
(ii) yields empirical not a priori evidence statements
(iii) does not require the truth of h
(iv) is related to the idea of a good reason for belief (in a non-veridical sense). Neither Salmon's theory nor Lewis's restrict or relativize evidence statements to some epistemic situation.[20]

However, as in the case of Carnap, the probability definitions of evidence yield relationships between evidence and belief that are different from that furnished by potential evidence. If e is potential evidence that h, then e is a good reason to believe h. Using the positive relevance definition of evidence, together with the theories of either Salmon or Lewis, we obtain this relationship: if e is evidence that h, given b, then e together with b is a reason to believe h to a higher degree than is b without e. (An analogous relationship between the high-probability definition and belief was given in the previous section.)

c. *Subjective theory.* The subjective theory of probability, as applied to the positive relevance and high-probability definitions, yields concepts of evidence most akin to what I have called subjective evidence. Like the latter, each of the concepts of evidence generated by the subjective theory of probability

(i) is subjective (in the sense that whether e is evidence that h depends on what people believe);
(ii) is empirical (what a person or community believes is an empirical matter);
(iii) does not require the truth of e or h (but only certain truths about what is believed about e and h).

Significant differences emerge, however, when we consider the relationship to belief. For one thing, as we saw in the previous two cases, positive relevance and high-probability definitions relate evidence that h to an increase in the degree of belief in h, and to a higher degree of belief in h than *not-h*, respectively.

---

20. Combining Lewis's propensity theory with the two probability definitions of evidence yields evidence statements that are relativized to a world and a time. But they do not require supposing that anyone knows anything about that world at that time. It is not that e is evidence for those in any particular epistemic situation.

This is different from the relationship between evidence and belief in chapter 2, where degrees of belief were not introduced.

There is another difference as well. A subjective evidence statement of the form

e is x's evidence that h at time t,

we said, is true if and only if at t, x believes that e is (veridical) evidence that h; x believes that h is true or probable; and x's reason for believing that h is true or probable is that e is true. On the subjective theory of probability, however, a subjective evidence statement of the form above is not to be interpreted simply as a statement about what x believes and why. It has an evaluative component as well. Using the positive relevance definition, not only does an evidence statement claim that x has a higher degree of belief in h with e than without e, but also that it is rational or reasonable for x to be in that belief-state. (An analogous point holds for the high-probability definition.) This claim of rationality or reasonableness on x's part is absent in the case of what in chapter 2 were called subjective evidence statements. Saying that in 1883 Hertz's experimental results constituted Hertz's evidence that cathode rays are electrically neutral carries no implication that Hertz's beliefs in this regard were rational or reasonable.

13. Conclusions

Two standard probability definitions of evidence—positive relevance and high probability—must be augmented by some interpretation of probability if a full theory of evidence is to be formulated and evaluated. Four popular theories of probability have been presented, each of which yields a different concept of evidence, even if we focus on just one of the two probability definitions of evidence. With Carnap's logical theory of probability we obtain a concept of evidence most akin to ES-evidence. With the frequency theory of Salmon and the propensity theory of Lewis we obtain concepts similar to potential evidence. Using a subjective theory of probability we generate a concept of evidence something like what I have called subjective evidence (or x's evidence). In none of these cases do we obtain veridical evidence.[21] However, even in the other cases the concepts yielded by the two probability definitions are different in significant ways from the ones described in chapter 2, particularly as concerns the relationship between evidence and belief. These differences have important consequences for our evaluation of the probability theories of evidence in the next chapter.

---

21. In certain restricted cases, Lewis's theory will yield veridical evidence, viz. when, with respect to an evidence statement of the form "in world w at time t, e is evidence that h," world w at time t includes a situation in which h is true.

4

What's Wrong with These Probabilistic Theories of Evidence?

In this chapter I will present objections in the form of counterexamples to the positive relevance and high-probability definitions of evidence. These counterexamples are supposed to work no matter what definition of probability is held. I will then discuss how probabilists can respond using their own interpretations of probability, and whether such responses are successful.

1. The Counterexamples

The first two counterexamples purport to show that the positive relevance definition fails to provide a sufficient condition for evidence.

(i) *First lottery counterexample*
- b: On Monday all 1000 tickets in a lottery were sold, of which John bought 100 and Bill bought 1. One ticket was drawn at random on Wednesday.
- e: On Tuesday all the lottery tickets except those of John and Bill were destroyed, and on Wednesday one of the remaining tickets was drawn at random.
- h: Bill won.

On any of the interpretations of probability in chapter 3, a reasonable assignment of probabilities in this case is this:

$p(h/b) = \frac{1}{1000}$, since given just b, Bill has just one of the 1000 tickets
$p(h/e\&b) = \frac{1}{101}$, since given the additional information e, Bill now has 1 of the 101 tickets remaining.[1]

---

[1]. For Lewis, who relativizes objective probability to a world and a time, these probabilities could be relativized to the imagined world prior to Monday. Similarly, on a subjective theory of probability, the probabilities could be relativized to one's knowledge prior to Monday. On a frequency theory, since probabilities are not assigned to sentences, "weights" would need to be substituted.

Since $p(h/e\&b) > p(h/b)$, on the positive relevance definition of evidence, e is evidence that h, given b.[2] But although Bill's probability of winning has increased almost tenfold from $\frac{1}{1000}$ to $\frac{1}{101}$, do we want to conclude that the destruction of all the tickets except for those of John and Bill is evidence that Bill won? On the positive relevance definition this fact about the ticket destruction is evidence that John won, and it is also evidence that Bill won. Isn't it much more plausible to conclude that this is evidence that John won, not Bill, since on Tuesday, as a result of the ticket destruction, John owned 100 of the 101 remaining tickets?

(ii) *A swimming counterexample*
- b: Steve is a member of the Olympic swimming team who was in fine shape Wednesday morning.
- e: On Wednesday, Steve was doing training laps in the water.
- h: On Wednesday, Steve drowned.

Let us suppose that h's probability is increased by e over what it is on b alone: $p(h/e\&b) > p(h/b)$. That is, swimming in water, even for Olympic swimmers, increases their probability of drowning. On the positive relevance definition we must conclude that, given that Steve is an Olympic swimmer in fine shape Wednesday morning, the fact that he was doing training laps in the water on Wednesday is evidence that he drowned on Wednesday. This claim seems extremely implausible.

These counterexamples are designed to show that the fact that e increases the probability that h is true is not *sufficient* to make e evidence that h. Now for two examples that purport to establish that increasing the probability of a hypothesis is not even *necessary* for evidence.

(iii) *Second lottery counterexample*
- e_1: The New York Times reports that Bill Clinton owns all but one of the 1000 lottery tickets sold in a lottery.
- e_2: The Washington Post reports that Bill Clinton owns all but one of the 1000 lottery tickets sold in that lottery.
- b: This is a fair lottery in which one ticket drawn at random will win.
- h: Bill Clinton will win the lottery.

Given $e_1\&b$, information e_2 concerning The Washington Post report is, it would seem, (strong) evidence that h. Yet, assuming the following plausible assignment of probabilities, e_2 fails to increase h's probability on $e_1\&b$:

$$p(h/e_1\&e_2\&b) = p(h/e_1\&b) = \tfrac{999}{1000}.$$

If this counterexample is accepted, then, contrary to the positive relevance definition, there can be evidence that h that does not increase h's probability.

The next counterexample purports to demonstrate something even stronger, viz., that there can be evidence that h that *decreases* h's probability!

(iv) *Intervening cause counterexample*
- e_1: On Monday at 10 A.M. David, who has symptoms S, takes medicine M to relieve S.

---

2. It is objective or subjective evidence depending on whether an objective or a subjective interpretation of probability is being employed.

e_2: On Monday at 10:15 A.M. David takes medicine M' to relieve S.
b: Medicine M is 95% effective in relieving S within 2 hours; medicine M' is 90% effective in relieving S within 2 hours, but has fewer side-effects. When taken within 20 minutes of having taken M medicine M' completely blocks the causal efficacy of M without affecting its own.
h: David's symptoms S are relieved by noon on Monday.

Given $e_1\&b$, information e_2 is, one might claim, (strong) evidence that h will be true, since medicine M' is 90% effective in relieving symptoms S. Yet on the following plausible assignment of probabilities, the positive relevance definition of evidence is violated here with a vengeance:

$p(h/e_1\&b) = .95$
$p(h/e_2\&e_1\&b) = .90$.

Not only does e_2 not increase h's probability on $e_1\&b$, it decreases it. Yet, one might claim that e_2 is evidence that h, given $e_1\&b$. If this counterexample to the positive relevance definition is accepted, then there can be evidence that a hypothesis is true that decreases its probability.

These examples are constructed to show that the positive relevance definition provides neither a necessary nor a sufficient condition for evidence. How a defender of positive relevance can reply is a question I shall consider after introducing a counterexample against the high-probability definition of evidence.

(v) *Irrelevant information counterexample.*
e: Michael Jordan eats Wheaties.
b: Michael Jordan is a male basketball star.
h: Michael Jordan will not become pregnant.

It is plausible to assume that $p(h/b)$ is very high, certainly greater than $\frac{1}{2}$. Suppose we add the information e to b. This addition has no effect one way or the other on the probability of h. That is, $p(h/e\&b) = p(h/b) > \frac{1}{2}$. Now on the high-probability definition of evidence, if h's probability on e is high, given b, that is sufficient for e to be evidence that h, given b. If so, we obtain an evidence claim that seems absurd, viz. that, given that Michael Jordan is a male basketball star, the fact that he eats Wheaties is evidence that he will not become pregnant. We may be willing to say that the fact that he is male is evidence that he will avoid pregnancy. But it seems counterintuitive to say that, given that he is a male basketball star (or even just a male), his eating Wheaties is evidence that this hypothesis is true. If so, the high-probability definition of evidence fails to provide a sufficient condition for evidence. (Later I will argue that it does provide a necessary condition.)

2. The Ambiguity Response

It is important to consider responses to these counterexamples, because they force us to take a critical look at fundamental assumptions we are making about evidence. The first response I call the "ambiguity response," since it derives from a claim of Carnap and Salmon that the concept of confirmation or confirming evidence is ambiguous.[3] Here is a passage from Salmon:

> As Carnap pointed out in *Logical Foundations of Probability*, the concept of confirmation is radically ambiguous. If we say, for example, that the special theory of relativity has been confirmed by experimental evidence, we might have either of two quite distinct meanings in mind. On the one hand, we may intend to say that the special theory has become an accepted part of scientific knowledge and that it is very nearly certain in the light of supporting evidence. If we admit that scientific hypotheses can have numerical degrees of confirmation, the sentence, on this construal, says that the degree of confirmation of the special theory on the available evidence is high. On the other hand, the same sentence might be used to make a very different statement. It may be taken to mean that some particular evidence—e.g., observations on the lifetimes of mesons—renders the special theory more acceptable or better founded than it was in the absence of this evidence. If numerical degrees of confirmation are again admitted, this latter construal of the sentence amounts to the claim that the special theory has a higher degree of confirmation on the basis of the new evidence than it had on the basis of the previous evidence alone.[4]

Accordingly, the present response to my counterexamples (i)–(iv) to the positive relevance definition is this. These counterexamples all presuppose Carnap's and Salmon's first sense of confirming evidence (which involves *high* confirmation and is to be understood as high probability). But they are directed against the second sense of confirming evidence (which involves *increase* in confirmation and is to be understood as increase in probability).

Take, for instance, my swimming counterexample (ii). The fact that *e*, on Wednesday Steve was doing training laps in the water, increases the probability of the hypothesis *h*, that on Wednesday he drowned. But, I claimed, it is extremely implausible to say that this fact is evidence that this hypothesis is true. Carnap and Salmon will agree with this conclusion *if* we are speaking of evidence in the sense of high confirmation. The fact that *e* is true does not make *h* highly confirmed. But it does increase its confirmation, and in that distinct sense it is evidence.

Similarly, Carnap and Salmon will say, my irrelevant information counterexample (v) fails to succeed because it presupposes Carnap's and Salmon's second sense of confirming evidence, viz. increase in confirmation. True, if we are using the latter concept, then the fact that Michael Jordan eats Wheaties is not evidence that he will not become pregnant, since it does not increase the latter's confirmation. But it is evidence in Carnap's and Salmon's first sense. It is information on the basis of which, together with the background information that Michael Jordan is a male basketball star, the hypothesis is highly confirmed.

---

3. This response to my counterexamples is explicitly made by Steven Gimbel, "Peirce Snatching: Toward a More Pragmatic View of Evidence," *Erkenntnis*, 51 (1999), 207–231. Cf. Frederick M. Kronz, "Carnap and Achinstein on Evidence," *Philosophical Studies*, 67 (1992), 151–167, and my reply, "The Evidence Against Kronz," *Ibid*, 169–175.
4. Wesley Salmon, "Confirmation and Relevance," reprinted in Peter Achinstein, ed., *The Concept of Evidence* (Oxford: Oxford University Press), 95–123; quote on p. 95.

Reply to the ambiguity response. The ambiguity response assumes that there are two different concepts of evidence, increase in confirmation and high confirmation, which are to be explicated as positive relevance and high probability, respectively. But this is just what is at issue. So let us confront the issue.

Is there a sense of evidence which might be explicated as positive relevance, that is, increase in probability? (In section 9 I will raise the corresponding question for the high-probability account.) Salmon in the passage quoted earlier claims that there is a sense of evidence or confirmation in accordance with which evidence is what renders a hypothesis "more acceptable or better founded than it was in the absence of this evidence." Carnap writes that

> The verb "to confirm" is ambiguous and has perhaps the connotation of "making firmer" even more often than that of "making firm."[5]

Patrick Maher, a defender of the positive relevance account, writes that there is (what he calls) a "basic idea" underlying the concept of evidence, viz.

> E is evidence for H if and only if E makes it rational to be more confident that H is true.[6]

So evidence or confirmation, in the present sense, is something that makes a hypothesis "more acceptable or better founded" (Salmon), "firmer" (Carnap), and rational to be "more confident of" (Maher). Let us consider the expressions in quotations: *more acceptable, better founded, firmer,* and *more confident of.* I want to make two claims about these and similar terms.

First, in general, if there is more x under one set of circumstances than another, then there is at least some x under the former circumstances, even if there is none in the latter. If there is more money in the bank, more cake on the table, or more water in the pot, at 1 P.M. than there was at noon, then at 1 P.M. there is at least some money in the bank, cake on the table, or water in the pot. Similarly, if information e makes a hypothesis h more acceptable (better founded, firmer), or if e makes me more confident of h than I was without e, then with e hypothesis h has at least some acceptability (foundation, firmness), and I have at least some confidence in it. To be sure, money, cake, and water are "stuffs," while acceptability, firmness, and confidence are not. The latter, we might say, are "states" that admit of differences in strength or degree. But a similar principle holds for each. If a greater quantity of the "stuff" is present in one circumstance than in another, then there is at least some of that "stuff" in the former circumstance; if the state is present to a greater degree in one situation than in another, then it is present to some degree in the former.

My second claim is that concepts such as acceptability, having some foundation, firmness, and confidence, in so far as these depend on probability, are "threshold" concepts. A necessary condition that must be satisfied for a hypothesis h to have any acceptability, foundation, or firmness, and before I have any

---

5. Carnap, *Logical Foundations of Probability*, 2nd ed., p. xviii.
6. Patrick Maher, "Subjective and Objective Confirmation," *Philosophy of Science,* 63 (1996), 149–174. Unlike Carnap and Salmon, Maher does not claim that the concept of evidence is ambiguous. Maher's positive relevance theory will be examined in sections 10 and 11 of this chapter.

confidence in it, is that h's probability exceed some threshold.[7] In this regard, these concepts (with respect to probability) are like the following concepts (with respect to the items in parentheses): crowds (with respect to number of people), electrically induced pain (with respect to voltage), and my fear of death due to playing a sport (with respect to the rate of death in that sport). Each of these concepts is related to the concept in parentheses. But there needs to be a certain threshold amount falling under the concept in parentheses in order that any amount of the concept dependent on it is realized. One person in the audience is not a crowd, even a small one. Two volts of electricity are not sufficient to produce any pain in me. A death rate of 1 per 500 million rounds of golf is not enough to produce any fear in me that I will die playing golf.[8]

Applying this to probability, assuming I buy 1 ticket in a 100 million ticket lottery, where one ticket will be selected at random to win, I increase the probability that I will win from 0 to 1/100 million. However, I increase my confidence in winning (firmness of the support of the hypothesis that I will win, etc.) only if I come to have at least some confidence in winning (the hypothesis has some firmness of support, etc.). But I have no such confidence (the hypothesis has no firmness of support). There is not enough probability here to exceed a threshold necessary for me to have any confidence (for the hypothesis to have any firmness, support, etc.). By analogy, although my fear of dying playing golf does depend on the death rate from golf, even if that death rate were quadrupled from 1 fatality in 500 million rounds to 4, that would not suffice to produce in me any fear of dying playing golf.

In chapter 2, two of the objective concepts of evidence, potential and veridical evidence, involve the idea of "providing a good reason to believe a hypothesis," while the third objective concept, ES-evidence, involves "providing a justification for belief." The concepts of "good reason for belief" and "justification for belief," in so far as these depend on probability, are threshold concepts. A probability of death from playing a round of golf equal to 1/500 million is not sufficient to provide a good reason to believe, or a justification for my believing, that I will die during my next round of golf. (In chapter 6 I will argue that for e to be evidence that h the three objective concepts of evidence require as a necessary condition that h's probability on e be greater than $\frac{1}{2}$.)

Accordingly, I reject the ambiguity response. Even if probabilists were correct in supposing that there is a sense of evidence that involves the idea of increase-in-strength-of-evidence, and even if the latter is related to probability, it does not follow, and indeed it is false, that any increase in probability is an increase in the strength of the evidence. Contrary to what probabilists claim, the concept of increase-in-strength-of-evidence (which they attempt to analyze as positive relevance) requires a prior notion of evidence (which they associate with high probability).

---

7. The probability in question will be subjective or objective according to whether the concept depending on it is construed subjectively or objectively. For example, in the case of my confidence, a subjective concept, the probability in question is the subjective probability I assign.
8. As reported in *The New York Times Magazine*, June 18, 1995, in 1993 24.5 million golfers played nearly 500 million rounds; one golfer died via lightning.

Those, such as Carnap and Salmon, who say that "evidence" (or "confirmation") is ambiguous, claim that when we assert that e is evidence that h (or that e confirms h) we can mean either that e increases the strength of the evidence or that e provides strong evidence that h. Although some piece of evidence might do both, it can be evidence without doing either. This is readily seen if we invoke the previous intervening-cause counterexample (iv) involving the medicines M and M' but change the figures a bit. Suppose that medicine M is 75% effective in relieving symptoms S and that M' is 65% effective. Again, suppose that when M' is taken after taking M, it destroys M's causal efficacy without losing any of its own. We also suppose that at 10 A.M. David took M to relieve S. Given this information, the fact that David took M' at 10:15 A.M. is evidence that his symptoms will be relieved, despite the fact that it is not particularly strong evidence,[9] nor does it increase the strength of the evidence from what it was before.

Finally, is there a distinct sense of "evidence" according to which it means "increase in strength of evidence (or confirmation)," as well as a distinct sense according to which it means "evidence (or confirmation) of a high degree of strength," as Carnap and Salmon claim? No more, I suggest, than there is a distinct "increase-in-degree-of-probability" sense of "probability," an "increase-in-firmness" sense of "firm," or an "increase-in-salary" sense of "salary." In each of these cases, including evidence, we are dealing with something that admits both of increases and of high degrees or amounts. The fact that we can distinguish claims about increases from claims about high degrees does not establish that the concept of evidence (or probability, or firmness, or salary) is ambiguous.

3. The "Slight Evidence" Response

This response of probabilists is addressed to my counterexamples (i) and (ii) that purport to show that increase in probability is not sufficient for evidence. The response is that in my counterexamples the reason it may seem implausible to say that e is evidence that h, even though e increases h's probability, is that e only *slightly* increases h's probability. In such cases the numerical difference between $p(h/e\&b)$ and $p(h/b)$ is very small. So when I deny that e is evidence that h in such cases I am doing so, probabilists may say, because I am contrasting these cases with ones in which e boosts h's probability considerably. In both types of cases, however, e is evidence that h, although where the boost is small so is the strength of the evidence.

Reply. Again I invoke the idea of threshold concepts. According to Carnap, Salmon, and Maher, the present concept of evidence is tied to the idea of "making h firmer or more acceptable," "making it rational to be more confident that h is true," "giving h more support." But if e makes h firmer (more acceptable, etc.) than it was without e, then, with e, h has at least some firmness. But firmness, and the other concepts invoked, are threshold concepts with respect to probabil-

---

9. To be sure, the probability here is .65. But recall Salmon's claim that, in the first sense of "confirmation," to say that the special theory of relativity has been confirmed means that "it is very nearly certain in the light of supporting evidence."

ity. It is necessary that some threshold of probability be exceeded if a hypothesis is to have any firmness at all.

The fact that I am now riding on an elevator increases the risk of my being injured in an elevator accident from 0 to 1 in 6 million.[10] But it gives no firmness (acceptability) at all to that hypothesis. The reason it seems (and is) implausible to speak of evidence in such a case has nothing to do with the fact that the *boost* in probability is so small. The reason is rather that the probability itself is so small.

4. The "Justification for Higher Degree of Belief" Response

This response of probability theorists is related to the previous ones but is worth treating separately. It begins by noting the relationship between evidence and belief on probability theories of evidence *as this relationship was explicated in chapter 3*. The response can be stated by focusing on Carnap's theory and the positive relevance definition.

According to Carnap, if e is evidence that h, given b, then anyone whose epistemic situation involves knowing $e \& b$ is justified in believing h to a higher degree than anyone whose epistemic situation involves knowing just b. Let's apply this to the swimming counterexample. Surely anyone whose epistemic situation involves knowing b (Steve is a member of the Olympic swimming team who was in fine shape Wednesday morning) together with e (on Wednesday Steve was doing training laps in the water) is justified in believing h (on Wednesday Steve drowned) to a higher degree than anyone whose epistemic situation involves knowing just b. Even if, given b, the information e does not justify (or make it reasonable to have) a belief in h, it does something sufficiently important to give e the status of evidence that h: *it provides a justification for believing h to a higher degree than does b alone*.

Reply. I deny the emphasized proposition by making three claims. The first is that if, given e and b, one believes h to a higher degree than one does given just b, then, given e and b, one believes h to some degree or other.[11] (Analogous claims were made in the previous sections.) The second claim is related. If e provides a justification for believing h to a higher degree than does b alone, then it provides a justification for believing h to some degree or other.

The third and most important claim is that the concept of belief itself is a threshold concept with respect to the probability the believer assigns. This does not imply that there are no degrees of belief, that belief is an all-or-nothing affair. Let us grant that belief (like fear and confidence) can have degrees. I can believe h more than h', or more than you do. But in so far as the degree or amount of my belief depends on what probability I assign, it is a threshold concept. There

---

10. *Discover*, May 1996, p. 82.
11. Isaac Levi, *Gambling with Truth* (New York: Alfred A. Knopf, 1967), makes a related claim, viz. "to believe that P to a positive degree (to believe positively that P) requires as a necessary and sufficient condition the belief that P" (p. 122). In chapter 6 I take up this claim and defend it.

must be a certain significant probability I attach to h for me to have *any* belief in h at all. A probability of 1 in 6 million that I will be injured on this elevator ride is not sufficient to give me even the slightest amount or degree of belief that I will be injured, though I recognize that its probability is not zero. Nor does the fact that I am entering the elevator provide a justification for any degree of belief in this hypothesis. The probability is just not high enough. Therefore, by the second claim above, the fact that I am entering the elevator fails to justify my believing this hypothesis to a higher degree than without this information. Accordingly, even if e increases h's probability on b, it does not follow that e provides a justification for believing h to a higher degree than does b alone. The relationship between increase in probability and a justified higher degree of belief of the sort probabilists infer does not exist.

5. Defending the Claim That Belief Is a Threshold Concept

The conclusion just reached depends on the assumption that belief is a threshold concept with respect to probability. This assumption will be questioned by some or all probabilists. It may be urged that once I agree that beliefs have different strengths that can vary with probabilities, then since probabilities take on all values between 0 and 1 so do beliefs. This, however, does not follow. A threshold concept (such as crowds or electrically induced pain) can have different degrees or strengths that vary with some quantity that takes on values between 0 and some higher number. But it does not follow that such threshold concepts are exemplified whenever that quantity is greater than 0. (One person is not a crowd; two volts of electricity will not produce pain.)

It may also be urged that although other concepts I have mentioned (including crowds and electrically induced pain) are threshold concepts, belief is not such a concept with respect to probability. If I agree that the probability is 1 in 6 million that I will be injured on this elevator ride, then, it may be argued, I have some degree of belief in this hypothesis, albeit a very tiny amount. If I own 1 ticket in a million ticket lottery and I assign the corresponding probability to my winning, then I have some degree of belief that I will win. If I have no degree of belief that this will occur, if I do not at all believe I will win, why did I buy a ticket?[12]

My response is that I bought a ticket not because I believe to some small degree that I will win, but because buying a ticket is a necessary condition for winning (it gives me 1 chance in a million of winning, whereas buying no ticket gives me no chance), and if I do win, which is extremely unlikely, I will become a millionaire. To be sure, even though I do not at all believe I will win, I am not 100% certain I won't, only .999999. But all this means is that I assign that probability to my not winning and .000001 to my winning. It does not mean that I believe, even slightly, that I will win.

---

12. Alvin Plantinga raises this very question with regard to a similar example and concludes that it is false to claim I believe I won't win. *Warrant and Proper Function* (New York: Oxford University Press, 1993), p. 166.

Normally, if one believes something to some degree this is expressed in some actual or potential behavior, including verbal behavior; usually one is, or can readily be made, aware of having some degree of belief. If I believe to some degree that I will be in an elevator accident, then I may try to avoid elevators, or be a little bit nervous when riding one, or when prompted say that I believe to some degree that an elevator accident will occur, or at least be aware, or be readily made aware, that I have this belief. To be sure, there may be factors preventing me from expressing a belief I have either verbally or behaviorally, such as my desire to prevent you from knowing about my belief; and there may be factors causing me to be unaware that I have such a belief, and causing it to be difficult to make me aware of this, such as my need to repress that belief for deep-seated psychological reasons. Suppose that knowing the general accident rates in elevators, I assign a probability of 1 in 6 million that I will be injured on the elevator I have just entered. It is perfectly conceivable that in the absence of factors that may prevent me from expressing or realizing any beliefs I have about elevator accidents, no behavior expressing a corresponding degree of belief of mine is forthcoming or can be elicited. I do not hesitate even slightly to get on the elevator. I show no nervousness at all during the ride. If pressed I will not say, even to myself, that I have some tiny degree of belief that there will be an accident.

A degree-of-belief theorist may respond by saying that in such a case there is a corresponding degree of belief, but the reason it is not expressed in my behavior, and the reason I am unaware of it, is that the degree is so small. Normally, a degree of belief is expressed in behavior, and one is aware of it, only when it is sufficiently large.

To this I respond with a question: If it is not expressed in behavior, even when prompted, and if one is not aware of it, how does the degree-of-belief theorist know it exists? One answer suggested is that the probability one assigns to the hypothesis is not zero. But that is not enough to show that my degree of belief is not zero, unless we already agree that probability measures degree of belief.

A more popular answer is that degrees of belief, even tiny ones, *are* expressed in behavior, viz. betting behavior. If I am willing to bet that I will win the lottery with maximum betting odds 1 to 999,999, then I have some degree of belief that I will win. If I have none, then I should be unwilling to bet on my winning. My response is that I am willing to bet on h with those odds not because I have a certain degree of belief that I will win, but because the probability I assign to my winning is 1/1 million. I would assign such a probability if I own 1 ticket of the 1 million sold. So even though I have a chance of winning, albeit a miniscule one, I do not believe, even a little bit, that I will win.

On a threshold notion of belief, then, degrees of belief are not probabilities one assigns, although they depend on such probabilities. I can have no degree of belief in a certain hypothesis while assigning it a probability greater than zero. In chapter 5 I will develop an objective epistemic theory of probability that does not tie probability to degree of belief or to degree of rational belief. If probabilities are construed as I will suggest there, then we need to distinguish three different ideas: (1) the objective epistemic probability of a hypothesis h; (2) what someone takes the objective epistemic probability of h to be; (3) what someone's degree of belief in h is. I am claiming here that the third idea is not identical with the second, although it does depend on it.

6. Further Probabilist Responses to a Threshold Concept of Belief

Probabilists may say that a threshold concept of belief is incompatible with betting behavior that is rational.[13] Consider again the lottery in which I buy one of the 1 million tickets sold. I assign a probability of 1/1 million to the hypothesis that I will win. Yet I deny that I have some degree of belief that I will win. If I have no degree of belief at all, not even a scintilla, then, it might be said, I ought to bet at any odds whatever that I will not win. I should bet all that I own and will ever own to gain $1 from you. But surely that would be irrational.

I agree that it would be irrational, if this conclusion followed. But it does not follow. Let us suppose that b is the maximum amount I am willing to bet on hypothesis h to gain an amount a, so that $b:a$ represents my betting odds on h. Then even if $b/a+b$ is the probability I assign to h, that probability does not necessarily represent my degree of belief in h, since I may have no degree of belief in h at all. On the threshold view, even if the maximum betting odds I am willing to accept on h indicate what probability I assign to h, and even if this probability is greater than zero, this does not necessarily mean that I have some degree of belief in h. My willingness to bet with the odds 1:999,999 that I will win the lottery (yielding a probability for me of 1/1 million) is perfectly compatible with my having no degree of belief whatever that I will win, even if I am not 100% certain that I won't win. Moreover, as long as my betting behavior is in accord with the probability rules, I am not subject to a system of bets I am bound to lose. Avoiding a "dutch book" does not require that my betting odds indicate degrees of belief.

Probabilists may admit that although the ordinary, everyday concept of belief is a threshold concept with respect to probability, they are introducing an idealization—a technical concept of belief that is not a threshold concept with respect to probability. On this notion one can have any degree of belief that corresponds to a probability. If I assign a probability of 1 in 6 million to my being injured on this elevator ride, then, in this technical sense, I do have some degree of belief I will be injured, albeit a tiny one. A probabilist will now reiterate that the fact that I am entering the elevator is evidence, albeit a tiny bit, that I will be injured.

My response is that this involves (at best) a technical, nonstandard concept of evidence. Ordinarily, if I know of evidence, however small, that I will be injured on this elevator, I will get a little worried. I may get off the elevator as soon as I can, or at least feel tempted to do so. That is because if there is some evidence for this hypothesis of which I am aware, then normally I will have at least some degree of belief in the hypothesis (in the ordinary, nontechnical sense). But when

13. I am grateful to Adam Goldstein for raising this issue.

the "evidence" is simply that the injury rate on elevators is 1 in 6 million, I have no inclination to leave the elevator. I have no reason to have any belief at all (in a nontechnical sense) that I will be injured. At best this is a technical sense of evidence that in this and many other cases has no effect whatever on my beliefs (in the ordinary sense) or my actions based on those beliefs.

Probabilists may respond by complaining that I have multiplied entities beyond necessity, since I admit not only that there are beliefs but degrees of belief as well. They admit only the latter. If this is meant as an ontological complaint, it is off the mark. If I admit that crowds exist and say that some are larger than others, I am not postulating two sorts of entities, crowds and "crowd degrees." I am simply saying that there are crowds, some of which are larger than others. Similarly, I am saying that there are beliefs, some of which are held more strongly than others.

Finally, probabilists may say that my claims about beliefs are subject to the "lottery paradox,"[14] and hence a contradiction. Suppose there is a lottery consisting of 1 million tickets, one of which will be selected at random. Now, on a threshold view of belief, the probability of any particular ticket's winning (1 in a million) is too low to give me any degree of belief that that ticket will be selected. So I do not believe it will be selected. Indeed, so the paradox goes, I believe it won't be selected. The same can be said for each of the million tickets. But if I believe $not\text{-}h_1$ and I believe $not\text{-}h_2$, then I believe the conjunction $not\text{-}h_1 \text{ \& } not\text{-}h_2$. Therefore, if we consider a conjunction in which each conjunct is of the form "ticket i will not win" for each of the 1 million tickets, then I believe that conjunction. This is tantamount to believing that none of the 1 million tickets will win. And this implies that I do not believe that one of the tickets will win. But I do believe that one of the tickets will win. (That probability is equal to 1.) So a contradiction is generated: I believe one ticket will win, and I don't believe that any ticket will win.

My response is to point out that the threshold idea of belief precludes the conjunction principle of belief. Suppose that the probability of h_1 is sufficiently high to allow me to have some belief that h_1 is true. Suppose the same is true of h_2. It does not follow that the probability of the conjunction $h_1 \text{ \& } h_2$ is sufficiently high to allow me to have some belief that it is true. This is exactly the situation in the lottery paradox. Even if the probability of any particular hypothesis of the form $not\text{-}h_i$ (ticket i will not win) is sufficiently high to give me some belief that it is true, it does not follow, and indeed it is false, that the entire conjunction of these hypotheses has a probability that is sufficiently high to give me some belief that it is true.[15]

7. Degrees of Acceptability: A Positive-Negative Scale

Positive relevance enthusiasts may now concede that if evidence needs to make a hypothesis "more acceptable" ("firmer," etc.), and if the latter is tied to proba-

---

14. Henry E. Kyburg, *Probability and Inductive Logic* (New York: Macmillan, 1970), p. 176.
15. In chapter 6, sections 12 and 13, I consider (and reject) an interesting defense of the conjunction principle offered by Mark Kaplan.

bility, then the hypothesis needs to surpass some threshold of probability. However, instead of changing their definition of evidence, these enthusiasts may withdraw their claim that evidence needs to make a hypothesis "more acceptable." Here is how they might proceed.

With states such as acceptability it is possible to consider a positive-negative scale that includes both the acceptable and the unacceptable. One might then speak of "degrees of acceptability" ranging from 0 to 1, without presupposing that something with nonzero acceptability is acceptable to some extent. One punishment, such as receiving a lethal injection, might have a higher "degree of acceptability" to a condemned prisoner than another, such as being hung, without either having any acceptability at all. It is simply that one punishment is less unacceptable than the other.

If we suppose that acceptability in the case of hypotheses is tied to probability but that it is a threshold concept with respect to probability, we might draw the following graph:

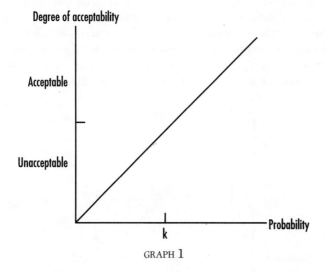

GRAPH 1

The x-axis represents probability, the y-axis degrees of acceptability. k represents a threshold of probability necessary for acceptability. That is, for h to have any acceptability at all, it must have a probability greater than k. Let us also suppose that exceeding k is not only necessary for acceptability but sufficient as well.[16] Finally, we suppose that below k a hypothesis has varying degrees of unacceptability. With a probability equal to k it is neither acceptable nor unacceptable. The present proposal then is that e is evidence that h, given b, if and only if either e makes h more acceptable on b than it is on b alone, or e makes h less unacceptable on b than it is on b alone.

This yields an extremely weak concept of evidence, especially where probabilities are below the threshold for acceptability. In this range evidence that h

---

16. In chapter 7 I will consider the possibility of having a precise number that is necessary but none that is both necessary and sufficient.

does not give h more acceptability but simply less unacceptability. In terms of a good reason to believe, which is supplied by potential and veridical evidence, where the probability threshold is not surpassed, evidence that h would not give a good reason to believe h but at best less reason to disbelieve h. So the fact that I am entering an elevator is evidence that I will be in an elevator accident, by virtue of the fact that my entering the elevator gives less reason to disbelieve that I will be in such an accident than without that fact. Such a notion of evidence, I think many (including the dean of chapter 1) would say, is too weak to be of scientific interest.

Moreover, graph 1 is considerably idealized. Consider an analogy between the relationship between the acceptability of a hypothesis and its probability, on the one hand, and the relationship between the acceptability of a job candidate and the candidate's grade-point average (GPA), on the other. Suppose that five levels are distinguished on the acceptable-unacceptable scale, denoted by numbers 0 through 4, with 0–2 being unacceptable and 3 and 4 being acceptable. Suppose that there is a certain threshold GPA, say 3.0, such that if the candidate's GPA is greater than this, the candidate satisfies minimum acceptability for the job (level 3). With a 3.5 GPA or above the candidate is highly acceptable (level 4). With a 2.0 GPA or less the candidate is unacceptable. Between 2.0 and 3.0, however, acceptability and unacceptability judgments do not depend simply on GPA. So there are no points on the graph between these grades. The graph looks like this:

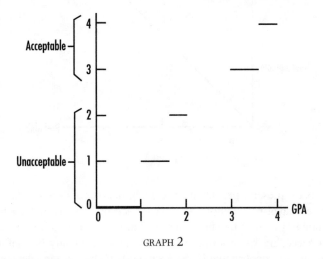

GRAPH 2

This is a step function. Although GPA takes on all possible values from 0 to 4, the degree of acceptability takes on only 5 possible values (0,1,2,3,4). On this graph, even though one candidate's GPA is higher than another's, his "degree of acceptability" may not be higher at all.

Acceptability in the case of hypotheses works in a similar way, if tied to probability. We make some finite number of distinctions among acceptable and unacceptable hypotheses. For example, if John owns 1 ticket in a million ticket lottery and Mary owns 2, the hypotheses "John will win" and "Mary will win" might be considered to have the same degree of unacceptability (say "complete unacceptability"), even though their probabilities are different. It is quite possible that there are more "degrees of acceptability" in the case of hypotheses than there are in our job candidate example, so that in the hypothesis case the graph might look like this:

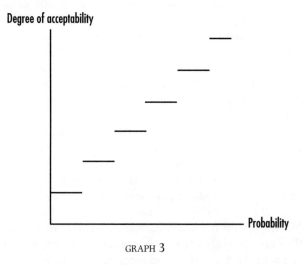

GRAPH 3

When we draw graph 1, we idealize by smoothing out the curve. We turn a step function into a continuous one by imagining more and more steps closer together. This is what is done by those giving an epistemic interpretation of probability in terms of concepts such as acceptability and believability. (In chapter 5, the epistemic concept of probability I introduce does just this.)

8. Must Evidence Increase Probability?

Defenders of the positive relevance definition of evidence might now agree that evidence requires that a certain threshold of probability be reached, but still insist that evidence (in any important sense given to that term) must always increase the probability of the hypothesis from what it is without the evidence. The modified view is this:

> e is evidence that h (in the present sense) if and only if $p(h/e) > p(h)$, and $p(h/e) > k$, where k is some threshold of probability required for evidence.

In effect, this would combine the two standard probability definitions of evidence into one.

I want to focus on the question of why evidence (in any sense), assuming that it depends on probability, must *increase* the probability of a hypothesis. So suppose that $p(h/e)$ surpasses the threshold required for e to be evidence that h. Why, in any sense of "evidence," must e also increase the probability of h over what it was without e?

Again, let us draw an analogy between evidence and two other concepts that can depend on probability: confidence and fear. Suppose that at noon I take drug

D to relieve pain. Drug D is known by me to relieve pain an hour later, but also to produce bad breath by that time. So at noon I have confidence that I will be free of pain around 1 P.M., but I also fear that I will have bad breath. Within a few seconds after taking D, my doctor gets me to take another drug D' which prevents D from acting, but, he suddenly remembers, is just as effective as D in relieving pain. Although it also produces bad breath, it has fewer potential long-term side effects than D. So taking D' makes me confident that my pain will be relieved by 1 P.M., and makes me fear that I will have bad breath by then. But my confidence and fear in the second case have not been *increased*. In short, something can make me confident that a certain event will occur, and can make me fear that another event will occur, without increasing my confidence or fear that these events will occur.

Indeed, we could alter the example so that something gives me confidence and fear that certain things will happen, even though the amounts of my confidence and fear have decreased. Just use the drug example above, and suppose that although drug D' cancels the power of D, it has a slightly lower probability of relieving pain and of producing bad breath than D does. I am still confident that my pain will be relieved by 1 P.M., and I still fear that I will have bad breath by then, even though my degrees of confidence and fear are slightly less.

Evidence works in exactly the same way, assuming it is tied to probability. Suppose that drug D is 98% effective in relieving pain, while D' is 90% effective. Again assume that when D' is taken after D it cancels D's power without losing any of its own. Given this background information, isn't the fact that I am taking D' after having taken D evidence that my pain will be relieved, even though the probability that it will be relieved has decreased from 98% to 90%? If so then increasing probability is not required for evidence.

Indeed, on the positive relevance view of chapter 3, if information decreases a hypothesis's probability, then that information is evidence that the hypothesis is false; it is evidence against the hypothesis. Shall we say, then, that the fact that I am taking drug D' after having taken D is evidence that the hypothesis that my pain will be relieved is false? Shall we say it is evidence against this hypothesis? That seems unacceptable. It is evidence that the hypothesis is true, it is evidence in favor of the hypothesis, although the strength of the evidence in favor of the hypothesis is slightly less than it was before. Assuming that the probability threshold for evidence has been reached, those who demand, in addition, that evidence increase the probability of the hypothesis h mistakenly equate information that constitutes evidence that h with information that strengthens the evidence that h.

9. Is High Probability Sufficient for Evidence?

Unlike positive relevance, the high-probability definition of evidence obviously satisfies the threshold condition. But is it sufficient, as some probabilists claim? Is there a sense of evidence for which h's having high probability on e (whatever the threshold for "high" is) suffices to make e evidence that h?

Let us recall how Carnap describes the alleged sense. He speaks of e as "making h firm" and of h "being firm on the basis of e." Similarly, to invoke other concepts mentioned, we might say that e makes h acceptable, or that e justifies a high degree of belief in h, or that e makes it rational to be confident that h is true. Let us draw an analogy between the concept of making firm (making acceptable, justifying, etc.) and the causal relation "causing it to be the case that," which, we might say, holds between events or between facts and events. A causal relation is *selective* in this sense. If some fact or event E_1 caused an event E, it will be false to say that E_1 and E_2 caused E, even if E_2 is a fact or is an event that occurred, unless E_2 is somehow causally involved, for instance, it is part of the cause of E, or it caused E_1 to cause E. Similarly, it will be false to say that E_2 caused E, given E_1, unless E_2 is somehow causally involved. If the fact that I took drug D caused my pain to be relieved, then it is false to say that the fact that I took drug D and that I have 10 fingers caused the relief of my pain, unless my 10 fingers, causally speaking, had something to do with this effect. And, I am supposing, it is false to say that my having 10 fingers caused this, given that I took drug D.

Similarly, if e makes h firm or acceptable, or if e justifies a belief in h, it will be false to say that e and e' together make h firm or acceptable, or that together they justify a belief in h, even if e' is true, unless e' contributes to making h firm or acceptable. And it will be false to say that e' makes h firm or acceptable given e.

Now let us return to the irrelevant information counterexample (v) involving Michael Jordan. The hypothesis h is that Michael Jordan will not become pregnant. The fact e that Michael Jordan eats Wheaties does not make the hypothesis h firm or acceptable, nor does it justify a belief in that hypothesis. Nor does this fact make h firm or acceptable given the additional fact b that Michael Jordan is a male basketball star. Nor, finally, do both facts together make h acceptable. The reason for the last two claims is that Michael Jordan's eating Wheaties plays no role in making hypothesis h firm or acceptable.

Probability, by contrast, is not selective in this way, at least not as conceived by probabilists. Given that Michael Jordan is a male basketball star (b), and therefore a male, it is highly probable that Michael Jordan will not become pregnant. That is,

$p(h/b)$ is close to or equal to 1.

The fact that Michael Jordan eats Wheaties (e) does not make it highly probable that Michael Jordan will not become pregnant. Yet e does not change the probability that h has on b; it is simply irrelevant information. So

$p(h/e\&b)$ is close to or equal to 1.

That is, for h's probability on both e and b to be high, it is not required that e contribute in some way toward that high probability. It suffices if e doesn't change the high probability that h has on b. This fact about probability is what allows h's probability on e and b together to be very high, even though e contributes nothing to that high probability (and even though h's probability on e alone is not very high).

This situation is very different from that with "making firm or acceptable." The fact b in the present case (or perhaps better, that part of b indicating that Michael Jordan is male) makes hypothesis h firm or acceptable. For both e and b

to make h firm or acceptable both must contribute. Since e makes no contribution toward making h firm or acceptable, it is false to say that e and b do together. And it is false to say that e makes h firm or acceptable, given b. If so, then if (as Carnap and others claim) "e is evidence that h, given b," or "$e\&b$ is evidence that h," is to be understood as "e makes h firm, given b," or as "$e\&b$ makes h firm," the latter two cannot be understood as

$p(h/e\&b) > k$.

This is because this probability statement is perfectly true, even though it is false that e makes h firm, given b, and it is false that $e\&b$ makes h firm. The truth of the probability statement $p(h/e\&b) > k$ may be a necessary condition for the truth of the statements in question involving firmness, but it is not sufficient (no matter what value is selected for k).

10. A Pragmatic Response: Maher's Theory

In this section and the next I will consider a reply to my counterexamples of section 1 and an interesting and powerful defense of the positive relevance conception of evidence offered by Patrick Maher.[17] I will examine how he responds to these counterexamples and how he modifies the positive relevance idea of evidence.

Maher agrees that "it would normally be inappropriate" to say that e is evidence that h in cases such as my swimming and elevator examples. He claims, however, that the inappropriateness has nothing to do with whether such evidence claims are true (which they are, according to him). Their inappropriateness derives from pragmatic rather than semantic features of language. Normally to claim that e is evidence that h in the swimming and elevator cases would not make a "relevant contribution to a conversation." They would do so, Maher insists, only "if there was some question about whether or not h was true."

Let me propose such a context for the elevator example. Suppose you see me getting on an elevator, and inform me that you have evidence that I will be injured on the ride. This claim of yours certainly raises the question for me about whether h is true (h is "I will be injured on this elevator ride"). Suppose, unbeknowst to me, you, being a worrywart, are continually (and unnecessarily) worried about elevator injuries (you were once in a freak elevator accident). So the question of whether I will be injured is also raised for you. Maher's pragmatic condition seems to be satisfied, since there is some question in the context about whether h is true. Unless other pragmatic conditions are violated (Maher mentions none), there should be nothing amiss with the statement that the fact that I am riding this elevator is evidence that I will be injured on this ride. But to me such a statement still seems "inappropriate." Despite the fact that you always worry about elevator injuries, and that the question of my being injured on this ride has been raised for both of us by your worry, it is "inappropriate" to claim

---

17. Patrick Maher, "Subjective and Objective Confirmation." See note 6. Where Maher speaks of "confirmation," I speak of "evidence."

that the fact that I am riding this elevator is evidence that I will be injured on this ride. It is inappropriate, I suggest, because it is false. No doubt, given what you believe, the fact that I am riding the elevator is *your* evidence that I will be injured (in the subjective sense of evidence of chapter 2). But it is not evidence that this is so (in any objective sense).

Maher offers a different pragmatic response to cases such as my second lottery and intervening cause counterexamples. When we consider whether *e* is evidence that *h* sometimes background information is given but usually, says Maher, it is not (p. 166). In the second lottery case if the conjunction $e_1 \& b$ is explicitly given as the background information, then, since $p(h/e_2 \& e_1 \& b) = p(h/e_1 \& b)$, e_2 is not evidence that *h*, given $e_1 \& b$. But suppose the claim is made that e_2 is evidence that *h*, where this is not explicitly relativized to any background information, although both e_1 and *b* are part of the total evidence available in addition to e_2. In such a case the context in which the evidence claim is made determines what background information is to be used for the relativization. But, unlike the situation in Carnap's account, Maher proposes that this background information should not be identified with the total evidence available. Rather,

> If $e_1, \ldots, e_n$ are all the propositions whose relevance to *h* is under discussion or otherwise salient in the context, then *ceteris paribus*, *b* is the total evidence available in that context *other than* $e_1, \ldots e_n$. (p. 167)

So, in my second lottery counterexample Maher will say that e_1 ("*The New York Times* reports that Bill Clinton owns all but one of the 1000 tickets sold in the lottery") is a proposition "whose relevance to *h* [Bill Clinton will win the lottery] is under discussion or otherwise salient in the context." Therefore, when we consider whether e_2 ("*The Washington Post* reports that Bill Clinton owns all but one of the 1000 tickets sold in the lottery") is evidence that *h* we should do so relative only to *b* ("this is a fair lottery in which one ticket drawn at random will win"), but not also to e_1. Accordingly, on Maher's positive relevance view, e_2 can be evidence that *h*, since $p(h/e_2 \& b) > p(h/b)$, despite the fact that $p(h/e_2 \& e_1 \& b) = p(h/e_1 \& b)$ and that $e_1 \& b$ comprises our total evidence available (other than e_2).

Similarly, in my intervening cause counterexample involving the two medicines *M* and *M'*, e_1 (mentioning *M*) is a proposition "whose relevance to *h* ["David's symptoms *S* are relieved by noon on Monday"] is under discussion or otherwise salient in the context." Therefore, when we consider whether e_2 (mentioning *M'*) is evidence that *h* we should do so relative only to *b* (which notes the rate of effectiveness of *M* and *M'*, and that *M'* cancels the effect of *M*), but not also to e_1. Accordingly, e_2 can be evidence that *h*, since $p(h/e_2 \& b) > p(h/b)$, despite the fact that $p(h/e_2 \& e_1 \& b) < p(h/e_1 \& b)$.

The pragmatic restriction on background information Maher offers renders his positive relevance definition of evidence different from the standard one offered by Carnap and Salmon. On the standard view, to determine the truth of

(i)   *e* is evidence that *h*,

where no background information is given, we determine the truth of

$p(h/e) > p(h)$.

On Maher's version this is not sufficient. To determine the truth of (i) we need information in addition to the putative evidence e and the hypothesis h. We need to know which, if any, propositions in b are ones "whose relevance to h is under discussion or otherwise salient in the context" in which (i) is asserted. This makes the truth of a statement of form (i) depend on the context in which it is uttered, where that context fixes the propositions that are under discussion or are otherwise salient. Accordingly, Maher's view imposes a condition that might be expressed as follows:

(ii) In context C, e is evidence that h only if $p(h/e\&b) > p(h/b)$, where b, which is fixed by the context C, contains no propositions whose relevance to h is under discussion or is otherwise salient in that context.

I state these as necessary but not sufficient conditions for Maher since he offers an additional condition that will be discussed in the next section.

Maher's proposal, which seems to be made solely to avoid counterexamples to the standard definition such as the second lottery and intervening cause cases, is, I think, vague. What is to count as "relevant to h," "under discussion," and "salient in the context." To invoke the second lottery case, let us suppose that e_1 (*The New York Times* report about Clinton's owning all but 1 of the 1000 lottery tickets) is part of the available information. Let us imagine a context in which *The New York Times* report is already known and it is accepted that this report is evidence that Bill Clinton will win the lottery. People in that context are asking whether there is evidence in other papers such as *The Washington Post*. In the context I am imagining it seems reasonable to say that the relevance (that is, the confirmatory relevance) of e_1 (*The New York Times*' report) to h is not under discussion; it is known and taken for granted. Is the confirmatory relevance of e_1 to h "otherwise salient" in the context? I am unsure what Maher has in mind here. It is *not* salient in the sense that e_1's confirmatory relevance to h is assumed and is not being questioned at the moment; only e_2's confirmatory relevance to h is being considered when it is asked whether e_2 is evidence that h, given b. It *is* salient in the sense that e_1 by itself (that is, without e_2) confirms, or is evidence that, h, and, I shall assume, this fact is known in the context.

If this is what Maher has in mind, then perhaps his positive relevance account should include the following condition rather than (ii):

(iii) In context C, e is evidence that h only if $p(h/e\&b) > p(h/b)$, where b, which is fixed by context C, contains no proposition e' which is such that $p(h/e'\&b') > p(h/b')$, where b' is formed from b by deleting e'.

(iii) precludes from the background information b any proposition e' that increases the probability of h from what it is without e'. This circumvents the second lottery and intervening-cause counterexamples by allowing e in those cases to count as evidence that h, provided that we suitably modify the background information to exclude propositions that increase h's probability. It also avoids certain vague features of (ii), since there is no appeal to what is under discussion or otherwise salient in the context. (Whether this is something desired by Maher is another matter.)

(iii), however, seems too strong, since it precludes the following. Let the background information b, which is fixed by the context, consist of two propositions:

b': Lottery L is a fair lottery in which 1000 tickets have been sold on Monday, one of which will win.

e': On Monday, John owns 100 of the tickets sold in lottery L, Bill owns 1.

The new piece of information is

e: On Tuesday, all tickets in lottery L except any owned by John and Bill have been destroyed.

The hypothesis is

h: John will win the lottery.

In this case, $p(h/e\&b) > p(h/b)$. (On a standard assignment of probabilities, $p(h/e\&b) = \frac{100}{101}$, while $p(h/b) = \frac{100}{1000}$.) But on condition (iii) we cannot conclude that e is evidence that h, since the special restriction is violated. This is because b contains a proposition, viz. e', which is such that $p(h/e'\&b') > p(h/b')$, where b' is formed from b (which is composed of b' and e') by deleting e'. (On a standard assignment of probabilities, $p(h/e'\&b') = \frac{100}{1000}$, and $p(h/b')$ should be less since b' contains no information about whether John owns any tickets or how many.)

Yet, I suggest, in a context that fixes b as background information we do want to say that e is evidence that h. The fact that (e), on Tuesday all the tickets in the lotttery except any owned by John and Bill have been destroyed, is evidence that John will win, since, in accordance with information in b, John owns 100 tickets and Bill only 1. Note that this example conforms with Maher's "basic idea" underlying the concept of evidence, viz. that e is evidence that h only if e makes it rational to be more confident that h is true (see section 2 above). The information that all the tickets except for any owned by John and Bill have been destroyed, in the light of b, makes it rational to be more confident that John will win than it was without that information. However, (iii) precludes the corresponding evidential claim. Accordingly, (iii) should not be defended by Maher, despite its advantage of eliminating references to propositions "whose relevance to h is under discussion or otherwise salient in the context." If, however, we retain the earlier (ii), we will keep these vague notions.

There is another feature of both (ii) and (iii) worth noting. According to Maher, sentences of the form "e is evidence that h" are incomplete, and hence lacking in truth-value, since no background information is explicit. Only sentences of the form "e is evidence that h, given b" are complete. However, Maher seems to be saying, although sentences of the first type are grammatically complete (they are not like "e is evidence that"), they are semantically incomplete. To complete them semantically one needs to specify a context in which "e is evidence that h" is uttered so that background information can be determined.

Now, let

e: John owns 999 of the 1000 tickets sold in the lottery.
b: This is a fair lottery, one ticket of which will be drawn at random.
h: John will win.

With this e, b, and h, consider the evidential claim

(A) e is evidence that h, given b

Maher regards this as semantically complete. But suppose we combine e and b and claim that

(B) $e \& b$ is evidence that h.

Does our claim suddenly and inexplicably become incomplete and lacking in truth-value, requiring a context to be specified so that background information can be fixed? On the contrary, the claim represented by (B) seems perfectly complete and indeed true. Or is Maher saying that (B) is complete because in the context in which it is uttered the b-component *is* the background information? But then in a sentence of the form

(C) e is evidence that h,

why can't e be taken to be the background information (Carnap's "total available evidence")? If it can then context is irrelevant for sentences of form (C).

11. Maher's Final Conditions

I turn finally to two more conditions Maher adds to the usual positive relevance definition. For e to be evidence that h, Maher requires that e itself be "evidence," which, for him, means that e must be known to be true "directly by experience." This condition is to be understood as relative to an individual or a community. For an individual it requires that the individual have "practical certainty" that e is true, which means, according to Maher, that "it is clear, without doing any calculations, that no such calculation is necessary" (pp. 159–160). For an individual "knowing e to be true directly by experience" also requires that e is "not inferred [by the individual] from any of the propositions under consideration" in the context (p. 162). For a community "knowing e to be true directly by experience" requires that some individual who knows e to be true directly by experience has made e, together with some justification of e, available to the community.

As noted, this condition relativizes evidence to an individual or community and a context containing propositions that are "under consideration." We might write it as follows:

(iv) e is evidence that h for X in context C only if e is evidence for X in context C, that is, only if X knows e to be true directly by experience in context C,

where what counts as "knowing e to be true directly by experience" is given in the previous paragraph. Condition (iv), like (ii), yields a concept of evidence that is relativized to a context which depends on the knowledge, beliefs, etc. of those in the context.

Maher's final condition makes reference to probability functions that he calls "rationally permissible." Many different probability functions are possible, not all of them rational. Maher claims that which of them it is rational to have in a situation depends on what is known "directly by experience" in that situation.

$R(b)$ denotes the set of probability functions that are rationally permissible to have in a situation in which b represents everything known directly by experience. (Maher calls b the "total evidence.") A probability function p is a member of the set $R(b)$ if and only if "there is a possible situation in which some person's total evidence is b, that person has probability function p, and that person does not violate any norm of rationality" (p. 163).

Maher's final condition, then, is that e is evidence that h, relative to the sort of b described above, only if $p(h/e) > p(h)$ for all probability functions that are members of the set $R(b)$—that is, for all probability functions that are rationally permissible in a situation in which b represents everything known directly by experience. More briefly,

(v) e is evidence that h, given b, only if $p(h/e) > p(h)$ for all $p \varepsilon R(b)$, where b comprises what is known directly by experience.

Condition (iv), in contrast to (v), invokes an individual or community X and a context C containing propositions that are "under consideration." Such relativizations, although not explicit in (v), are implicit in the relativization to b. The latter contains the propositions known "directly by experience," which requires for some individual X that such propositions are not inferred by X from any propositions under consideration in the context. But in view of the fact that condition (iv) is imposed, which relativizes evidence to an individual (or community) X, in (v) the propositions b known directly by experience must be understood as ones known *by* X.

Putting these various conditions together, we might express Maher's theory as follows, with respect to an individual person X:

(vi)   For person X, in context C, e is evidence that h, given b, if and only if
    (a) $p(h/e) > p(h)$, for all $p \varepsilon R(b)$, where b comprises what X knows directly by experience in C;
    (b) X knows e to be true directly by experience in C;
    (c)   Neither e nor b is inferred by X from any propositions under consideration in C.[18]

Although Maher does not spell this out sufficiently, for a community (rather than an individual person), e is evidence that h, given b, if and only if there is some person X satisfying the above conditions who has made e, together with some justification of e, available to that community.

In addition, we recall, Maher claims there are "pragmatic" conditions which do not affect the truth of an evidence claim, but only its appropriateness on a given occasion. A pragmatic condition he mentions is that a claim that e is evidence that h should make a relevant contribution to a conversation. It will do so only if there is some question about whether or not h is true.

Proposition (vi) above looks a bit different from Maher's own final formulation, which is this (p. 163):

---

18. Condition (c) follows from (a) and (b) in virtue of what Maher means by "knowing directly by experience." I add it here because I want to express fully the contextual features of the proposed definition.

(m) e is evidence that h, given b, if and only if (1) e is evidence, and (2) for all $p \in R(b)$, $p(h/e) > p(h)$.

Proposition (vi) spells out Maher's proposals more explicitly. In (vi) Maher's condition (1) in (m) is given by (b) and (c), while his condition (2) is given by (a) and (c). Despite the simple formulation in (m), Maher's concept of evidence, expressed in (vi), is relativized to an individual person (or community), and to a context in which certain propositions are under consideration. Accordingly, if (vi) adequately reflects his ideas, then, despite his claim that he "will offer an analysis of confirmation in the objective sense" (p. 158), I suggest that he is committed to a subjective concept, viz. evidence for some person (or community) in a given context in which certain things are known by observation and certain matters are or are not under consideration.

Alternatively, perhaps, Maher is relativizing his concept not to an individual person or community but simply to a context or (as in the case of ES-evidence) to a (type of) epistemic situation. He is saying that e is evidence that h relative to a context or epistemic situation in which certain propositions are under consideration and certain things are known directly by experience, whether or not anyone is in that context or epistemic situation. Another possibility is that Maher is not relativizing his concept of evidence at all but saying that e is evidence that h if and only if *there is* at least one person X and at least one context C that satisfy condition (vi). Both of these interpretations, however, seem to be inconsistent with what he says about his requirement (in (b)) that for e to be evidence that h, e must be "evidence," that is, it must be known directly by experience. He writes (p. 162):

> A necessary condition for E to confirm anything [i.e., for E to be evidence for some h] is that E be evidence. The notion of evidence is relative to an individual or a community; I have not specified this explicitly but have left it to be determined by the context. For an individual X, E is evidence if and only if X knows E directly by experience.

This strongly suggests, if it does not imply, a relativization of the sort in (vi).

Assuming that (vi) represents his position, what concept of evidence is Maher trying to explicate? The conditions in (vi) are clearly not meant to be sufficient ones for the subjective concept of evidence of chapter 2. For *that* concept, in order that e be X's evidence that h, it is necessary and sufficient that X believe that e is (probably) veridical evidence that h, that X believe that h is (probably) true; and that X's reason for believing that h is (probably) true is that e is true. None of these three conditions is required by (vi). Nor, of course, are the conditions in (vi) meant to explicate potential evidence or veridical evidence. In these cases, by contrast with (vi), there is no relativization to an individual X or to a specific context in which certain propositions are under consideration.

Intuitively, what Maher may be getting at in (vi) is this: given b, e is evidence that h that is *available to* X in context C, even if X does not realize that it is evidence that h or take it to be so. Of the concepts in chapter 2 the one that comes closest is ES-evidence—evidence that is relativized to some type of epistemic situation. Maher's background information b represents everything known by X

directly by experience. What is needed to add to the idea of ES-evidence is relativization to a person X and to a specific context C which are such that

(a) the background information b represents X's epistemic situation in C (it contains propositions known by X "directly by experience" in C—and hence not inferred from any propositions under consideration in C;
(b) X knows e to be true directly by experience in C.

Let us say that

e is evidence that h, given b, that is available to X in context C

if e is ES-evidence that h, given b, and if conditions (a) and (b) are satisfied. This does not require X to believe that e is evidence that h, given b, or X to have e as a reason for believing h (as does the idea of subjective evidence in chapter 2). It does not require h to be true (as does veridical evidence). Like ES-evidence, and unlike potential evidence, it relativizes evidence to an epistemic situation; but what it adds to ES-evidence is the idea, explicated in (a) and (b), that this ES-evidence is available to a particular person who is in some particular context.

If this is what Maher is after, then my comments are these, in ascending order of importance:

1. Maher claims to be analyzing an objective notion. Unlike ES-evidence, however, this concept is subjective in the sense that it relativizes evidence to a particular individual and a particular context in which certain propositions are under consideration.

2. This concept of evidence is derivative. It depends on a prior account of ES-evidence.

3. The first condition in (vi)—the positive relevance condition—would be the one meant to express the idea that e is ES-evidence that h, given b. But, as I have argued earlier, positive relevance is neither necessary nor sufficient for ES-evidence. A basic idea underlying ES-evidence is that e provides a justification for believing h. But increase in probability is neither necessary nor sufficient for this (as the earlier counterexamples are meant to show). Having a justification (or a good reason) for belief, in so far as this depends on probability, is a threshold concept. There needs to be a certain significant threshold of probability for h before h can be said to have a justification. Increase in h's probability is neither necessary nor sufficient to achieve that threshold.

Accordingly, Maher must reject the claim that his concept of evidence is a species of ES-evidence. Indeed he will do so because the "basic idea" underlying his concept of evidence is not that e provides a justification for believing h, but that e provides a justification for a higher degree of belief than without e, or, in his words, that "e makes it rational to be more confident that h is true." My reply here is similar to ones earlier in this chapter. If with e it is rational to be more confident that h is true than without e, then with e at least some confidence in h is rational. But the concept of rational confidence, in so far as it depends on probability, is a threshold concept. Increase in probability is neither necessary nor sufficient to reach that threshold. So even if we substitute Maher's "basic idea" underlying evidence (rationality of more confidence) for the basic idea underlying ES-evidence (providing a justification for belief), Maher's most important condition—increase in probability—is neither necesssary nor sufficient for representing this basic idea.

12. Conclusions

In this chapter I have presented counterexamples to the positive relevance and high-probability definitions of evidence. These counterexamples purport to show that positive relevance is neither necessary nor sufficient for evidence and that high probability is not sufficient.

Various responses by probabilists are considered. These include the idea that "evidence" is ambiguous and can mean either positive relevance or high probability, that counterexamples involving small increases in probability (albeit low probabilities) do constitute evidence (in the positive relevance sense), however slight, and that they do so because they provide an increased justification for believing the hypothesis. I reject these responses because they involve a mistaken idea of the relationship between belief (justification, acceptability, firmness, etc.) and probability. I defend the claim that belief is a threshold concept with respect to probability (likewise for the other concepts mentioned). The relationship between positive relevance and higher degrees of belief of the sort probabilists take for granted does not exist.

My claims about belief are not subject to the lottery paradox since the threshold concept precludes the conjunction principle required by that paradox. If the positive relevance theorist modifies his position by saying that evidence is not required to make a hypothesis more acceptable but only less unacceptable, an extremely weak concept of evidence emerges. Indeed, I present an argument to show that evidence need not increase the probability of a hypothesis at all and may even decrease it. What is required is not increase in probability but high probability. However, the latter is not sufficient, since evidence, by contrast to probability, is a selective concept. A hypothesis can have high probability on information that is evidentially irrelevant to it. Finally, I discuss a pragmatic response to my counterexamples, and a novel version of the positive relevance theory, offered by Patrick Maher.

5

OBJECTIVE EPISTEMIC PROBABILITY

If the positive relevance and high-probability accounts of evidence do not suffice, must we abandon attempts to provide a probabilistic definition of evidence? On the contrary, I believe such a definition is possible. But before developing this, an interpretation of probability different from the four standard ones of chapter 3 will be given in the present chapter. This concept of probability will be utilized in the definitions of potential and veridical evidence in later chapters. The latter in turn will be used to define ES- and subjective evidence.

According to the present theory, probability, understood as a concept satisfying the probability rules of chapter 3, can be construed as a measure of how reasonable it is to believe a proposition. On this view, reasonableness of belief in a proposition admits of degrees and is subject to the formal rules of probability. So, for example, according to an addition rule, if h_1 and h_2 are mutually exclusive, then the degree of reasonableness of believing the disjunction h_1 or h_2 is the sum of the degree of reasonableness of believing each disjunct. Since, in accordance with the rules, probability is measured between 0 and 1, the claim that the probability of a hypothesis h is 1 is understood as saying that believing h has the highest degree of reasonableness. At the other extreme, if the probability of h is 0, then believing that h is false has the highest degree of reasonableness. At the midpoint, saying that the probability of h is $\frac{1}{2}$ means that it is just as reasonable to believe h true as to believe h false, in which case it is reasonable to have no belief either way, that is, it is reasonable to suspend belief. The claim that the probability of h is $\frac{3}{4}$ entails that believing h true is 3 times more reasonable than believing h false. More generally, if the probability of h is r, so that the probability of not-h is $1-r$ then believing h is $r/1-r$ as (or more) reasonable as (than) believing not-h.

What sort of claim is one that says that the degree of reasonableness of believing h is r? To answer, let me start with a more general, nonquantitative concept: being reasonable to believe something.

1. Two Concepts of Reasonableness of Belief

The question of whether something is reasonable to believe can be considered in one of two senses: in a sense that depends on what (other) beliefs are held by some individual (or type of individual), group, or generally by people, and in a sense that does not. For example, although eating a pound of arsenic is fatal within 24 hours, suppose that the authorities in the community believe that arsenic, even a pound of it, promotes health when put on food. And suppose that people know that Ann ate a pound of arsenic 24 hours ago, and also know what the authorities believe and have no reason to question them. Is it, or is it not, reasonable to believe that Ann is dead or dying? Both answers are possible.

We can say that it is not reasonable, at least for members of this community, to believe this, since the authorities preach the benefits of arsenic. We can also say that it is reasonable, even for members of this community, to believe that Ann is dead or dying, since (unbeknownst to the community and its authorities) arsenic is a deadly poison. The latter normative fact about what it is reasonable (even for this community) to believe is determined by, or "supervenes" on, physical facts about arsenic and the human body. It is not determined or affected by facts about what people believe about arsenic and the body. Nor is it relativized to an epistemic situation. The claim is not that it is reasonable for someone in such and such an epistemic situation to believe that Ann is dead or dying.[1]

In the sense that depends on beliefs, determining whether it is reasonable to believe h requires identifying a person, or group, and/or an epistemic situation, for whom (or which) it is reasonable to believe h. In this sense, whether it is reasonable to believe h depends on who or what type of person is doing the believing, since it depends on (other) beliefs of that person. In the sense of reasonableness of belief that does not depend on other beliefs, determining whether it is reasonable to believe h does not require identifying a person, or group, or epistemic situation for whom, or with respect to which, it is reasonable to believe h. We can put the same point by saying that, in this sense, if it is reasonable to believe h, it is reasonable for anyone to do so. This sense of "reasonableness of belief" can be called abstract, since it abstracts the question of whether h is reasonable to believe from the set of (other) beliefs of actual or potential agents.

It might be objected that such an abstract sense of "reasonableness of belief" is impossible, since there are only two alternatives. One is that "reasonableness of belief," as I am using it here, is just an epistemic- and normative-sounding name for the physical facts or states of affairs that (as I have been putting it) make something reasonable to believe (such as the fact that Ann ate a pound of arsenic and that arsenic is a deadly poison). Another is that "reasonableness of belief" is always, at least implicitly, relativized to a particular person, group, or epistemic situation; it is never really abstract.

As for the first alternative, I am not denying that in the sorts of cases I have in mind physical facts underlie abstract facts about what it is reasonable to believe. My claim is that these physical facts are not identical with the fact about

---

1. In chapter 2, section 6, analogous points were made about the related concept of a "good reason to believe."

what it is reasonable to believe. If they were identical, then whatever causes one to be the case should cause the other. However, what caused Ann to eat a pound of arsenic (say it was a drug-induced desire to try something very different) is not what caused it to be reasonable to believe she is dead or dying. What caused the latter to obtain (if "cause" is appropriate here) is the fact that she ate the arsenic, which, of course, is not what caused her to eat the arsenic.

As for the second alternative, I am not denying that there is a (nonabstract) sense of what is reasonable to believe for someone in a given epistemic situation. But if this were the only sense of reasonableness of belief, then in explaining why, or in determining that, it is reasonable to believe that Ann is dead or dying, one could not simply appeal to the fact that she ate a pound of arsenic. One would need also to determine an epistemic situation. The fact that Ann ate the arsenic does not explain why, or determine that, it is reasonable *for you* to believe she is dead or dying if you are in the epistemic position imagined above in which the authorities claim that arsenic promotes health.

Nor am I denying that if it is reasonable to believe h, then there will be some epistemic situation such that it will be reasonable for anyone in it to believe h. What I am denying is that a nonrelativized abstract claim must always be understood as implicitly relativized to some particular epistemic situation. Can a weaker claim be made to the effect that, in the abstract sense, it is reasonable to believe h if and only if *there exists* some (type of) epistemic situation that is such that it is reasonable for anyone in it to believe h? No, it cannot. Otherwise the existence of an epistemic situation in which the known authorities believe that arsenic promotes health would make it reasonable to believe (in the abstract sense) that Ann is alive and well, since she has eaten the arsenic. But it is not reasonable to believe this, since arsenic is lethal.

Analogous points apply to the concept of "degree of reasonableness of belief." This can be understood in a sense that depends on the beliefs of particular individuals or on epistemic situations, as well as in an abstract sense that does not. It is the latter that I am invoking as an interpretation of probability. Certain physical and/or mathematical facts or states of affairs may make it reasonable to degree r to believe something. Suppose that this coin, which has two sides marked heads and tails, is balanced; it will be tossed randomly three times; there are 8 possible outcomes in 3 tosses, 3 of which involve exactly 2 heads. These facts make it reasonable to the degree $\frac{3}{8}$ to believe that the coin will land heads exactly twice. That there is this degree of reasonablenesss is an objective, nonphysical, normative fact determined by the former physical and mathematical facts. It does not depend on what anyone knows or believes. It is abstract in the sense that it is divorced from particular individuals and types of epistemic situations. It is not to be understood as relativized to the epistemic situation of any actual or type of person or group. If the mathematical experts in your community inform you (incorrectly) that there are 10 possible outcomes in 3 tosses, 2 of which involve exactly 2 heads, and you have no reason to question this, then, in the nonabstract sense, the degree of reasonableness *for you* to believe the hypothesis in question is $\frac{1}{5}$ not $\frac{3}{8}$. Despite what these authorities say, in the abstract sense, the degree of reasonableness of believing that the coin will land heads exactly twice is $\frac{3}{8}$ not $\frac{1}{5}$. Finally, if the degree of reasonableness of believing this is $\frac{3}{8}$, then there will be some epistemic situation such that for anyone in it, the degree of reasonableness of believing this is $\frac{3}{8}$.

I call this concept epistemic because it has to do with a certain fact about believing something (that is, the state of believing, not just the content of the belief), viz. how reasonable it is to believe. But it is an abstract epistemic concept. The claim is that one can ask how reasonable certain facts make it to believe some proposition p, without considering the knowledge and beliefs of persons, if any, who may be in the position of believing that p. In this respect it is similar to inferring and concluding. We can ask how reasonable it is to infer or to conclude that p from some fact, without considering the knowledge and beliefs of persons, if any, who may be in the position of inferring or concluding that p.[2]

2. A Contrast with Other Probability Views

This interpretation of probability is different in two important respects from the subjective theory of probability described in chapter 3. First, although both theories interpret probability epistemically, on the present epistemic view, probability is not a measure of *how much* belief one has or ought to have or *how strong* that belief is or ought to be. It does not state or presuppose that beliefs come in degrees of firmness. Rather, it is committed to the idea that how reasonable it is to believe something is subject to differences of degree. However, it is not committed to the idea that reasonableness of belief is only a matter of degree. The degree to which it is reasonable to believe some proposition may make it reasonable to believe that proposition. Tautologies would be one extreme example. It is reasonable to believe them, and the degree to which it is reasonable is 1. On the other hand, it is reasonable to the degree $\frac{1}{2}$ to believe that a fair coin will land heads, but it is not reasonable to believe that the coin will land heads, since it is equally reasonable to believe it won't. In this case it is reasonable to suspend belief. Accordingly, I would claim, reasonableness of belief is a threshold concept with respect to probability: some probability significantly greater than 0 is necessary for it to be reasonable to believe a proposition. But *degree* of reasonableness is not a threshold concept with respect to probability.

A second important difference between the present view and the subjective theory of probability is that, in accordance with the latter but not the former, a probability statement is relativized to an actual person (or group). For subjec-

---

2. Bernard Williams (*Moral Luck* [Cambridge: Cambridge University Press, 1981], 101–113) draws a distinction between "internal" and "external" reasons for doing something. Both reasons are relativized to an agent, with the difference being that an external reason is independent of the motives or desires of the agent, whereas an internal reason is not. Williams argues that external reason statements, if isolated from internal ones, are false, or incoherent, or misleadingly expressed (p. 111). I don't know whether Williams wants to apply his distinction to reasons for believing something, and, if he does, whether he wants to say that such reasons cannot be divorced from an agent's beliefs. If he does, his view is indeed incompatible with the one I defend here.

In this section I have benefited from discussions with Josh Gert.

tivists a probability statement must be understood as a probability *for someone*. On the present epistemic view, there is no such relativization. When I speak of the probability that John's symptoms will be relieved, I am not speaking of how reasonable it is for some particular person or group with a given set of beliefs to believe this. In accordance with my epistemic view, it may be very reasonable to believe that his symptoms will be relieved, since he is taking medicine M, which is 95% effective, even if it is not particularly reasonable for you or anyone else to believe this, since neither you nor anyone else knows that he is taking M or how effective it is.

How does the present epistemic view differ from Carnap's theory of probability (outlined in chapter 3)? Although Carnap regards his concept of probability as (what he calls) a "logical" one defined by reference to state-descriptions, he does provide an epistemic interpretation for the concept defined: if the probability of h is equal to r then one is justified in believing h to the degree r. So one way his view differs from the present one is that for Carnap probability is a measure of how much belief or how strong a belief one is justified in having. His view, like that of the subjectivists, is that beliefs come only in degrees; the present one is not committed to this.

Second, on Carnap's theory, but not the present objective one, a probability statement is to be understood in terms of justification of belief. For Carnap, whether one is justified in believing a proposition to a certain degree depends on what one knows and does not know to be true. It is justification for someone (real or imagined) in a certain type of epistemic situation. Carnap's probabilities, construed epistemically, are more akin to ones required for ES-evidence (evidence that provides a justification for belief for someone in a certain type of epistemic situation). By contrast, objective epistemic probabilities reflect a degree of reasonableness of belief that is not tied to epistemic situations.

There is a third difference between the present view and Carnap's. In *The Continuum of Inductive Methods* Carnap shows how an infinite number of probability functions can be defined, all of them satisfying what he regards as rational objective necessary conditions for probability.[3] These objective conditions do not in general yield a unique value for the probability of a given hypothesis. Different persons can legitimately choose different probability functions based on criteria some of which, such as familiarity and ease of use, are subjective. As in the case of the subjective theory, the result will be the existence of different probabilities for the same hypothesis, depending on the probability function in question. By contrast, objective epistemic probability is not subject to subjective factors. If a numerical probability for a given hypothesis exists, it is unique.

The objective epistemic view is also different in fundamental respects from the frequency and propensity theories outlined in chapter 3. Although both of these views are objective, and both insist on unique probabilities, they are not epistemic, at least not directly so. According to both of them a probability statement is to be understood as describing a physical fact about the world, not a fact about belief or reasonableness of belief. On the relative frequency view, "$p^S(H) =$

---

3. See chapter 3, footnote 6.

r" means that in the sequence S of events, the limit of the relative frequency of events of type-H is r. Loosely speaking, such a statement describes what the long-run relative frequency of type-H events will be or would be if the sequence were extended indefinitely. On the propensity theory, a probability statement gives a disposition for some specified system to yield a type-H event (which might or might not be given by the same number as the long-run relative frequency of H-type events in some sequence). On Lewis' version, a probability statement, construed in propensity terms, gives the propensity of a proposition to be true at a certain time in a certain world.

Frequency and propensity theorists may assert that a relationship exists between probability and belief. (Lewis does so for propensities via his Principal Principle, and in chapter 3, sections 5–7, I suggested how frequentists might relate statements about "weights" to beliefs.) But on these views, a probability statement itself is not about beliefs but relative frequencies or propensities in nature. According to the objective epistemic view, a probability statement is not to be construed in either of these ways. It does not attribute a relative frequency or a propensity to anything. The fact that the limit of the relative frequency of heads with this coin is .8, or that this is the propensity of the coin to land heads next time, may make the degree of reasonableness of believing this coin will land heads next time equal to .8. If it does, then why the degree of reasonableness of believing this coin will land heads next time is .8 can be correctly explained by appeal to the limit of the relative frequency of heads with this coin, or to the propensity of the coin to land heads next time. But why the coin has this limiting frequency or this propensity cannot be correctly explained by appeal to the fact that the degree of reasonableness of believing this coin will land heads next time is .8.

Even if objective epistemic probability statements do not mean the same as frequency or propensity statements, do they assign the same probability values? This question will be addressed in section 7 after introducing the idea of relativized epistemic probability.

Of the four theories of probability discussed in chapter 3, one is clearly epistemic (subjective probability), one has an epistemic "interpretation" (as Carnap puts it), two are not epistemic (frequency and propensity), although their defenders may assert that there is some relationship between their conceptions of probability and belief. As noted earlier, I classify the present theory as epistemic because the probabilities in question represent degrees to which it is reasonable to believe something rather than relative frequencies, propensities, ratios of weights assigned to state-descriptions, and so forth. However, the concept of belief involved is abstract, since probability statements of the sort in question are not relativized to epistemic situations or to what in fact anyone knows or doesn't know.

There is another way of dividing the four probability theories of chapter 3. Two of them (Carnap's and frequency) propose formal definitions of probability (Carnap in terms of state-descriptions, frequentists in terms of limits) from which the probability rules can be derived; two of them do not (subjective and propensity). In the latter two cases it is simply asserted that the subjective and propensity probabilities are to be construed as ones satisfying these rules. The objective epistemic interpretation falls into this latter category. It construes probability as a measure of how reasonable it is to believe a proposition—a measure that satisfies the probability rules. But like the subjective and propensity theories, it does not propose a more basic definition from which these rules follow.

What advantage accrues for characterizing degrees of reasonableness of belief as subject to the rules of probability? Subjectivists supply one type of a priori answer. If a system of bets is made in accordance with degrees of reasonableness of belief, where these satisfy the probability rules, then no system of bets based on these degrees is bound to lose (no "dutch book" is possible). No doubt this is an advantage, at least in typical betting situations. Whether, as subjectivists insist, a betting criterion supplies a necessary and sufficient condition for rationality in general, or even just for rationality in betting, is a controversial topic that will not be explored here.[4] The justification I would offer for characterizing degrees of reasonableness of belief as subject to the rules of probability is also a priori, but it does not appeal to betting. It is simply that the rules of probability provide defining necessary conditions for degrees of reasonableness of belief. This a priori justification for requiring that the probability rules be satisfied is similar to Carnap's, although the semantical interpretation of such rules offered here is different from his.[5]

3. Are Objective Epistemic Probabilities Empirical? Idealized?

Some objective epistemic probability statements are empirical, some are not. The claim that it is reasonable to the degree $\frac{1}{2}$ to believe that this coin will land heads next time it is tossed can be defended empirically by appeal to the fact that the coin is a fair one, that it will be tossed in a random way, and that there are no interfering conditions. The claim that it is reasonable to the degree $\frac{1}{2}$ to believe that this fair coin will land heads next time when tossed in a random way with no interfering conditions can be defended a priori.

This represents another important difference between the present epistemic probabilities and Carnap's. For Carnap all epistemic probability statements are a priori: whether one is justified in believing h on e to the degree r is to be determined entirely by linguistic and mathematical "computation" by considering no

---

4. For a discussion and references, see Colin Howson and Peter Urbach, *Scientific Reasoning*, 2nd ed. (La Salle, IL: Open Court, 1993), 89ff.
5. Carnap speaks of "explicating a concept," by which he means a priori "transforming a given more or less inexact concept into an exact one" (*Logical Foundations of Probability*, p. 3). Carnap requires that the resulting concept be similar to the original, exact, fruitful, and simple (pp. 3–8). As indicated, in his later work Carnap shows how the concept of probability can be "explicated" in an infinite number of ways (that is, he defines an infinite class of probability functions). All of them, however, satisfy the rules of probability. These he regards as a priori necessary for probability.

empirical facts whatever, just the a priori relationship between e and h and the chosen linguistic system in which these sentences occur (which includes the chosen probability function). On the present view, there are epistemic probability statements, such as the first one in the previous paragraph, that can be defended only empirically.

I have been assuming that probability has unique values that include all real numbers from 0 to 1. This is a standard assumption of most probability theories. Some probabilists, however, defend the idea that epistemic probabilities are imprecise.[6] On such a view, a probability can be represented by an interval, rather than a point, between 0 and 1. For example, on a subjective view, an agent's probability interval for a hypothesis h might be determined at the extremes by the least upper bound of all betting quotients on h at which the agent takes a bet on h to be advantageous, and the greatest lower bound at which the agent thinks it is advantageous to bet against h. In the case of intermediate values, the agent has no beliefs about advantages of either side of the bet.[7]

The idea of imprecise probabilities (without the introduction of betting odds) can also be incorporated into objective epistemic interpretations. On either interpretation, suppose the y-axis of a graph represents some ordered set of propositions and the x-axis represents the real numbers from 0 to 1. For simplicity consider 5 propositions, numbered 1 through 5. The following graph depicts a case with point probabilities:

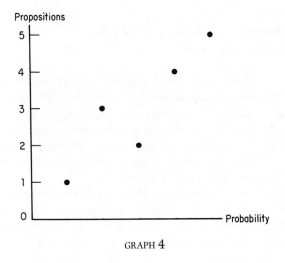

GRAPH 4

The points represent the probability of the propositions in question. A corresponding interval graph might look like this:

---

6. See, for example, Peter Walley, *Statistical Reasoning with Imprecise Probabilities* (London: Chapman and Hall, 1991); Isaac Levi, "On Indeterminate Probabilities," *Journal of Philosophy*, 71 (1974), 391–418; Richard Jeffrey, *Probability and the Art of Judgment* (Cambridge: Cambridge University Press, 1992), ch. 5.
7. See Howson and Urbach, p. 87.

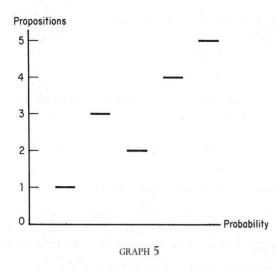

GRAPH 5

The intervals in Graph 5 represent intervals of probability. On this conception, it is not that each proposition has a precise probability which is unknown. Rather, probabilities, whether construed as subjective degrees of belief or as objective degrees of reasonableness of belief, are imprecise; they are "smeared out."

Even if this represents a more realistic view of probabilities, I will continue to use the simpler, more precise, concept of point probabilities employed by most probabilists. I will idealize by supposing that in general point probabilities exist, and that if a hypothesis has a probability, it has a unique point probability. By analogy, a physicist might idealize by supposing that a three-dimensional rod has a unique length that can be represented by a real number. The physicist is treating the rod as having perfectly smooth, straight edges, and as being perfectly rigid. To be sure, there are different methods for measuring the length of the rod that, when employed, may yield slightly different values for the length. Which method is chosen may vary with the interests and capabilities of the measurer. But the idealizing assumption is that the rod has some unique real-valued length. This fact is to be distinguished from the method(s) for ascertaining what that length is. Similarly, in the case of probability the idealizing assumption is that a hypothesis has a unique point probability—a fact that is to be distinguished from method(s) for determining what that probability is. This assumption is different from the Carnapian idea (in *The Continuum of Inductive Methods*) that, in effect, denies the existence of a unique point probability even as an idealizing assumption. For Carnap a hypothesis has different point probabilities depending on the probability function being used.

It might be noted that subjectivists, who define probability in terms of subjective degrees of belief, usually understand the latter concept in terms of betting odds which are based on discrete units of money. Probability, so construed, is not a continuous function on the real numbers. If the x-axis of a graph represents all real numbers from 0 to 1, and the y-axis represents probability (that is, degree of belief) measured in terms of betting odds based on discrete units of money, the result will be a step function. Subjective probabilists idealize by imagining more and more steps resulting in a point probability that can take any real number from 0 to 1 as a value. They idealize by supposing that for an individual at a given time such a probability exists.[8]

4. Degrees of Reasonableness of Belief versus Reasonable Degrees of Belief

On the present epistemic theory, a probability statement of the form "$p(h) = r$" is understood to mean that

(a) The degree of reasonableness of believing h is equal to r.

Now, as indicated above, the degree of reasonableness of believing h can be some number r (such as $\frac{1}{2}$) without its being reasonable to believe h. If so, instead of construing "$p(h) = r$" as (a) above, why not understand it to mean that

(b) It is reasonable to believe h to the degree r.

That is, instead of speaking of degrees of reasonableness of belief, why not speak of the reasonableness of having certain degrees of belief?

We could do so if we do not construe having some degree of belief in h as implying having a belief in h, which is how in chapter 6 I will construe degrees of belief.[9] I will claim that believing, like speaking and running, is something we do that admits of degrees along one or more dimensions. One can speak with different degrees of loudness and run with different degrees of speed. But even if one is speaking very softly or running very slowly, one is still speaking or running. Similarly, I will claim, if one believes something with some degree of firmness or strength, one still believes.

Suppose, then, that the probability of h is some value r, say $\frac{1}{2}$. Then on interpretation (b) it is reasonable to believe h to the degree $\frac{1}{2}$, which, as I will construe belief, entails that it is reasonable to believe h (to some degree or other). But it is not. If the probability of my getting heads on the next toss with this coin is $\frac{1}{2}$, then it is not reasonable to have some degree of belief that I will get heads and some that I will get tails. One should not believe even a little bit that it will land heads (or tails). One should suspend belief.

A defender of interpretation (b) might modify the position, as follows. Instead of speaking of degrees of belief, we will speak of degrees of belief and degrees of disbelief. If the probability threshold for belief is $\frac{1}{2}$, then any number greater than $\frac{1}{2}$ represents some degree of belief it is reasonable to have, and any number less than $\frac{1}{2}$ represents some degree of disbelief it is reasonable to have. The value $\frac{1}{2}$ represents the probability at which it is reasonable to suspend belief. On this interpretation, we have a scale in which the x-axis represents probability and the y-axis represents degrees of belief it is reasonable to have, degrees of disbelief, and one degree of "suspension of belief."

---

8. See Howson and Urbach, pp. 88–89.
9. This is related to, but not identical with, claims in chapter 4, section 4, that belief is a threshold concept with respect to probability, and that if, given e and b, one believes h to a higher degree than one does, given just b, then, given e and b, one believes h to some degree or other.

If the probability threshold for belief is greater than $\frac{1}{2}$, we need not suppose that it is also an upper bound for disbelief. Suppose that a probability significantly greater than $\frac{1}{2}$ is required for belief, while a probability significantly less than $\frac{1}{2}$ is required for disbelief. With probabilities between these thresholds one neither believes nor disbelieves but suspends belief. This proposal requires a scale where the x-axis represents probability and the y-axis represents degrees of belief it is reasonable to have, degrees of disbelief, and (more obscurely) varying degrees of "suspension of belief." On this reinterpretation of probability statements of the form "$p(h/e) = r$," (b) becomes disjunctive:

(b)′ It is reasonable either to believe h to the degree r, to disbelieve h to the degree r, or (where r is within a certain range) to suspend belief in h.

"Degree of reasonableness" can be construed much more easily than "degree of belief" as representing a "positive-negative" scale that includes both the reasonable and the unreasonable. In this respect it is like concepts such as "degree of intelligence" and "degree of strength." "How intelligent or how strong is Irving?" does not presuppose or suggest that Irving has any intelligence or strength (Irving might be a moribund human or a live worm). Similarly, "how reasonable is it to do x?" does not presuppose or suggest that it is at all reasonable to do x. By contrast, "how fast is Bill running?" and "how firmly does John believe h?" presuppose or suggest that Bill is running and that John does believe h. No doubt, one can answer these questions by saying that Bill is not running at all, and that John completely disbelieves h. But although this answers the question, it does so by rejecting a presupposition or a suggestion of that question. One can answer the question "how reasonable is it to do x?" by saying "it is not reasonable at all." But this answer, unlike the previous ones, does not reject any presupposition or suggestion of the question. Accordingly, I prefer interpretation (a) to (b) or (b)′.

As noted in chapter 4, a threshold concept of belief does not imply that a dutch book can be made against a bettor. Nor will the adoption of the present notion of epistemic probability. If one's betting behavior is in accord with degrees of reasonableness of belief, a dutch book is avoided. Doing so does not require that the probability of h, that is, the degree of reasonableness of believing h, match one's degree of belief in h. I can accept that the degree of reasonableness of belief in getting 100 heads in a row with a fair coin is $(\frac{1}{2})^{100}$ without having any degree of belief at all that I will get 100 heads in a row.

Accordingly, as indicated in chapter 4, three concepts need to be distinguished. First, there is the degree of reasonableness of believing h. Second, there is what someone takes this degree of reasonableness to be. Third, there is someone's actual degree of belief in h. The first is an objective epistemic probability. The second is what someone takes this probability to be (which may or may not be equal to the objective epistemic probability). The third is not itself a probability or what someone takes to be a probability, although it depends on the latter. Whether I believe to any degree at all that I will be in an elevator accident depends on whether I take the probability of such an accident to be sufficiently high. I may take this probability to be greater than zero without having any degree of belief that I will be in such an accident. We may call the second concept a "subjective" probability. However, unlike standard subjective interpretations of probability, my subjective probability for hypothesis h—the probability I assign to h—is not the same thing as my degree of belief in h.

5. Relativizations

How should relativized probability statements of the form "the probability of h given e is equal to r" be understood? Such statements are ambiguous. "Given e" can mean (and frequently does) the same as "since it is the case that e." If so e is (supposed to be) true. It can also mean the same as "supposing that e," where e need not be true. We can write $p(h/e) = r$ in the latter case, which involves no commitment to the truth of e, and $p_e(h) = r$ in the former, which has such a commitment. In terms of my epistemic probabilities, statements of the form "the probability of h given e is r" can be understood as saying "the degree of reasonableness of believing h, on the assumption of e, is r" or as "since e is the case, the degree of reasonableness of believing h is r."

There are three special relativizations, one or more of which are frequently made when probability statements are asserted:

1. *A temporal condition.* Suppose that at time $t1$ Sam owns no tickets in the lottery, while at $t2$ he purchases 80% of all the tickets. Let h be that Sam wins. How reasonable it is to believe h is different at $t1$ and $t2$. We might write:

$p(h/\text{the time is } t1) \neq p(h/\text{the time is } t2)$,

or

$p_{t1}(h) \neq p_{t2}(h)$.

2. *No interference condition.* Let e be that this coin, which is balanced, is tossed once in a random way. Let h be the hypothesis that it will land heads on the next toss. In determining a probability for h given e, we usually make an assumption that no interference will occur, for example, the coin is not under the control of a magnet, it will not be destroyed in midair, etc. Let $n.i.$ stand for the condition that there is no interference. $n.i.$ can but need not spell out interference conditions. In the former case we can write $n.i.(\text{-})$, where the blank is filled with a specific or general type of interference condition. In the latter case we can simply write $n.i.$, and formulate probability claims such as

$p(h/e \& n.i.) = \frac{1}{2}$,

and

$p_{e \& n.i.}(h) = \frac{1}{2}$.

In the second case, if there is interference the probability claim is false.

Recall Hertz's experimental result (e) that no electrical deflection was produced in his experiments. His hypothesis (h) is that cathode rays are not electrically charged. Hertz claimed that

$p_{e \& n.i.}(h)$ is very high.

Thomson showed that $n.i.$ is false: there is interference with the electrical charge caused by the fact that the cathode tubes Hertz used were not sufficiently evacuated. Accordingly, the probability claim above is false.[10]

3. *The disregarding condition.* Consider once again the coin example, in which *e* is that this coin, which is balanced, is tossed once in a random way, and *h* is that the coin will land heads next time. Let us suppose that in fact this coin will be tossed next time under certain "microconditions" M (pertaining to the force applied, its direction, the state of the air, and so on) which are such that it follows from laws of nature that under conditions M the coin will always land heads. Let *m* be that microconditions M obtain. Then the following is true:

(1) $p_{e\&m}(h) = 1$.

Establishing its truth, however, for some specific *m*, would be extremely difficult, since it would be extremely difficult to determine the microconditions under which this coin will be tossed and whether it follows from a law of nature that the coin's being tossed under such conditions necessitates its landing heads. What we normally do when determining probabilities in coin tossing and other gambling cases is to disregard microconditions, since we cannot determine what they are.

The disregarding condition does not involve supposing counterfactually that some different state of the world obtains. Disregarding microconditions in the case of coin tossing is not the same as assuming that no forces exist when the coin is tossed (whatever that would entail) or that the ones that do exist are different. It is to presuppose that they do exist, but to ignore them for the purpose of determining a probability. By analogy, if the college admissions committee disregards age and SAT scores in determining whether a particular applicant is eligible for admission, the committee members are not supposing that the applicant has no age or SAT scores, or that she has different ones from those she actually has. Nor are they asserting that these factors are irrelevant in determining admission. They are simply suspending them in this case. They are saying that in this case, age and SAT scores are being ignored for the purpose of determining admission.

Let us write $d(mi)$ for "microconditions are being disregarded." We may simply indicate the type of microconditions in question without specifying specific ones. More generally, we can write $d(\text{——})$ for "conditions of type —— are disregarded." In the coin tossing case we might write

$$p_{e\&d(mi)}(h) = \tfrac{1}{2}$$

for a case in which we are disregarding microconditions of the sort in question. Since we are also assuming that no interfering conditions (such as magnets) are present, we can write the resulting probability as follows:

(2) $p_{e\&n.i.\&d(mi)}(h) = \tfrac{1}{2}$.

---

10. In section 1, I cited the arsenic example in which *e* is that Ann ate a pound of arsenic at time *T*, and *h* is that Ann died within 24 hours of *T*. Suppose, however, that Ann died not from the arsenic but from an unrelated accident. If we relativize the claim that *e* is evidence that *h* to a no interference assumption, which is claimed to be true, then our evidential claim is falsified if in fact Ann died from an unrelated accident.

This can also be conditionalized in other ways, such as,

(3) $p_{n.i.\&d(mi)}(h/e) = \frac{1}{2}$

(4) $p(h/e\&n.i.\&d(mi)) = \frac{1}{2}$.

Claim (2) is to be understood as saying that since this coin is balanced and tossed randomly, and there are no interfering conditions, and microconditions are being disregarded, the degree of reasonableness of believing the coin will land heads on the next toss is $\frac{1}{2}$. Claim (4) is to be understood by changing the "since" to "on the assumption that" in (2).

Claims (2), (3), and (4) are not refuted by the truth of claim

(1) $p_{m\&e}(h) = 1$.

In general, when determining probabilities in coin tossing and other gambling cases we have in mind probabilities such as (2), (3), and (4), rather than (1). Our concern is with how reasonable it is to believe that a certain gambling outcome will occur, since, or on the assumption that, there are no interfering conditions (such as cheating) and that microconditions are disregarded. If it turns out that there is interference or that we are able to learn the microconditions and do not disregard them, then claims (2) and (3) will turn out to be false. But if there is no interference, and if microconditions are being disregarded, then claims (1), (2), and (3) could all be true. These relativizations (like any others) may yield different correct probabilities for the same hypothesis.

These are not relativizations to epistemic situations. The claim is not that interference conditions and microconditions are unknown by some (type of) person or group. It is simply that they are being ignored in determining a probability. By analogy, suppose the admissions committee, in deciding to admit a particular applicant, assumes there are no "interference conditions" (for example, it assumes the applicant will not die before matriculating), and disregards age and SAT scores in determining admission for this applicant. In doing so the committee is not, or need not be, claiming that these things are unknown, but only that in this case they are being ignored in determining whether to admit.

6. Choosing a Probability Statement

Suppose that (1)–(4) are all true. Which of these statements one chooses to make does, and ought to, depend on one's epistemic situation. Since people tossing coins generally are not in a position to determine microconditions and try to arrange the situation to avoid interfering conditions, a statement such as (1) is not usually made, but ones of the sort (2), (3), and (4) are. This is a corollary of the idea that, other things being equal, one should assert statements that one can justify.

What probability statement one chooses to make also depends, and ought to depend, on one's interests, including planned actions, if any, to be taken on the basis of the probability statement. Coin tossers interested in betting will use statements such as (2), (3), and (4) in fixing betting odds, not (1), since they cannot determine the truth of the latter. However, the fact that what probability statement one chooses to assert and employ is, and ought to be, influenced by one's epistemic situation and one's interests does not mean that the probability statement itself is to be understood as relativized to one's epistemic situation and interests. (Indeed, the same holds for any statement whatever.)

Suppose that this coin, which has two sides marked differently, is balanced (e_1), and that, having been tossed randomly many times, it has landed heads not half the time but one-third of the time (e_2). Each of the following may be true:

(5) $p_{e1 \& d(e2 \& mi)}(h) = \frac{1}{2}$
(6) $p_{e1 \& e2 \& d(mi)}(h) < \frac{1}{2}$.

In (5), the results of the tosses are disregarded, in (6), they are not. The truth of each of these statements does not depend on anyone's epistemic situation or interests. However, which of these statements, if either, one chooses to assert and use for purposes of action, such as betting, does and ought to depend on one's epistemic situation and interests. If I know that e_2 is true, then I should employ (6) rather than (5). This is a corollary of the idea that, other things being equal, one should utilize all the relevant information one possesses in determining what statements to assert and use for purposes of action.[11] Even if both (5) and (6) are true, if I know that e_2 is true and I wish to place a bet, I should utilize (6), not (5). Given my epistemic situation and my interests, I have no reason to disregard e_2.

Can one disregard anything? Yes, but, depending on the case, doing so may mean that the probability does not exist. In the previous example, suppose one disregards both e_1 (that the coin, which has two sides marked differently, is balanced) and e_2 (that it has landed heads $\frac{1}{3}$ of the time in the past). Disregarding these things, the question of how reasonable it is to believe it will land heads next time has no answer. By contrast, I am claiming, given that e_2 is disregarded but e_1 is not, e_1 makes it just as reasonable to believe the coin will land heads as it is to believe it will land tails.

At the beginning of this chapter I claimed that certain facts or states of affairs can make it reasonable (to some degree r) to believe something. Its being reasonable (to this degree) is, I said, determined by, or "supervenient" on, the facts in virtue of which it is reasonable. With relativizations the same is true. Whether, and to what degree, it is reasonable to believe something is determined, in part, by what relativizations, if any, are introduced. Assuming that the results of the tosses are being disregarded, the fact that the coin is a fair one makes it reasonable to the degree $\frac{1}{2}$ to believe the coin will land heads next time. Assuming these results are not being disregarded, the fact that the coin is a fair one does not make it reasonable to the degree $\frac{1}{2}$ to believe the coin will land heads next time.

Let me now make the following claim about unrelativized probability statements of the form $p(h) = r$. On the present epistemic theory a statement of this

---

11. This is analogous to (although not the same as) Carnap's "requirement of total evidence," according to which "in the application of inductive logic to a given knowledge situation, the total evidence available must be taken as a basis for determining the degree of confirmation." Rudolf Carnap, *Logical Foundations of Probability*, p. 211.

form, when asserted, will usually be construed as elliptical for one of the form $p\_\_\_(h) = r$, where the blank is filled by some claim(s) (including possibly one or more of the special relativizations) whose truth is being assumed. So, for example, where e is that this balanced coin is tossed randomly and h is that this coin will land heads on the next toss, if I say

$$p(h) = \tfrac{1}{2},$$

my claim might be understood as

(2) $p_{e\&n.i.\&d(mi)}(h) = \tfrac{1}{2}.$

This is not the same as saying that some propositions have no objective epistemic probabilities. It is to say that objective epistemic probability claims may be incomplete as stated. Statements of form (2) are not conditional probability statements in the traditional sense. They are not of the form $p(h/-) = r$. But they are relativized to something that is being claimed to be the case.

When is an objective epistemic probability claim established as true (or false)? If it is a priori, the matter can be settled with finality by computation. By a priori calculation we can conclusively establish the claim that it is reasonable to the degree $\tfrac{3}{8}$ to believe that I will get exactly 2 heads with this coin, given that the coin is fair, that I toss the coin randomly 3 times, that there is no interference, and that microconditions are disregarded. If an objective epistemic probability claim is empirical, its truth or falsity cannot be settled with such finality. (This holds true for nonepistemic probability theories, such as the relative frequency and propensity views, according to which probability statements are empirical.) Suppose my hygrometer, which has been accurate 99% of the time in the past, registers 45% humidity. And suppose I assert that

(7) It is reasonable to the degree .99 to believe that (h) the relative humidity is 45%, since (e) my hygrometer registers that number, and ($n.i.$) there are no interference conditions. (This will be represented as $p_{e\&n.i.}(h) = .99$.)

I might defend (7) empirically by appeal to the past accuracy of the hygrometer. Suppose, however, that unbeknownst to me, there were interference conditions resulting in the hygrometer's being broken and stuck on 45%. Claim (7) is then false. Given the past accuracy of my hygrometer, and given that I have no reason now to suspect its accuracy, someone in my epistemic situation may be justified (to the degree .99) in believing that the relative humidity is 45%, since that is what the hygrometer registers. In fact, however, it is not reasonable to believe this, since the hygrometer is broken. As with any empirical statement, one might be justified in believing (7), given what one knows, even though (7) turns out to be false.

7. Epistemic Probability versus Propensity

I now return to a claim made in section 2 that objective epistemic probability is different from other objective interpretations. I will concentrate on the propensity theory. Earlier I raised the following question: Even if propensity and objective epistemic probability statements do not mean the same, do they assign the same probability values? My answer is: sometimes yes, sometimes no.

Suppose that this coin will be tossed next time under microconditions M which are such that under these conditions the coin always lands heads. On a propensity view, the propensity of this coin to land heads next time, given that such microconditions obtain, is 1. Similarly, the objective epistemic probability of heads next time, given that such microconditions obtain, is 1.

Now suppose that the following is true:

e: This coin, which has been tossed randomly a large number of times, has landed heads 80% of the time.

Let

h: This coin will land heads next time.

Using Lewis' version of the propensity theory, since (we are assuming) deterministic microconditions M obtain in world w at time t,

$C_{w,t}(h/e) = 1$.

In this world, because microconditions M obtain, the propensity or chance of this coin to land heads next time is 1, despite the fact that the coin has landed heads only 80% of the time in the past. Now suppose microconditions are disregarded. Then, using objective epistemic probability, $p_{d(mi)}(h/e)$ may well be equal to .8. What happens on a propensity view? The problem for a propensity view is that microconditions always exist in the world, whether M or something else, and they determine what the propensity of the coin is to land heads next time. Assuming that microconditions for coin tossing are deterministic, the propensity of heads next toss will always be either 0 or 1. What could it mean, on the propensity theory, to disregard all microconditions? Here are three possibilities.

1. It might mean that the propensity of heads next time is being considered under the assumption that no microconditions exist. If $w(n)$ is some world in which no microconditions for coin tossing exist (for example, no forces, or at least no forces applied to coins), then we are considering $C_{w(n),t}$(heads next toss). But what propensity the coin has to land heads next time depends on what microconditions exist; they determine this propensity. If no microconditions are assumed to exist—if no force of any particular magnitude or direction is applied to the coin, etc.—then the propensity of the coin to land heads is either 0 or undefined.

2. It might mean that the propensity of heads next time is being considered under the assumption that some different set of microconditions obtain, not the actual ones. If w' is a world in which the force on this coin at t is different from what it is in world w at t, then we are considering $C_{w',t}$(heads next toss). But again this propensity will either be 0 or 1, if microconditions for coin tosses are deterministic.

3. Most plausibly, it might mean simply that whatever microconditions exist for the next toss are being ignored. But, on the propensity theory, the fact that they are being ignored will not change the propensity of the coin to land heads

next time. If such microconditions exist and are deterministic, then, even if they are being ignored, the propensity will either be 0 or 1. That is, for any world w in which coin tossing is deterministic, $C_{w,t}$(heads next toss/microconditions are being ignored) = $C_{w,t}$(heads next toss), which is 0 or 1.

On the epistemic view, however, we can readily consider the degree of reasonableness of believing that heads will occur next time, given that microconditions are being ignored. In ignoring them, we are not supposing that none exist or that some microconditions other than the actual ones obtain. We are simply asking how reasonable it would be to believe in heads next time if the question of microconditions is put aside. More generally, it is possible to consider how reasonable it is to do something (whether to believe or take some action) putting aside some type of question that might otherwise be relevant. If the question of microconditions is put aside in the coin tossing case, there is no reason to suppose that the degree of reasonableness of believing in heads next toss is either maximal or minimal.

Earlier I claimed that physical facts about the world can generate a nonphysical objective fact about whether it is reasonable to believe something and how reasonable it is to believe it. Deterministic microconditions may make the degree of reasonableness of believing in heads next toss equal to 1 (or 0). With a "disregarding condition," however, a new situation emerges. In disregarding such microconditions we generate a different degree of reasonableness of belief without changing the microconditions or assuming they are different. In disregarding microconditions we do not alter the physical propensity of the coin to land heads next time, nor do we assume such a physical propensity is altered.

Accordingly, propensities and epistemic probabilities can yield different values. It may be the case that

$p_t(h/e \, \& \, d(mi)) = .8$,

while

$C_{w,t}(h/e \, \& \, d(mi)) = 1$.

Does this violate Lewis' Principal Principle? In chapter 3, section 9, this principle was stated as follows:

> *Principal Principle:* Let X be the proposition that $C_{w,t}(h/e) = r$. Let p be some rationally corrected subjective probability function. Then $p(h/X\&e) = r$.

Where h and e are given above, this principle requires that if

$C(h/e \, \& \, d(mi)) = 1$,

then

$p(h/C(h/e \, \& \, d(mi)) = 1 \, \& \, e) = 1$.

But the latter is not violated if

$p(h/e \, \& \, d(mi)) = .8$,

whether p is a rationally corrected subjective probability function or an objective epistemic one.

Generalizing, if no "disregarding condition" is imposed, then if a propensity probability for h exists and has a certain value, there will be a corresponding epistemic probability for h with the same value. That is, if $C_{w,t}(h) = r$, then $p_t(h) = r$. But a "disregarding condition" can change the situation. If $C_{w,t}(h/d(-)) = r$, it is not necessarily the case that $p_t(h/d(-)) = r$.

6

EVIDENCE, HIGH PROBABILITY, AND BELIEF

I propose to argue that potential and veridical evidence require probability greater than $\frac{1}{2}$. "Probability" here will be construed as objective epistemic probability. The requirement that the probability exceed $\frac{1}{2}$ (which will be taken to be a necessary but not a sufficient condition for evidence) allows us to avoid various counterexamples introduced in chapter 4 against the positive relevance (increase-in-probability) account. More to the point, I will argue that a probability greater than $\frac{1}{2}$ is required by the fundamental idea underlying both potential and veridical evidence, viz. that of providing a good reason for belief (as well as by the fundamental idea underlying ES-evidence, viz. providing a justification for belief). In the course of defending this claim, I will examine several theories of evidence and belief, including two that deny any relationship between evidence and belief (or high probability), one that denies the usefulness of the concept of belief, insisting on replacing it with degrees of belief, and one that defines belief as high subjective probability.

1. The Counterexamples and High Probability

On the positive relevance account, e's increasing h's probability is sufficient to make e evidence that h. In the first lottery counterexample of chapter 4 the positive relevance definition counts the fact that (e) all the tickets except for those of John and Bill were destroyed as evidence that (h) Bill won, despite the fact that Bill had only 1 of the remaining tickets and John had 100. If we insist that a hypothesis must have probability greater than $\frac{1}{2}$ on the putative evidence then this counterexample is avoided. In the present case the probability assigned to h on e is $\frac{1}{101}$.

In the swimming counterexample the positive relevance definition counts the fact that (e) Steve was doing training laps in the water on Wednesday as evidence that (h) he drowned, despite the fact that he is a member of the Olympic swimming team who was in fine shape Wednesday morning. If the hypothesis h must have probability greater than $\frac{1}{2}$ on the putative evidence e, then, in this example, as in the previous one, e will not be evidence that h, and the counterexample is avoided.

According to the positive relevance account, not only is e's increasing h's probability sufficient for evidence, it is also necessary. The second lottery and intervening cause counterexamples present cases in which, intuitively, e is evidence that h but e does not increase h's probability. In the second lottery case, given the background b (that the lottery is a fair one in which one ticket drawn at random will win) and given the information e_1 (that *The New York Times* reports that Bill Clinton owns all but one of the 1000 lottery tickets sold), the information e_2 (that *The Washington Post* reports the same thing) is evidence that h (Bill Clinton will win)—or so I claimed. The positive relevance account disallows this claim, since e_2 does not increase h's probability over what it is on e_1 and b alone. In the intervening cause counterexample involving the medicine's relieving the symptoms exactly the same thing occurs: something that intuitively looks like evidence that a hypothesis is true is precluded from that category since it fails to increase the probability of the hypothesis. However, these two counterexamples may be avoided if we relinquish the idea that evidence must increase a hypothesis' probability in favor of the idea that the hypothesis must have high probability.

2. A Good Reason for Belief Requires Probability Greater Than One-Half

My argument for this claim makes two assumptions:

1. For any hypothesis h and putative evidence e there is some number k greater than or equal to zero such that if e is a good reason to believe h, then $p(h/e) > k$.

The concept of "a good reason for belief" employed here is one involved in potential evidence (where e can be a good reason to believe h even though h is false). The concept is objective, in the sense that whether e is a good reason to believe h does not depend on whether anyone in fact knows or believes e, h, or that e is a good reason to believe h. Therefore, the idea of probability in this assumption is objective. Since it is, and since assumption 1 is concerned with a good reason for belief, the concept of probability employed in assumption 1 (and throughout this chapter, except where otherwise indicated) will be the objective epistemic one of chapter 5, viz. degree of reasonableness of belief.

The assumption above is weak in two respects. First, it does not require (or disallow) that k be the same for all hypotheses. Second, the assumption places no restriction on k. (As far as this assumption is concerned, k could be zero.) The strength of the assumption is to tie "good reason to believe" to probability. For e to be a good reason to believe h, it is required that there be some probability of h on e, and that this probability be greater than 0, the lowest probability possible.

(The present assumption, although weak, is not without its critics. I will consider one of the most challenging in sections 12 and 13.)

2. For any e and h, if e is a good reason to believe h, then e cannot be a good reason to believe the negation of h ($-h$).

This is the central assumption of the argument for the claim that e is a good reason to believe h only if $p(h/e) > \frac{1}{2}$. I will return to assumption 2 after presenting the argument.

Suppose that e is a good reason to believe h, so that, in accordance with assumption 1, $p(h/e) > k$, for some k greater than or equal to 0. Suppose $k < \frac{1}{2}$. Then it could be the case that $p(-h/e) > k$. If so, $-h$ and e would satisfy a necessary condition for e's being a good reason to believe $-h$. So unless other necessary conditions for "good reason to believe" are imposed that preclude the possibilities imagined, for some e and h if $k < \frac{1}{2}$ it could be the case that e is a good reason to believe h and also a good reason to believe $-h$. This violates assumption 2.

Now consider the "other necessary conditions" that might be imposed on "good reason to believe" to form a set of sufficient, as well as necessary, conditions. If this set requires that k be greater than or equal to $\frac{1}{2}$, then a good reason to believe requires a probability greater than $\frac{1}{2}$. By contrast, suppose this set of conditions allows something to be a good reason to believe even when k is less than $\frac{1}{2}$. Then for some e and h, where $p(h/e) > k$ and $p(-h/e) > k$, it could be that e is a good reason to believe h and to believe $-h$, in violation of assumption 2. If k is greater than or equal to $\frac{1}{2}$, then such a possibility is precluded. If k is less than $\frac{1}{2}$, it is not precluded.

Providing a good reason to believe is a concept underlying both potential and veridical evidence. Accordingly, by the argument of this section, each of the latter requires probability greater than $\frac{1}{2}$. What I have said can be applied *mutatis mutandis* to the concept of providing a justification for belief, which underlies ES-evidence.

In the sections that follow I will be considering objections to this argument, and, more generally, to the claim that e is a good reason to believe h only if $p(h/e) > \frac{1}{2}$. Before doing so, however, one must be very clear that the claim I am defending provides only a necessary and not a sufficient condition for being a good reason for belief. It is not part of the claim that any probability greater than $\frac{1}{2}$ suffices for such a reason. We may want to require that the probability be significantly greater than $\frac{1}{2}$. And we may want to say that whatever the probability of h on e is, this by itself is not sufficient for e's being a good reason to believe h. Both issues will be addressed in chapter 7.

3. First Objection

One might be tempted to reject the previous argument, and, in particular, assumption 2 on which it is based, by invoking the following example. Suppose medicine M relieves symptoms S in half the cases and in the other half makes them worse. Then, it might be said, the fact that (e) Sam is taking M to relieve symptoms S is a good reason to believe (h) that his symptoms will be relieved, since M relieves S in half the cases. And it is a good reason to believe ($-h$) that his symptoms won't be relieved (but made worse), since that is what happens in half the cases. If so, e is a good reason to believe h and a good reason to believe $-h$, in violation of assumption 2.

My response is to admit that in such a case e is *just as good* a reason for believing h as for believing $-h$. But the fact that two things are equally good does not by itself make either of them good. (Two equally good essays can both be quite bad.) Suppose that Sam does not take medicine M to relieve his symptoms S. And suppose that 90% of those with S who do not take M (or any other medicine) get worse, while 10% get better. Then, taking M is clearly the better course of action than not taking M. Taking M has a better chance of relieving S (50%) than does not taking M (10%). But this is not enough to make the fact that Sam is taking M a good reason for believing that his symptoms will be relieved.

The claim that some e is a good reason for believing h is usually an empirical claim. It can be defended or criticized on empirical grounds. For example, the claim that

R: The fact that Sam is taking medicine M to relieve symptoms S is a good reason to believe that Sam's symptoms S will be relieved

might be defended by appeal to the following empirical association (here I change the percentages from the ones originally given):

A1: Symptoms S are relieved by M in 90% of the cases, while without M there is relief of S in only 5% of the cases.

Similarly, the claim R can be criticized and indeed falsified by showing A1 to be false and the following to be true:

A2: Symptoms S are relieved by M in 1% of the cases, while without M there is relief in 5% of the cases.

More generally, the weaker the association between taking M and relief of S the less one can expect relief of S from M, and the less good is the reason for believing S will be relieved by M.

One way to criticize a good-reason-to-believe claim is to show that the association is such that it provides as good a reason to believe h as it does to believe $-h$. Suppose you claim that the fact that this fair coin will be tossed once is a good reason to believe it will land heads. I am *criticizing* your claim by saying

Not really, since this fact provides an equally good reason for believing it will land tails.

I am saying that the fair coin's being tossed is not a good reason to believe it will land heads *because* it is an equally good reason to believe that it won't.

Similarly in the symptom case, suppose that A1 and A2 are both false, but the following is true:

A3: Symptoms S are relieved by M in 50% of the cases, and made worse in the other 50%, while without M there is relief only in 5% of the cases.

I am criticizing claim R above by saying that, in view of A3, the fact that Sam is taking M to relieve S provides an equally good reason to believe that Sam's symptoms will not be relieved, but indeed will be made worse. I am saying that because of this, the fact that Sam is taking M is not really a good reason to believe his symptoms will be relieved. This is true despite the fact that Sam's chances for relief are considerably improved by taking M, just as the coin's chances for landing heads are considerably improved by tossing it. But because the facts leading to such improvements are equally good reasons for believing that the outcomes in question will not occur, they do not provide a good reason for believing they will occur.

Now we need to consider some other fundamental objections.

4. Belief and Degrees of Belief

It might be held that there are no beliefs as such, only degrees of belief. This is the view of some Bayesians, who seek to define evidence (and related concepts) entirely by reference to probability. Since the concepts of evidence I have distinguished involve the idea of providing a good reason (or justification) for believing, they presuppose that there are beliefs, not simply degrees of belief. If there are only degrees of belief, not beliefs as such, the objective concepts of evidence, or at least my characterizations of them, seem flawed.[1]

On the Bayesian view, belief is like temperature. Both come in degrees. Anything that has a temperature has some degree of temperature. This view has two versions. On the stronger one, noted above, there are no beliefs at all, only degrees of belief. The assertion that one believes a proposition is either false or lacking in truth-value. The assertion that one believes it to some particular degree has a truth-value. On the weaker version there are beliefs but these are to be understood in terms of degrees of belief. One might take the state of affairs of X's believing h to be identical with the state of affairs of X's believing h to a degree greater than some specified number k.[2] Or one might take the state of affairs of X's believing h to be identical with the state of affairs of X's believing h to the particular degree X does, provided that this is greater than k.

Despite appearances, neither the stronger nor the weaker version of the degree-of-belief view is particularly damaging to the idea that evidence provides a good reason (or justification) for belief. Bayesians understand degrees of belief as probabilities. So, the view that evidence provides a good reason to believe can be understood as saying that evidence furnishes a high probability. The relationship between evidence and belief can be expressed in a way that will satisfy even the more militant Bayesians who deny the existence of beliefs in favor of degrees of belief.

1. These claims about belief refer not to the content of belief (the proposition believed) but to the state or act of believing. It is the latter, rather than the former, that is being claimed to be subject to degrees.
2. This version seems to be one offered by Daniel Hunter, "On the Relation between Categorical and Probabilistic Belief," *Nous*, 30 (1996), 75–98. On Hunter's view, "to believe a proposition is simply to assign that proposition a high subjective probability, where what counts as sufficiently high probability is vague" (p. 87).

One argument brought against both versions of the Bayesian view is empirical. It is that if degrees of belief exist, but not beliefs, or if beliefs are simply sufficiently high degrees of belief, we would be, and need to be, overwhelmed with information that we could not process. But in fact we are not overwhelmed. One defender of this argument is Richard Foley, who writes:

> If all of the information provided to us by others were finely qualified with respect to the provider's degree of confidence in it, we would soon be overwhelmed. It is no different with our private deliberations. We normally don't have finely qualified degrees of confidence in a wide variety of propositions . . . but even if we did, we would soon find ourselves overwhelmed if we tried to deliberate about complicated issues on the basis of them.[3]

I will not pursue this here. In this book I will assume that there are states of believing and that these are not identical with having degrees of belief. If they are not, how are they related? Believing is something we do. In this respect it is like speaking, hitting, or running (even though obviously in other respects it is different). These doings admit of degrees, indeed along one or more dimensions. One can speak with different degrees of speed and loudness. One can hit with different degrees of frequency and intensity. And one can believe with different degrees of firmness or strength. The fundamental idea is the doing, the dimensions represent ways it can be done, and the degrees represent how much is done along a given dimension. So the fundamental idea is speaking, which can be qualified along the dimension of loudness, which admits of varying degrees. With whatever degree of loudness one is speaking, one is still speaking. Similarly, believing is the fundamental idea; it can be qualified along the dimension of firmness or strength, which admits of degrees. With whatever degree of strength or firmness one believes, one is still believing.[4]

It might be objected that analogous claims can be made for temperature, which comes in degrees. Although "having a temperature" is not to "do" anything, one might say that it is the fundamental idea. Whatever degree of temperature something has, it has a temperature. Nevertheless, having a temperature is always having a temperature of some degree. Indeed, a standard dictionary offers this definition of temperature:

> The degree of hotness or coldness of anything, usually as measured on a thermometer.[5]

In the kinetic theory of gases temperature is understood as a quantity proportional to the mean translational kinetic energy of the molecules. Even if this is not taken as a definition, nevertheless temperature is by definition a quantity that admits of degrees (where this quantity is proportional to the mean molecular kinetic energy). Now, just as having a temperature is by definition the same

---

3. Richard Foley, "The Epistemology of Belief and the Epistemology of Degrees of Belief," *American Philosophical Quarterly*, 29 (1992), 111–124; quote on p. 122. See Gilbert Harman, *Change in View* (Cambridge, MA: MIT Press, 1986), ch. 3.
4. See Isaac Levi, *Gambling with Truth*, p.122.
5. *Webster's New World Dictionary*.

as having a temperature of some degree, so, it may be said, believing is by definition believing to some degree.

My response is to say that the analogy has not been drawn correctly. To be sure, since temperature is by definition a quantity admitting of degrees, having a temperature is by definition having a temperature of some degree. The analogy with belief should be this. By definition believing is believing something (a proposition, a person). Believing, by definition, requires an object of belief. Necessarily one cannot simply believe, just as necessarily an object cannot have a temperature without having some specific degree of temperature.

Proponents of the original analogy may now argue that believing is not *by definition* believing to some degree, but just that this is always so, even necessarily so. Even if it is not true by definition that one cannot believe without believing to some degree, it is nevertheless true (just as one cannot speak without speaking with some degree of loudness). So, suppose that someone who believes h does so to the degree r. Then, it might be claimed, the state of affairs of his believing h is identical to the state of affairs of his believing h to the degree r (just as the state of affairs of a gas's having a certain temperature is identical to the state of affairs of its having the corresponding mean molecular kinetic energy). This identity thesis suffices for a degree-of-belief view (so long perhaps as the degree of belief exceeds some threshold k).

This version of the degree-of-belief view I would reject by invoking the causal principle that if states of affairs are identical, they have the same causes and effects. Even if it is true that if you believe h you must do so to some degree or other, your believing h and your believing h to the degree you do (where this exceeds k) can have different causes and effects. Therefore, they are not identical. For example, suppose that you believe very strongly (say to the degree .99) that O. J. Simpson is guilty of murder, and that your believing it to this degree causes me to be annoyed. What causes me to be annoyed, however, is not your believing it (something that I too believe), but your believing it to the degree you do (how can you be so damn sure?). But if your believing that O. J. Simpson is guilty is the same state as your believing this to the degree you do, then, by the causal principle, they must have the same effects. Since they do not, these states are not identical.

Accordingly, a degree-of-belief theorist must either reject beliefs altogether in favor of degrees of belief or hold that the state of affairs of believing some h is identical with the state of affairs of believing h to a degree greater than some specified k. As I said earlier, however, neither of these views requires abandoning the idea that evidence provides a good reason for belief. The latter will simply be understood as a high probability.

5. Must Evidence Supply a Good Reason for Belief?

Suppose it is granted that a good reason for belief requires high objective epistemic probability (greater than $\frac{1}{2}$) on that reason. Let us confine our attention to potential and veridical evidence, both of which provide reasons for belief (rather than to ES-evidence and subjective evidence which do not). Why must (potential and veridical) evidence provide a *good* reason for belief? Won't it suffice for such evidence to provide just *some* reason to believe a hypothesis, but not necessarily a good one, or perhaps a *better* reason for belief than without it, but one that is not necessarily a good reason?

Suppose, according to the background information b, there is a coin that is precariously balanced on its edge. The new information e is that this coin is being subjected to a force. Let h be the hypothesis that this coin will soon be lying heads up. The fact that the precariously balanced coin is being subjected to a force is, I think, *some* reason to believe it will soon be lying heads up, and also *some* reason to believe it will soon be lying tails up and hence not heads up. It is not, however, a *good* reason for believing either, since (as I argued in sections 2 and 3) if something is a good reason to believe h it cannot also be a good reason to believe $-h$. One way of criticizing the claim that e is a good reason to believe h is to show that it is an equally good reason to believe $-h$.

However, suppose we were to drop the "good reason to believe" requirement. Can we do so in favor of a "some reason to believe" requirement? Can we say that since the fact that this precariously balanced coin is being subjected to a force is some reason to believe it will soon be lying heads up and some reason to believe it will soon be lying tails up, this fact is potential evidence that each claim is true?[6] I reject this idea because of what I will call the "canceling" effect to which evidence is subject.

Potential (as well as veridical) evidence claims of the form "e is evidence that h" are usually empirically incomplete. They can be falsified or at least have doubt cast upon them by additional empirical information. So if you claim that the fact that this precariously balanced coin is being subjected to a force is evidence that it will land heads, I can cast doubt on your claim by pointing out the empirical fact that this coin has two sides, one of which is not heads, and that the force is being applied so as to make the coin spin randomly, so that there is no more reason to expect it to land heads rather than tails. If this is so, then your evidence claim has been seriously weakened. Has it been canceled altogether? Has it been falsified?

I think it has. In this respect potential evidence is like net force in physics. A book sitting on the table is being subjected to two forces. There is the force of gravity directed toward the center of the earth; and there is the force exerted by the table on the book in the opposite direction. The net force in this case is zero, since the two independent forces are equal and opposite. The net force is determined by the balance of all the forces. Whether some particular force acting on a body is the net force depends upon whether there are other forces acting on it. The claim that this force is the net force is defeated by the existence of other forces. In this respect evidence is like net force. Some fact might pull in one direction, while another might pull equally in the opposite direction and prevent the first from being (potential or veridical) evidence. Or (unlike the case of force) some fact might pull equally in opposite directions, with a net pull in neither. (This is so with our precariously balanced coin.)

---

6. Obviously it cannot be veridical evidence, since this requires the hypothesis to be true; and both hypotheses cannot be true.

It might be objected that in many situations (including scientific ones) the evidence pulls equally in opposite directions, and it is still evidence. For example, suppose that the effectiveness of a new drug is being tested with regard to its capacity to reduce certain symptoms. Two trials are run. In the first there are 2000 patients with the symptoms, 1000 of which are given the drug and 1000 a placebo. Of those taking the drug, 950 have their symptoms relieved, while none in the control group does. Call this information e_1. The second trial is run just like the first, except that the result is quite different (call it e_2): of the 1000 patients taking the drug only 50 have their symptoms relieved, while none in the control group does. Consider the hypothesis h that the drug is more than 50% effective. The results of the two trials pull equally in opposite directions with respect to h. Yet we might say that the combined result $e_1 \& e_2$ constitutes some evidence that h and some evidence that $-h$.

My response is that what is being claimed is that part of the combined result, viz. e_1, is evidence that h, and a different part, viz. e_2, is evidence that $-h$. The concept of evidence here is ES-evidence. Where the epistemic situation does not contain knowledge of the result e_2, e_1 can be ES-evidence that h (analogously for e_2 with respect to $-h$). Someone in an epistemic situation in which result e_2 is unknown might be justified in believing h on the basis of e_1, just as someone in an epistemic situation in which result e_1 is unknown might be justified in believing $-h$. But where we are not separating and relativizing evidence in this way, the combined results pulling in opposite directions cancel. The combined results are not evidence that h is true, and they are not evidence that $-h$ is true. They are puzzling and inconclusive.

A somewhat similar analysis applies to the case of the precariously balanced coin. Where h is that the coin will soon be lying heads up, and e is that a force is being applied to the coin, e can be evidence that h and also evidence that $-h$, if we mean ES-evidence and if we have in mind different ES's for each claim. One ES includes knowledge of the fact that one side of the coin shows heads, but includes no knowledge about the other side. The other ES includes the analogous situation involving tails. Someone who knew that at least one side of the coin is marked heads, and who was not in a position to know that the two sides of the coin are different, might be justified in believing that the coin will soon be lying heads up; similarly, for someone else with this knowledge concerning tails. But where ES includes the fact that the two sides of the coin are marked differently, then, with respect to such an ES, one would not be justified in believing the coin will soon be lying heads up. And the fact that a force will be exerted on the precariously balanced coin is not ES-evidence that the coin will soon be heads up, nor is it ES-evidence that the coin will soon be tails up. It is evidence that the coin will soon be *either* heads up *or* tails up.

6. How Scientists Deal with Conflicting Studies

Multiple studies with conflicting results frequently occur in the sciences. Bemoaning this fact, here is how one pair of authors describe the situation in social science research:

The sequence of events that led to this state of affairs has been much the same in one research area after another. First, there is initial optimism about using social science research to answer socially important questions that arise.... Next, several studies on the question are conducted, but the results are conflicting. There is some disappointment that the question has not been answered, but policy makers—and people in general—are still optimistic. They, along with the researchers, conclude that more research is needed to identify the interactions (moderators) that have caused the conflicting findings....

In the third phase, a large number of research studies are funded and conducted to test these moderator hypotheses. When they are completed, there is now a large body of studies, but instead of being resolved, the number of conflicts *increases*. The moderator hypotheses from the initial studies are not borne out. No one can make much sense out of the conflicting findings. Researchers conclude that the phenomenon that was selected for study in this particular case has turned out to be hopelessly complex, and turn to the investigation of another question, hoping that this time the question will turn out to be more tractable.[7]

What these authors advocate, and what their book expounds, is *meta-analysis*, a method or set of methods for combining the results of disparate and even conflicting studies to determine whether new statistical information can be gleaned from these studies that will support or disconfirm a hypothesis of interest. Even before the term "meta-analysis" was introduced in 1976 by Gene V Glass, other investigators had decried conflicting results of studies in various areas. In 1971 Light and Smith, who introduced a precursor to meta-analysis which they called the "cluster approach," quoted the following passage from a 1966 U.S. Department of Health, Education, and Welfare survey regarding the benefits of ability-grouping in schools:

> Many studies throughout the years have compared academic progress of children grouped according to ability with progress made when grouped heterogeneously. Conclusive and definite anwers to questions commonly asked are difficult to get. Some studies show gains favoring ability grouping, some favoring heterogeneous grouping. Others show little or no significant difference between procedures used. The evidence against or in favor of ability grouping remains vague in spite of a rather persistent belief that learning problems would be greatly alleviated if children on similar levels of ability or achievement could be grouped together for instructional purposes.[8]

In a recent book on meta-analysis for the more general reader, Morton Hunt documents the problem of conflicting studies in many different fields of science.[9] These include several examples of medical studies yielding conflicting conclusions. For instance,

---

7. John E. Hunter and Frank L. Schmidt, *Methods of Meta-Analysis* (Newbury Park, CA: Sage Publications, 1990), p. 36.
8. Quoted in Richard J. Light and Paul V. Smith, "Accumulating Evidence: Procedures for Resolving Contradictions among Different Research Studies," *Harvard Educational Review,* 41 (1971), 429–471. Quote on p. 438.
9. Morton Hunt, *How Science Takes Stock* (New York: Russell Sage Foundation, 1997); see pp. 2–4.

Twenty-one studies of the use of fluorouracil against advanced colon cancer all find it beneficial, but findings of its effectiveness vary so widely—from a high of 85% to a low of 8%—as to be meaningless and useless to clinicians. (p. 2)

When there are studies with conflicting results what do researchers in the field generally do besides try to conduct new studies or give up and tackle a new problem? Several strategies are noted by the authors cited above:

1. Write a review article that summarizes the different studies and results, without attempting to resolve the issue.
2. Choose a single, favorite study from the set and agree with its conclusions.
3. Compute overall averages for relevant statistics across the entire set of studies, independently of the sizes of the sample in each study or the conditions under which the samples were taken.
4. Take a vote. If a majority of the studies favor one conclusion, then that is the conclusion supported by the studies.
5. Employ meta-analysis, which its proponents regard as a much more sophisticated and reliable set of methods than 3 and 4 above.

When different studies yield conflicting results strategies 2–5 may generate the claim that, despite these conflicting results, the evidence, considered as a whole, does support (or disconfirm) a certain hypothesis. But it need not generate such a claim at all. Various conflicting studies may seem equally good, so no favorite emerges using strategy 2. Overall averages for the studies computed using strategy 3 may not favor a hypothesis over its negation. The studies may be evenly divided with respect to a given conclusion, so the vote in strategy 4 results in no selection. And even a sophisticated meta-analysis of strategy 5 may not yield a conclusion favoring one hypothesis over another. Indeed the use of meta-analyses is controversial.[10] Different meta-analyses by different analysts have scored the same studies differently, resulting in different conclusions. And the results of a meta-analysis of a fairly large set of small trials may conflict with the results of a very large controlled trial.[11]

Accordingly, even strategies 2–5, which can lead to some resolution, need not do so. If not, scientists are in the same boat as Hunter and Schmidt described earlier: "No one can make much sense out of the conflicting findings." They cancel. There is (ES-)evidence that h is true (the results of some of the studies). There is (ES-)evidence that h is false (results of other studies). But considered as a whole, the studies are not both evidence that h is true and evidence that $-h$ is true. They are not evidence that either hypothesis is true, despite the fact that they contain parts that are (ES-)evidence that each is true. That is why the government report, quoted by Light and Smith, concerning ability-grouping in school, claims that while "some studies show gains favoring ability grouping,

---

10. A summary of criticisms is contained in a review of Hunt, *How Science Takes Stock*, by John C. Bailar III, *Science*, 227 (1997), July 25, 1997, 528–529. See also Bailar, "The Practice of Meta-Analysis," *Journal of Clinical Epidemiology*, 48 (1995), 149–157.
11. See, for example, Jacques LeLorier et al., "Discrepancies between Meta-Analyses and Subsequent Large Randomized, Controlled Trials," *The New England Journal of Medicine*, August 21, 1997, 536–542.

some favoring heterogeneous grouping . . . the evidence against or in favor of ability grouping remains vague." That is also why Hunter and Schmidt say that in such cases of conflict "researchers conclude that the phenomenon that was selected for study . . . has turned out to be hopelessly complex, and turn to the investigation of another question, hoping that this time the question will turn out to be more tractable."

From the discussion in this and the previous section, I conclude that if e is potential or veridical evidence that h, then e provides a good reason to believe h, not simply some reason. And, I have argued in section 2, e provides a good reason to believe h only if h's (objective epistemic) probability on e is greater than $-h$'s probability on e, that is, only if h's probability on e is greater than $\frac{1}{2}$. So if e is potential or veridical evidence that h, then h's probability on e is greater than $\frac{1}{2}$. In the sections that follow I propose to consider two new objections to these conclusions. The first divorces evidence from belief; the second divorces belief from probability.

7. The Likelihood View of Evidence

On this view, evidence is unrelated to belief, or at least, it is not related in any of the ways I have outlined. The idea has been defended by Ian Hacking,[12] A. W. F. Edwards,[13] and most recently, Richard Royall, a biostatistician.[14] I will consider the latter in what follows.

Royall defines evidence by formulating what he calls the

Law of likelihood: If hypothesis A implies that the probability that a random variable X takes the value x is $p_A(x)$, while hypothesis B implies that the probability is $p_B(x)$, then the observation $X = x$ is evidence supporting A over B if and only if $p_A(x) > p_B(x)$, and the likelihood ratio $p_A(x)/p_B(x)$ measures the strength of that evidence.[15]

An example will be helpful, and I will introduce one used by Royall himself. There is a diagnostic test for a disease D which is such that when D is present the test detects it 95% of the time and gives an erroneous reading 5% of the time. When D is absent the test correctly yields a negative result 98% of the time and incorrectly yields a false positive 2% of the time. The probabilities, then, are represented as follows:

$p(+/D) = .95$ (the probability that the test will yield a positive result given the presence of D is .95)
$p(-/D) = .05$
$p(-/not\text{-}D) = .98$
$p(+/not\text{-}D) = .02$

---

12. Ian Hacking, *Logic of Statistical Inference* (Cambridge: Cambridge University Press, 1965).
13. A. W. F. Edwards, *Likelihood* (Baltimore: Johns Hopkins Press, 1992).
14. Richard Royall, *Statistical Evidence* (London: Chapman and Hall, 1997).
15. Royall, p. 3. Although he does not provide an interpretation for probability, Royall is talking about some concept of objective probability.

Now we consider two hypotheses regarding a certain patient.

A: The patient has D
B: The patient does not have D

On Royall's view, hypothesis A implies that the probability of a positive test result for this patient is .95. We can write this as $p(+/A) = .95$, or $p_A(+) = .95$. Similarly, we can write $p(+/B) = .02$, or $p_B(+) = .02$, for the probability of a positive test result on hypothesis B. $p(+/A)$ is called the *likelihood* of hypothesis A on the test result +. The *random variable* X in this case is the test result, which can take two possible values: positive (+) and negative (−). Let us suppose that our patient is given the test, and that it yields a positive result (+). This result is what Royall denotes by the lower case variable x. So, in this example, "the observation $X = x$" is understood to mean that the test yields a positive result. I shall refer to this outcome simply as e.

The question of interest to Royall is whether e is evidence supporting hypothesis A (that the patient has disease D) over hypothesis B (that the patient does not have D). On his view, it is if and only if

(1) $p(e/A) > p(e/B)$,

that is, if and only if the likelihood of A on e is greater than that of B on e.[16] In the present case,

$p(e/A) = p(+/D) = .95$
$p(e/B) = p(+/not\text{-}D) = .02$.

So (1) is satisfied, and e is evidence supporting A over B. The strength of the evidence is given by the ratio $p(e/A)/p(e/B) = .95/.02 = 47.5$.

Royall makes a number of claims about this view, which help to clarify it and highlight its advantages.

1. One advantage he claims is that it is a plausible extension of reasoning in deterministic cases to probabilistic ones. He writes:

> One favorable point is that it seems to be the natural extension, to probabilistic phenomena, of scientists' established forms of reasoning in deterministic situations. If A implies that under specified conditions x will be observed, while B implies that under the same conditions something else, *not x*, will be observed, and if those conditions are created and x is seen, then this observation is evidence supporting A versus B. This is the law of likelihood in the extreme case of $p_A(x) = 1$ and $p_B(x) = 0$. The law simply extends this way of reasoning to say that if x is more probable under hypothesis A than under B, then the occurrence of x is evidence supporting A over B, and the strength of that evidence is determined by how much greater the probability is under A. This seems both objective and fair—the hypothesis that assigned the greater probability to the observation did the better job of predicting what actually happened, so it is better supported by that observation. (p. 5)

16. As mentioned in footnote 15, Royall is speaking of some type of objective probability, although he does not specify any particular one. In what follows, in order to contrast his view of evidence with the one I am developing for potential evidence, I will construe his probabilities as objective epistemic ones. However, the criticisms I develop will be applicable no matter what objective interpretation is given for his probabilities.

2. A second advantage claimed by Royall is that, unlike the positive relevance definition of evidence, the likelihood definition does not require the determination of a prior probability for a hypothesis h, which Royall regards as "generally unknown and/or personal." Obviously, then, Royall is regarding the likelihood probability $p(e/h)$, and therefore evidence itself, as objective rather than subjective. Immediately following his rejection of prior probabilities as generally unknown and/or personal, he writes:

> Although you and I agree (on the basis of the law of likelihood) that given evidence supports A over B, and C over both A and B, we might disagree about whether it is evidence supporting A (on the basis of the law of changing probability) purely on the basis of our different judgements of the a priori [prior] probabilities of A, B, and C. (p. 11)

3. Royall emphasizes that the definition is comparative. It is intended to explicate the idea that some set of observations is evidence for one hypothesis over another. It

> represents a concept of evidence that is essentially relative, one that does not apply to a single hypothesis, taken alone. Thus it explains how observations should be interpreted as evidence for A vis-a-vis B, but it makes no mention of how those observations should be interpreted as evidence in relation to A alone. (p. 8)

So from the fact that e is evidence for h_1 over h_2, or that e supports h_1 more than h_2, we cannot conclude that e is evidence for h_1, or that e supports h_1.

4. Finally, let me mention a claim made by Edwards, another defender of the likelihood account. He writes:

> It is sometimes objected that the [likelihood] measure of support does not have any "meaning," by which is usually meant any "probability interpretation." It is indeed true that a statement of support, though derived from probabilities, does not make any assertion about the probability of a hypothesis being correct. And for good reason: the [likelihood] method of support has been developed by people who explicitly deny that any such statement is generally meaningful in the context of a statistical hypothesis....
>
> There is, however, a perfectly simple "operational interpretation" of a likelihood ratio for two hypotheses on some data. It is, of course, the ratio of the frequencies with which, in the long run, the two hypotheses will deliver the observed data.[17]

What Edwards seems to have in mind by the last statement can be illustrated by means of the following example. We have two weighted coins. One is weighted so that in the long run 60% of the tosses yield heads. The other is weighted so as to yield 40% heads. We cannot tell which coin is which simply by looking, only by tossing. We choose one of the two coins at random and consider these hypotheses:

h_1: This coin has the 40% weight.
h_2: This coin has the 60% weight.

17. Edwards, *Likelihood*, p. 33.

Now we toss the coin 10 times, resulting in the following outcome:

e: The coin landed heads 3 times.

The likelihoods are these:

$p(e/h_1) = .2150$
$p(e/h_2) = .0425$

The likelihood ratio $p(e/h_1)/p(e/h_2) = .2150/.0425$, which is approximately 5. So on the likelihood account, e supports h_1 over h_2, and the strength of the evidence, as measured by the likelihood ratio, is approximately 5.

Now what Edwards is claiming is that the "operational interpretation" of this likelihood ratio can be given as a long-run relative frequency: if we continue to toss this coin indefinitely, then on the hypothesis h_1 (that the coin is the 40% coin) the relative frequency of sequences of 10 tosses in which 3 yield heads is 5 times greater than it would be on the hypothesis h_2 (that the coin is the 60% coin). More generally, if the likelihood ratio $p(e/h_1)/p(e/h_2) = r > 1$, then if we continue to run the test as we have been doing, we will obtain results like e, the results we have actually obtained, r times more if hypothesis h_1 is true than if h_2 is true.

8. Likelihood and Belief

Is there a likelihood concept of evidence? Do scientists and others actually employ such a concept? If not, should they?

To begin with, Royall explicitly acknowledges that the concept of evidence he introduces is to be distinguished from any concept that concerns whether, or to what degree, one should believe the hypothesis on the basis of the evidence. Suppose one has made an observation. Royall distinguishes what he regards as three separate questions:

1. What do I believe, now that I have this observation?[18]
2. What should I do, now that I have this observation?
3. What does this observation tell me about A versus B? (How should I interpret this observation as evidence regarding A versus B?) (p. 4)

An answer to the third question, he makes clear, does not provide an answer to the first two.

Indeed, assuming, as I have been doing, that if e supplies a good reason to believe h then $p(h/e)$ must be high ($> \frac{1}{2}$), we can readily see why the likelihood definition does not guarantee that evidence supplies a good reason for belief.[19] On

---

18. In the context of his discussion, it seems pretty clear that Royall intends this question to be interpreted as "What should I believe?"
19. This is perfectly consistent with construing Royall's probabilities as objective epistemic ones measuring degrees of reasonableness of belief. Royall's concept of evidence, understood in terms of such probabilities, requires the degree of reasonableness of believing e given h to be greater than the degree of reasonableness of believing e given h'. It does not require the degree of reasonableness of believing h given e to be greater than $\frac{1}{2}$, or than h' given e.

the likelihood definition, e can be evidence for hypothesis A over hypothesis B even if $p(A/e)$ is extremely low. Suppose, for example, that the presence of a certain gene G1 reduces the chance of getting a certain disease D from .05 to .02, while the presence of gene G2 reduces it from .05 to .01. That is, $p(D) = .05$, $p(D/G1) = .02$, and $p(D/G2) = .01$. And suppose that in the general population gene G1 is on the average present in one out of 1000 people, while G2 is present in one out of 100 people. That is, $p(G1) = .001$ and $p(G2) = .01$. Now, using Bayes' theorem, we have

$$p(G1/D) = \frac{p(G1) \times p(D/G1)}{p(D)} = \frac{.001 \times .02}{.05} = .0004$$

$$p(G2/D) = \frac{p(G2) \times p(D/G2)}{p(D)} = \frac{.01 \times .01}{.05} = .002$$

Since $p(D/G1) > p(D/G2)$, on the likelihood definition, the fact that a certain patient has disease D is evidence for the hypothesis that the patient has gene G1 over the hypothesis that he has G2. (The likelihood ratio is 2.) Yet in this case, the probability that a patient has gene G1 given that he has disease D, $p(G1/D)$, is very low (.0004). If $p(G1/D)$ must be greater than $\frac{1}{2}$ for the presence of disease D to be a good reason to believe the patient has gene G1, the fact that the patient has D in this case is not a good reason for such a belief. Yet on the likelihood view, this fact is supporting evidence that the patient has G1 over the hypothesis that the patient has G2. As noted, likelihood evidence for one hypothesis over another need not provide a good reason to believe the better supported hypothesis.

If likelihood evidence does not necessarily provide a *good* reason to believe a hypothesis, does it provide at least *some* reason? Since Royall distinguishes his concept of evidence for h from a concept of whether, or to what degree, to believe h, the answer would seem to be no. This can be confirmed by reference once again to probability. If e is at least *some* reason to believe h (though not necessarily a good reason), then, we might suppose, there is some threshold of probability (say, less than $\frac{1}{2}$) that h's probability on e must exceed. For example, in section 5, in the case of the coin precariously balanced on its edge, I claimed that the fact that it is being subjected to a force is not a good reason to believe it will soon be lying heads up. Nevertheless, I claimed it was *some* reason to believe this (and equally some reason to believe it will soon be lying tails up). The probability we assign to each of these outcomes is $\frac{1}{2}$. Without assuming that for e to provide some reason to believe h, $p(h/e)$ must be this high, let us suppose that it must exceed some threshold value $k > 0$. Whatever that value we can devise a case in which it is not exceeded, but e is likelihood evidence for h_1 over h_2.

For example, suppose the probability threshold value for "some reason" is .01. That is, for e to be at least some reason to believe h, the probability of h on e must exceed .01. Yet in our gene example, $p(G1/D) = .0004$, which is only 4% of this threshold value. If .01 is the threshold value, then the fact that the patient has disease D is not a reason to believe he has gene G1, even though, on the likelihood definition, the fact that the patient has D is evidence for his having G1 over his having G2. Could we say at least that if e is likelihood evidence for h_1 over h_2, then e is a better reason for believing h_1 than h_2, or a reason for believing h_1 more than h_2? This also seems precluded. If e is a better reason for believing h_1

than h_2, or a reason to believe h_1 more than h_2, then h_1 should be more probable on e than is h_2. Yet in the gene example, $p(G1/D) < p(G2/D)$. But since $p(D/G1) > p(D/G2)$, D is likelihood evidence for G1 over G2.

Indeed the presence of disease D lowers the probability that a person has gene G1 (from .001 to .0004). Yet it is likelihood evidence that a patient has gene G1 over gene G2. (The presence of D also lowers the probability that a person has gene G2, from .01 to .002.)

Accordingly, in this example, the presence of disease D is likelihood evidence for the hypothesis that the patient has G1 over the hypothesis that the patient has gene G2, even though

1. $p(G1/D) < p(G2/D)$, that is, even though the probability that the patient has G1 is less (not more) on D than the probability that the patient has G2.
2. $p(G1/D) = .0004$, that is, the probability that the patient has G1, given that he has D, is *very* low.
3. $p(G1/D) < p(G1)$, that is, the probability that the patient has G1 is lowered by his having D.

It is for these reasons that Royall wants to divorce support from belief.

If likelihood evidence does not provide a good reason, or indeed some reason, to believe a hypothesis, or a better reason to believe one hypothesis than another, what does it provide? What of value do we get when we get likelihood evidence? Let us grant that we get something that does not require a determination of prior probabilities, and that we get something for which objective probabilities can be used. Of what value is the product? What can we do with it, if it does not provide a basis for believing a proposition or (on Royall's view) a basis for action? Why should we be interested in such a concept? Why is this a concept of *support* or *evidence*?

In its defense, we recall, Royall claims that the likelihood concept of evidence is a natural extension of what he calls "deterministic" reasoning. He construes the latter as involving two hypotheses h_1 and h_2, where h_1 implies that e is the case and h_2 implies that e is not the case, and e turns out to be the case. If so, he claims, e is evidence supporting h_1 over h_2 (p. 5). There is, however, an important difference between the deterministic case and the nondeterministic ones of interest to Royall. In the former, if h_1 implies e, and h_2 implies $-e$, then $p(e/h_1) = 1$ and $p(e/h_2) = 0$. In the deterministic cases, unlike the nondeterministic ones, it is not possible that $p(h_1/e) < p(h_2/e)$, since the latter is equal to zero. Unless $p(h_1/e) = 0$, in the deterministic cases we are guaranteed that $p(h_1/e)$ will always be greater than $p(h_2/e)$. Let us assume that if $p(h_1/e) > p(h_2/e)$, then, given e, it is more reasonable to believe h_1 than h_2, or at least it is less unreasonable to believe h_1 than h_2. In the deterministic cases, then, that is, in cases in which $p(e/h_1) = 1$ and $p(e/h_2) = 0$, there will be at least some relationship between Royall's likelihood evidence and belief. But in nondeterministic cases there is no such guarantee, since even though $p(e/h_1) > p(e/h_2)$, in such cases it is possible that $p(h_1/e) < p(h_2/e)$ (and that $p(h_1/e)$ is very small, and that $p(h_1/e) < p(h_1)$). In short, even if the deterministic cases have some (minimal) connection with belief, no such connection is guaranteed for the nondeterministic ones. The question remains, then, why in the nondeterministic cases (which are, after all, the ones Royall is concerned with) we should be interested in his likelihood concept of evidence. What makes it a concept of evidence?

Another claim made by Royall is that since on the likelihood definition "the hypothesis that assigned the greater probability to the observation did the better job of predicting what actually happened, so it is better supported by that observation" (p. 5). (Edwards makes a related claim about the continued predictive sucesses of the hypotheses.) This reply, I believe, begs the question. What justification is there for saying that of two hypotheses, the one that assigned the greater probability to the event that actually occurred is the *better supported* by that event? Such a justification seems necessary since we cannot conclude that the event provides a good, or any, reason for believing the preferred hypothesis, or that it provides a better reason for believing the preferred hypothesis than for believing its competitor.

Ian Hacking makes a claim that might provide some justification for taking likelihood to be a measure of support. After formulating a principle of likelihood which is essentially the one propounded by Royall,[20] Hacking writes:

> This I venture as the explication of the thesis, "if p implies that something happens which happens rarely, while q implies that something happens which happens less rarely, then, lacking other information, q is better supported than p."[21]

I want to focus here on Hacking's phrase "lacking other information." The idea is that if the likelihood of h_1 on e is greater than that of h_2 on e, then, *in the absence of any other information*, e supports h_1 better than h_2. Perhaps the thought is that if one is in a situation where one can determine only whether the likelihood of h_1 on e, that is, $p(e/h_1)$, is greater than that of h_2 on e, that is, $p(e/h_2)$, and where one cannot compare prior probabilities, $p(h_1)$ and $p(h_2)$, and where one cannot compare posterior probabilities, $p(h_1/e)$ and $p(h_2/e)$, then at least one can judge support based on likelihoods.

If this is the idea, then again the question is begged, since it needs to be shown that in the absence of other information about these probabilities, judgments about support can be made at all. By analogy, suppose we want to determine whether one ship on the ocean is closer to the Baltimore harbor than another. Suppose that latitudes can readily be determined, and we can say that one ship is on a latitude closer to that of Baltimore than the other ship. But suppose that (as was historically the case until the invention of a gyroscopic clock) longitudes are very difficult to determine. It could be disastrous to say that, in the absence of longitudinal information, one ship is closer to Baltimore than the other because its latitude is closer to that of Baltimore than the other. Both latitude and longitude are needed to determine which is closer. What Hacking needs to demonstrate is that there is some intuitive sense of support for which information regarding prior and posterior probabilities is not needed. He needs to show that in that sense of support, in the absence of such information, the hypothesis with the higher likelihood is the better supported.

---

20. Indeed, Royall cites Hacking when he formulates his principle.
21. Hacking, *Logic of Statistical Inference*, p. 65.

9. Does Evidence Require Prior or Posterior Probabilities? Mayo's Error-Statistical Account

Deborah Mayo develops a novel probabilistic approach to evidence that forcefully rejects the positive relevance and high probability accounts (as well as a likelihood view of the sort just discussed).[22] She rejects positive relevance and high probability because both require a determination of the posterior probability $p(h/e)$, and, in addition, the former requires a determination of a prior probability $p(h)$. In general, she believes, it is not possible or necessary to assign probabilities to hypotheses, whether these are prior or posterior probabilities. I will not discuss Mayo's rejection of these standard views, but only her own positive view, which she calls an error-statistical approach.

On her view, to have evidence for a hypothesis, in a sense that is important to science, is to have something that passes a good test for that hypothesis:

> Following a practice common to testing approaches, I identify "having good evidence (or just having evidence) for H" and "having a good test of H." That is, to ask whether e counts as good evidence for H, in the present account is to ask whether H has passed a good test with e. (p. 179)

She divides this idea into two parts: (a) passing a test, and (b) the test's being a good one. The first idea she describes by saying that the test result e must "fit" the hypothesis. Her criterion of "fit" is rather open-ended and contextual. She does not require that the putative evidence e be entailed by the hypothesis h, or even that e be highly probable on h. Minimally, she says, e must not be improbable on h (p. 179).

It is the second idea, that of a good test, that is the most important in her account. For this purpose she proposes a

> *Severity requirement:* Passing a test T (with [result] e) counts as a good test of or good evidence for H just to the extent that H fits e and T is a *severe test* of H (p. 180).

The notion of a severe test in this requirement is to be understood by reference to what she calls a severity criterion, for which she offers two formulations:

> *Severity criterion 1:* There is a very high probability that test procedure T would not yield such a passing result [e] if H is false. (Alternatively, there is a very low probability that the test procedure T would yield such a passing result, if H is false.)

> *Severity criterion 2:* There is a very high probability that the test procedure T would yield a worse fit if H is false. (Alternatively, there is a very low probability that test procedure T would yield so good a fit if H is false.)

---

22. Deborah Mayo, *Error and the Growth of Experimental Knowledge* (Chicago: University of Chicago Press, 1996). I am very much indebted to her for her patience and impatience in explaining her very interesting but complex views to me. Her patience allowed me to correct mistakes and improve my formulations; her impatience motivated me to do so. In any case, induction tells me that she will not completely sanction what I say here.

Mayo's criteria for "evidence" (or a "good test") invoke a concept of probability. However, unlike the two standard probability accounts, which she rejects, her account does not appeal to a posterior probability for hypothesis h, that is, to $p(h/e)$, or to a prior probability for h, that is, to $p(h)$. The only probabilities she invokes are these:

(a) the probability that the test T will yield the putative evidence e, given the hypothesis h. I will write this as $p(e(T)/h)$. This probability is not supposed to be low, thus satisfying her minimal condition for e "fitting" h.

(b) the probability that the test T will yield the putative evidence e, given that the hypothesis is false, that is, $p(e(T)/-h)$. This probability is supposed to be low, thus satisfying her severity requirement.

There is no requirement, however, that the following probabilities be determined—$p(h/e(T))$ or $p(h)$—or that the former be high or that it be higher than the latter, in order that $e(T)$ be evidence that (or a good test of) h. Finally, the probabilities that Mayo invokes in (a) and (b) are objective, not subjective. Her examples employ probabilities construed in the relative frequency sense, and it is clear that these are the kinds of probabilities she has in mind for her concept of evidence.

Let's consider an example. There is a very large set of bags, all of which contain many balls. Each bag in the set is one of two types. In the first type, 60% of the balls are white and 40% are red. In the second type, 40% are white and 60% are red. We will proceed with the following experiment. We will choose one bag at random and select a sample of balls from it, with replacement, determining how many are white. (We will assume that the probability of selecting a white ball is the same each time.) We consider two hypotheses. The first, h, is that the bag selected contains 40% white balls. The second, h', is that it contains 60% white balls.[23]

The particular test T we run involves selecting 100 balls, with replacement, from the chosen bag. We need a test rule for deciding whether a result passes the test T with respect to the hypotheses h and h'. Here is one:

Test Rule: In 100 selections of balls from the chosen bag, if the observed proportion of white balls is less than .5 (that is, if less than 50 out of 100 are white), then the resulting selection passes the test T for hypothesis h as against hypothesis h'.

Now we run test T. We select 100 balls, with replacement, and obtain the result $e(T)$ that 40 are white. Since 40 of the 100 are white, the hypothesis h (that the bag selected is 40% white) "fits" this result (as precisely as you like). The question is how severe the test is. According to Mayo's severity criterion 1, there must be a very high probability that our test procedure would not yield this pass-

---

23. Mayo prefers to state hypotheses of these sorts to be making an assertion about the value of a parameter p in a binomial distribution. So if we consider drawing balls from the chosen bag, p represents the percentage of white balls drawn in n trials. The hypotheses of concern to her would then be $h: p = .4$, $h': p = .6$, in n trials (for sufficiently large n). In what follows I simplify by letting h and h' be the hypotheses given in the text.

ing result $e(T)$ if hypothesis h were not true but h' were. The probability that our test procedure would yield this passing result if h were false and h' were true,

p (we obtain this passing result for h (40 white out of 100)/h' is true),

is .00002. So the probability that our test procedure would *not* yield this passing result if h were not true but h' were is $1 - .00002 = .99998$, which is indeed a very high probability. Accordingly, Mayo's severity requirement is satisfied. On her view, then, the result $e(T)$ of the test (40 of the 100 balls selected are white) is good evidence for the hypothesis h.

Finally, we note that determining whether this test result $e(T)$ is evidence that h does not involve determining $p(h)$, the prior probability of h, or $p(h/e(T))$, the posterior probability of h given the test result $e(T)$.

10. Does Mayo's Account Give Us What We Want?

Is the result $e(T)$ described in the previous section really a good test of, and hence evidence for, hypothesis h, as Mayo's account implies? I suggest that it depends on whether certain additional conditions obtain. In the example we stipulated that the bag in question is chosen at random from a very large set of bags. Suppose that half of these bags are ones in which 40% of the balls are white and 60% are red, while the other half are ones in which 60% are white and 40% are red. If this is so, then, where a bag is selected at random, the probability of selecting a bag in which 40% of the balls are white is $\frac{1}{2}$, that is,

(1) $p(h) = \frac{1}{2}$.

Now, the probability of obtaining 40 white balls in a sample of 100, $e(T)$, on the assumption that h is true, is

(2) $p(e(T)/h) = .08$.

And the probability of obtaining the result $e(T)$, assuming that h is false (and that h' is true) is

(3) $p(e(T)/-h) = .00002.$[24]

According to Bayes' theorem,

$$p(h/e(T)) = \frac{p(h) \mathrm{x} p(e(T)/h)}{p(h) \mathrm{x} p(e(T)/h) + p(-h) \mathrm{x} p(e(T)/-h)}.$$

From (1), (2), and (3) above, using Bayes' theorem, the posterior probability

$$p(h/e(T)) = \frac{.5 \times .08}{(.5 \times .08) + (.5 \times .00002)}$$

24. Following Mayo, probabilities in (1)–(3) can be construed in a relative frequency sense. To simplify notation, however, I use h both for the type of event "selecting a bag in which 40% of the balls are white" (e.g., in (1)) and for the hypothesis "the bag selected has 40% white balls" (e.g., in (2)). Similar interpretations of h can be given for Mayo's "parameter" idea in note 23.

is approximately 1, and $p(-h/e(T))$, which is equal to $p(h'/e(T))$, is approximately 0. Note that in this case, since $p(h/e(T)) > .5$, obtaining the result $e(T)$ comports with my high-probability condition on evidence. So far so good.

By contrast, let us consider a very different scenario. Suppose that instead of half of the bags being 40% white, only 1 in 100,000 are, so that the probability of selecting a bag in which 40% of the balls are white is .00001, that is,

(4) $p(h) = .00001$.

Using this value in Bayes' theorem, and keeping the other values the same, we obtain

(5) $p(h/e(T)) = .004$ and $p(h'/e(T)) = .996$.

Accordingly, the probability that the bag selected is 40% white, *even assuming that in the sample of 100 examined 40 are white*, is extremely low, considerably less than 1%.

In what sense, if any, does this count as *passing a good test of, or evidence for, h?* This raises a fundamental question of what Mayo means by these italicized terms. I find the following two different claims about evidence (or "passing a good test") in her work.

(a) Sometimes she claims that "evidence indicates the correctness of hypothesis H, when H passes a severe test" (p. 64). This seems to imply that evidence that H provides a good reason to believe that H is *true*. But if this is what evidence is supposed to do, then, as I argued earlier in this chapter, it will do so only if the probability of the hypothesis, on the putative evidence, is greater than $\frac{1}{2}$. Clearly if (5) is the case, obtaining the experimental result $e(T)$ from test T is not evidence that h, in the sense required.

(b) Sometimes she claims that evidence that H indicates that H is *reliable*, which for her means that "what H says about certain experimental results will often be close to the results actually produced—that H will or would often succeed in specified experimental applications" (p. 10).

There are hypotheses that are extremely reliable, with reference to certain types of experimental results, yet very improbable, and indeed false. Consider the following hypothesis about a certain coin:

H = The devil made this coin fair, that is, he made it so that it will land heads approximately half the time.

If we consider experimental results pertaining to how the coin actually lands when tossed, this hypothesis may be very reliable. It may accurately predict what happens in the long run when the coin is tossed. Yet the hypothesis itself is extremely improbable. So, Mayo may say, tossing the coin 1000 times and obtaining heads approximately half the time is evidence for the reliability of the devil hypothesis H in experiments in which the coin is tossed and the side it lands on is noted. It is not evidence for the truth of the devil hypothesis H.

Mayo seems to allow evidential claims of both types (a) and (b). Which one someone is justified in making will depend on the hypothesis and the putative evidence. I do not at all reject claims of type (b) since they simply amount to saying that e is evidence (in a "good reason to believe" sense) not for the truth of H, but for the truth of the claim "H is experimentally reliable (in a specified range of experiments)." I would reject the claim that this is the only, or the main, question about evidence of concern to scientists and others. With respect to a hypothesis such as the devil hypothesis H, one may seek evidence not only of its "reliability" but of its truth as well. And, if $p(H/e)$ is extremely low, and hence e is not a good reason to believe H, then e is not evidence that H is true, even if it is evidence that H is "reliable" in experiments in which the coin is tossed and the side it lands on is noted.[25] Moreover, as I will argue in chapter 8, any plausible attempt to define instrumental evidence—evidence that H is experimentally reliable—will need to appeal to the probability that H is true.

What about my example involving bags of balls? Here Mayo may claim that the evidence obtained (40 of the 100 selected are white) is evidence that hypothesis h (the bag selected contains 40% white) is "experimentally reliable" (with respect to experiments that involve selecting balls from the bag). Or she may claim that it is evidence for the truth of h. Or she may claim that the truth of h in such a case amounts to its experimental reliability. Any of these alternatives results in problems with her account, I believe. The evidence obtained (that 40 of the 100 are white) is not evidence that h is true or "experimentally reliable" (with respect to experiments involving selecting balls from the bag). If only 1 in 100,000 of the bags contains 40% white balls, that is, if $p(h) = .00001$, then, as we said, $p(h/e(T)) = .004$. More generally, in the type of case envisaged, the probability is extremely low that the hypothesis h is true or "experimentally reliable," no matter what the test results $e(T)$ show. Yet on her view, the specific test result is a good test of, and hence evidence for, hypothesis h. On the contrary, I would claim, it is not.

11. Mayo Responses

Mayo considers an example that is analogous in certain respects to my bags-of-balls example. Her response, tailored to my own example, is this.[26] Selecting a sample from the chosen bag, is, or at least can be, a severe test of the hypothesis h, that 40% of the balls in the bag are white, the larger the sample the more severe the test. If the bag contains such a large number of balls that counting them all is infeasible, then sampling is surely the best means of detecting error. So, in a suitable sample, if approximately 40% of the balls are white, then obviously we cannot infer that 60% of all the balls in the bag are white. Our sample

---

25. Complicating the picture is the fact that for the purposes of testing, Mayo wants to divide a hypothesis such as the devil hypothesis H into two separate ones, each of which answers a different question. H_1: the devil made this coin; H_2: the coin is fair. These hypotheses would be tested in different ways. Perhaps, then, the way in which each would be tested (for example, by tossing the coin in the case of the second hypothesis but not the first) would be such that passing a severe test would indicate not only the "reliability" of the hypothesis but its "correctness" as well. So it might be the case, on her view, that when testable hypotheses are suitably isolated, then passing a severe test indicates both "reliability" and "correctness."
26. Deborah G. Mayo, "Response to Howson and Laudan," *Philosophy of Science*, 64 (1997), 323–333.

does not indicate that, despite the fact that in the example the probability that the bag has 60% white balls, given that 40% in the sample are white, is .996.[27]

My response to this contains two points.

1. The fact that 40% of the balls in the sample are white ($e(T)$) does not "indicate," in the sense of "is evidence," that 60% of the balls in the bag are white (h'). Although $p(h'/e)$ is .996 (so that my high probability condition is satisfied), my second (explanatory) condition of evidence (to be developed in chapter 7) is not satisfied. Given the truth of h' and e, the probability is not high that there is an explanatory connection between the two. For example, it is not highly probable that in the sample 40% of the balls are white *because* in the bag approximately 60% are white.

2. Perhaps what tempts Mayo to conclude that finding 40% white balls in the sample is evidence for the hypothesis that 40% in the bag are white is that there is *conflicting* evidence in this case. The fact that we have a sample from the bag in which 40% are white, *considered by itself*, is, or seems to be, evidence that 40% of the balls in the bag are white. On the other hand, the fact that the bag has been randomly chosen from a large set of bags only 1 in 100,000 of which contain 40% white balls, the remaining being 60% white, is, or seems to be, evidence that 60% of the balls in the bag are white. Earlier we spoke of cases of this sort, some of which, at least, can be understood in terms of ES-evidence. If one is in an epistemic situation in which all one knows is that 40% of a suitable sample of balls in this bag are white, but one has no idea that this bag was randomly selected from a large set of bags only 1 in 100,000 of which contains 40% white balls, then one may be justified in believing that 40% of all balls in the bag are white. And if one is in a situation in which all one knows is that the bag has been randomly selected from a large set of bags, .99999 of which contains 60% white balls, then one may be justified in believing that 60% of the balls in the bag are white.

With both pieces of information, what can we conclude? Consider just potential evidence. Can we conclude that the fact that 40% in the sample are white ($e(T)$) is evidence that the bag is 60% white (h')? No, despite the fact that

---

27. In Mayo's example, which is derived from one of Colin Howson, a randomly chosen student passes a college admissions test (e). There are two hypotheses. H = the student is ready for college (the student is not deficient); H' = the student is not ready for college. Now we suppose that college-ready students pass the test with probability 1 ($p(e/H) = 1$), while college-deficient students pass only 5% of the time ($p(e/H') = .05$). In the general population .999 are college-deficient, $p(H') = .999$, while $p(e) = .051$. Using Bayes' theorem, we calculate that $p(H'/e) = .98$, that is, the probability that the randomly chosen student is college-deficient, given that the student has *passed* the test, is .98, which, according to Mayo, Howson would take to indicate that the student is college-deficient, given that he has passed the test! Mayo responds that "scoring the passing grade e hardly shows *a lack* of readiness—and on these grounds, the error statistician denies that e indicates H'" (p. 328).

If by "indicates" Mayo means "is evidence that," then Howson would reject Mayo's claim. For Howson (who holds a positive relevance view of evidence), e is not evidence that H' (the student is college-deficient), but that H (the student is college-ready), since $p(H'/e)$ = .98 and $p(H') = .999$, so $p(H'/e) < p(H')$. On the other hand, $p(H/e) = 1 - p(H'/e) = .02$, $p(H) = .001$, so that $p(H/e) > p(H)$, from which it follows, according to Howson, that e is evidence that ("confirms") H, while e "disconfirms" H'. So, contrary to Mayo, for Howson "scoring the passing grade" indicates "readiness," not a lack thereof.

$p(h'/e\&b)$ is approximately .996 (where b indicates that only 1 in 100,000 of the bags is 40% white). The reason is the one given in point 1 above. The explanation condition is violated. Can we conclude that the fact that 40% in the sample are white ($e(T)$) is evidence that 40% in the bag are white (h)? No, because, as we saw, $p(h/e(T)\&b)$ is .004. So, on my view, in the light of background information, $e(T)$ is not evidence that h or that h'.

I turn now to a second possible Mayo response.[28] The suggestion is to reformulate the test rule to reflect the fact that the sample of balls selected occurred in two stages: first, a bag was selected randomly from a set of bags with respect to which the probability of selecting a 40% white bag is .00001; second, 100 balls were randomly selected from the chosen bag, of which 40 are white. In view of this two-stage procedure, the test rule, given in section 9, should be reformulated as follows:

> *New test rule:* In 100 selections of balls from a bag chosen at random from a set of bags with respect to which the probability of selecting a 40% white bag is .00001, if the observed proportion of white balls is less than .5, then the resulting selection passes the test T' for hypothesis h as against hypothesis h'.

Hypothesis h is that the bag selected in this way contains 40% white balls; h' is that it contains 60% white balls. Now we run the new test T' by selecting a bag at random from the set in question and selecting 100 balls from the chosen bag. We obtain the result $e(T')$ that 40 of the 100 are white.

Does the new test rule make a difference? Does it prevent the result $e(T')$ from counting as good evidence for the hypothesis h, on Mayo's view? No, it does not. First, the result $e(T')$ passes the new test rule for h as opposed to h', since the observed proportion of white balls (.4) is less than .5. Second, Mayo's severity criterion is satisfied, since the probability of obtaining the result $e(T')$, assuming that h is false, is very low:

$$p(e(T')/-h) = p(e(T)/-h) = .0002.$$

Using Bayes' theorem, and the given assumption that $p(h) = .00001$, we get

$$p(h/e(T')) = .004, \text{ and } p(h'/e(T')) = .996.$$

In short, we get the same results using the new test rule as we do using the original one. Moreover, the new test rule has the prior probability $p(h)$ built into its formulation. This is something Mayo seeks to avoid.

Here is a third possible Mayo response. She may reply in the same manner she does when a hypothesis is constructed to fit the data, where the procedure for constructing it would probably yield a good fit even if the hypothesis is false. She calls this Gellerization. For example, let an experimental result consist in outcomes of tossing a coin, where a heads outcome is a "success" and tails a "failure." We are interested in hypotheses giving the probability of "success" (heads) on each coin toss. Suppose the experimental result e of tossing the coin 4 times is: s,f,f,s (s = success, f = failure). Let the Gellerized hypothesis $G(e)$ be that the

---

28. This possibility was raised for me by Kent Staley.

probability of success is 1 on trials 1 and 4, and 0 on trials 2 and 3.[29] Then $G(e)$ maximally fits e: $p(e/G(e)) = 1$.

The test T Mayo imagines is this: observe the series of coin tossing outcomes and "find a hypothesis $G(e)$ that makes the result e maximally probable, and then pass that hypothesis" (p. 202). Now, claims Mayo, in determining whether test T is a severe test for some such $G(e)$, you cannot just determine that $p(e/G(e))$ is maximal. You also need to consider "as part of the testing procedure, the particular rule that is used to determine which hypothesis to test" (p. 202). The following severity condition must be satisfied:

> There is a very high probability that the test procedure T would *not* pass the hypothesis it tests, given that the hypothesis is false. (p. 202)

With the test procedure T just noted for a Gellerized hypothesis $G(e)$, the probability that the test procedure would not pass the hypothesis it tests, given that the hypothesis is false, is not high but zero.

The question now is whether the hypothesis h in the previous section ("the bag selected is 40% white") is Gellerized in Mayo's sense, so that the procedure for constructing it would probably yield a good fit with observed data even if the hypothesis is false. It does not seem to be Gellerized. There is no procedure for constructing h analogous to that for the Gellerized coin tossing hypothesis $G(e)$. We are simply told in advance that any bag in the set is either 40% or 60% white. So we are considering each hypothesis in turn. Moreover, if the hypothesis h were false, so that the bag selected was 60% white not 40%, then the probability would not be high that the test T consisting of selecting 100 balls and noting colors would yield the result it did, viz. 40 white ($p(e(T)/-h) = .00002$). So the severity condition just noted is satisfied.

Let me briefly mention a final possible response by Mayo. It is to restrict her theory of evidence to cases in which prior and posterior probabilities of hypotheses are unobtainable.[30] Mayo admits that there are cases in which prior and posterior probabilities, in a frequency sense, can be assigned. She writes:

> *Except* for such contexts, however, the prior probabilities of the hypotheses are problematic. Given that logical probabilities will not do, the only thing left is subjective probabilities. For many [including Mayo], these are unwelcome in scientific inquiry. (p. 82)

On this proposal, Mayo's concept of evidence is applicable when and only when hypothesis h is such that a frequency interpretation of $p(h)$ and $p(h/e)$ cannot reasonably be applied, yet where frequency interpretations can be given to $p(e/h)$ and $p(e/-h)$.

What concept of evidence, then, are we to use when (as in my bags-of-balls example) a frequency interpretation can readily be assigned to $p(h)$ and $p(h/e)$? If we choose one according to which $p(h/e)$ should be high, we obtain a very different concept of evidence from the one proposed by Mayo. In any case, why

---

29. See Mayo, p. 201, footnote 17.
30. I thank Thomas Hood for this idea.

should our concept of evidence be different when frequency probabilities can be assigned to hypotheses? Why should e's passing a severe test with respect to h (in Mayo's sense) count as evidence that h when $p(h)$ and $p(h/e)$ cannot be assigned by frequency means but not when they can?

12. Is Belief Related to Probability? Mark Kaplan's View

I turn now to the final objection to the view that for e to be potential or veridical evidence that h, and hence for e to provide a good reason to believe h, h's probability on e must be greater than $\frac{1}{2}$. The objection derives from a theory that completely divorces belief from probability. I will examine an interesting version due to Mark Kaplan.[31]

On Kaplan's theory, there is no relationship between whether you believe something and your degree of confidence in it. For Kaplan, your degree of confidence is something that satisfies, or ought to satisfy, the probability rules. So on his view you can believe something even though you have very little confidence in it, that is, even though you assign very low probability to it. And you can fail to believe it, even though you have a great deal of confidence in it, that is, even though you assign very high probability to it. If this is right, then if you believe h and do so for a good reason, that reason need not be one that you take to confer a high probability on h. Kaplan is concerned here with subjective probability. But one who wants to invoke objective probability might extend Kaplan's argument and say that a good reason to believe something, and therefore potential and veridical evidence, need not confer a high degree of (objective) probability.

Let us consider Kaplan's reasons for the first part of this argument, namely, the claim that there is no relationship between your believing something and having some degree of confidence in it. On what he calls the "confidence threshold view" (p. 94), you believe h if your degree of confidence in h exceeds some threshold value k.[32] The confidence threshold view, Kaplan claims, violates a fundamental principle of rationality of beliefs, viz. "deductive cogency," which states that the set of hypotheses you believe includes the consequences of that set but no contradictions. (Among the consequences of a set of propositions is the conjunction of members of the set.) That this principle is violated can be shown by appeal to the lottery paradox (discussed in chapter 4). Suppose there is a fair lottery consisting of 1000 tickets, one of which will win. Suppose your confidence level for belief is .9, so that you believe a proposition p if your confidence in p exceeds .9. With respect to any particular ticket in the lottery your confidence level that it will not win is, let us suppose, greater than .9. Accordingly, you believe it won't win. Conjoining all these beliefs results in the belief that no ticket will win. But you are extremely confident (>.9) that some ticket will win. Therefore, your beliefs are contradictory.

---

31. Mark Kaplan, *Decision Theory as Philosophy* (Cambridge: Cambridge University Press, 1996).
32. This is a simplification of the version he considers.

A defender of the confidence-threshold view will not be persuaded, but will reject "deductive cogency." If you believe h_1 and you believe h_2, it does not follow that you believe the conjunction $h_1 \& h_2$, since the probability you assign to the latter may be less than the threshold required for beliefs, even if the probabilities you assign to h_1 and h_2 separately exceed this threshold.

Kaplan is aware of this response by confidence-threshold theorists, and he offers a reply. It is that if you reject "deductive cogency" then you must reject *reductio ad absurdum* arguments generally (p. 96). In a *reductio* argument, Kaplan writes, "a critic derives a contradiction from the conjunction of a set of hypotheses which an investigator purports to believe," thereby demonstrating a defect in this set.

Is Kaplan right? Must a confidence-threshold theorist reject *reductio* arguments? In the lottery paradox the confidence-threshold theorist claims to believe each proposition in the set consisting of propositions of the form "ticket *i* will not win." But he does *not* believe the conjunction of propositions in this set. So we have the following situation. The *set* of beliefs which the confidence-threshold theorist believes is inconsistent: the beliefs can't all be true. Does the confidence-threshold theorist believe they are all true? Yes and No. He believes that *each* of them is true, but he does not believe that they are all true together; he does not believe that the *conjunction* of these beliefs is true, even just the conjunction of beliefs of the form "ticket *i* will not win." So even though the set of his beliefs can't all be true, you cannot derive a contradiction of the form "p and $-p$" from this set, because the conjunction of beliefs is not a member of the set of his beliefs. You cannot derive the proposition "no ticket will win, and some ticket will win," or even just the first conjunct, from his set of beliefs, unless, that is, you accept the principle of "deductive cogency," which he does not.[33]

This does not commit the confidence-threshold theorist to abandoning *reductio* arguments. In a *reductio* argument the critic takes the investigator to believe not just each proposition in the set of premises but the conjunction of these propositions as well. The critic then derives a contradiction of the form "p and $-p$" from that conjunction. In the lottery situation the confidence-threshold theorist does not believe the conjunction of propositions of the form "ticket *i* will not win."

Kaplan has a second reply for confidence-threshold theorists (p. 98). It is that even if we do not require "deductive cogency," it is a *possible* course of action. That is, it is possible to believe in such a way that the set of hypotheses you believe includes the consequences of that set but no contradictions. But, says Kaplan, this is not even possible on the confidence-threshold view. That view necessarily leads to the lottery paradox and the situation in which the set of your beliefs includes a contradiction.

To this a confidence-threshold theorist will respond that if the confidence-threshold view is correct, then Kaplan is just mistaken in his claim that "deduc-

---

33. Here is an analogy. Consider situations that I am willing to have be the case. These might include my seeing a movie tonight and my not seeing a movie tonight. But it doesn't follow from this that I am willing to have it be the case that I see a movie tonight and that I don't see a movie tonight. Although I am willing to have each of the two states of affairs obtain that cannot obtain together, I am not willing to have both of them obtain together.

tive cogency" is a possible course of action. It is impossible. If confidence beyond a certain threshold entails belief, then necessarily in a lottery situation of the sort in question one will believe each member of a set of propositions not all of which can be true, which is a violation of "deductive cogency." Kaplan must argue independently that confidence beyond a certain threshold does not entail belief.

This leads to Kaplan's own view about belief, to which I now turn.

13. Kaplan's "Assertion View" of Belief

According to Kaplan,

> You count as believing P just if, were your sole aim to assert the truth (as it pertains to P), and your only options were to assert that P, assert that $-P$, or make neither assertion, you would prefer to assert that P. (p. 109)

By having the "sole aim to assert the truth" Kaplan makes it clear that he means the truth and only the truth, so that avoiding error is part of this aim. "The truth," he writes

> is just an error-free, comprehensive story of the world: for every hypothesis h, it either entails h or entails $-h$, and it entails nothing false. (p. 111)

In pursuing the aim of truth,

> the desire for comprehensiveness and the desire to avoid error are bound to conflict.... We would expect that different individuals would differ as to how much risk of error they were willing to tolerate in order to satisfy the desire for comprehensiveness.... It is compatible with the Assertion View that you tolerate very great risk of error, in which case you might well believe hypotheses in whose truth you have very little confidence. (p. 111)

So this view divorces belief from confidence.

An immediate objection is that this leads to a bizarre consequence. You can believe something and be very confident that it is false. Kaplan admits that this is a consequence of this view.[34] His response is this. The reason it sounds bizarre is that we are operating with the "ordinary notion of belief," which "construes belief as a state of confidence short of certainty" (p. 142). The problem, he says, is that this ordinary notion of belief also "takes consistency of belief to be something that is at least possible, and perhaps even desirable." But he claims to have shown that "no coherent notion of belief can do both" (p. 142). So he is offering the assertion concept of belief as an alternative to the ordinary concept.

Kaplan's new concept of belief will also lead to a bizarre consequence for ev-

---

34. He writes: "Suppose we ask someone whether hypothesis h is true. It is easy to understand how she might reply 'Yes, h is true,' or 'No, h is not true,' or how she might be unwilling to commit herself either to h's truth or to its falsehood. But it is hard to see what can be made of the responses, 'Yes, h is true, but I am extremely confident it is false,' and 'I believe h is true, but I'd give you 20 to 1 odds that I'm mistaken.' Yet, not only does the Assertion View countenance these apparently bizarre responses..., it deems them entirely reasonable ones to make in the context of inquiry." (p. 142)

idence, if the latter is construed as something that provides a good reason, or a justification, for believing.[35] For, on Kaplan's assertion concept of belief, it is possible that e is evidence that h, and hence that e provides a good reason (or justification) to believe h, even if it offers no basis for any confidence in h, and indeed even if it offers a basis for confidence that h is false. This, I think, departs considerably from any ordinary notions of evidence.

The reason that Kaplan offers his assertion concept of belief is that the ordinary one, which, he claims, is tied to confidence, is "incoherent," since it violates deductive cogency (p. 142). But, I suggest, this charge of "incoherence" does not stick. Rejection of "deductive cogency" by confidence-threshold theorists does not lead to believing contradictions of the form p and $-p$. It leads only to believing each member of a set of beliefs not all of which can be true.

Indeed, an inconsistent set of beliefs is possible on Kaplan's own "assertion view." Consider two propositions, P and Q. Suppose my sole aim is to assert the truth as it pertains to each, and my only options (in each case) are to assert that $P(Q)$, assert that $-P(-Q)$, or make neither assertion. And suppose I prefer to assert that $P(Q)$. Then, on the assertion view, I believe P and I believe Q. But this could happen even if P and Q form an inconsistent set.

Kaplan may reply that the lottery paradox shows that the confidence-threshold view of belief *necessarily* leads to an inconsistent set of beliefs, whereas his own assertion view does not entail that an inconsistent set of beliefs is necessary, only that it is *possible*. Since both views allow for the possibility of such inconsistency, and since Kaplan does not take this as a defect for his assertion view, I'm not sure why the necessary consequence of certain inconsistent sets of belief is particularly damaging for the confidence-threshold view (which Kaplan later calls the ordinary view of belief).

14. Conclusions

In this chapter I have defended the following claims.

1. A high-probability condition on evidence enables us to avoid counterexamples to the positive relevance view.

2. If e provides a good reason to believe h, then h's objective epistemic probability on e must be greater than $\frac{1}{2}$. Moreover, if e is potential or veridical evidence that h, then e must provide a good reason to believe h. It does not suffice for e to provide just some reason. Conflicting studies may furnish some reason to believe h and some reason to believe $-h$, but not a good reason to believe both. Taken separately, they may provide ES-evidence for each hypothesis. But taken together, they cancel and do not provide potential or veridical evidence for both.

3. Neither the strong Bayesian view that there are no beliefs only degrees of belief, nor the weaker one that beliefs are degrees of belief greater than some

---

35. Kaplan does not construe evidence in this way, but suggests a view according to which e, if true, is evidence that h for you if you have more confidence in h given e than you do in h without e, where you are not certain of the truth-values of e or h (p. 48). Since confidence for him is a probabilistic notion, Kaplan is suggesting a subjective positive relevance concept of evidence.

number, is incompatible with the idea that evidence provides a good reason to believe a hypothesis, and hence that it confers a probability greater than $\frac{1}{2}$.

4. The view that evidence provides a good reason to believe, and hence furnishes a probability greater than $\frac{1}{2}$, is defended against three views. The first, the likelihood view, completely divorces evidence that h from whether, or to what extent, one should believe h. So, in effect, does the second, Mayo's error-statistical account, since even "experimental reliability" does not guarantee a good reason to believe the hypothesis, or that its probability is greater than $\frac{1}{2}$. The third, due to Mark Kaplan, completely divorces belief from probability. I argue that the likelihood and error-statistical views fail to provide useful concepts of evidence, notwithstanding arguments to the contrary by their defenders. Also I argue that Kaplan is mistaken when he claims that only a divorce between belief and probability will save *reductio* arguments. Finally, I discuss unintuitive consequences of Kaplan's own assertion view of belief, which allows you to believe something to which you assign very low probability.

7

THE EXPLANATORY CONNECTION

High probability of h given e, by which is meant that h's probability is greater than that of its negation, that is, greater than $\frac{1}{2}$, is necessary for e to be potential (and hence veridical) evidence that h. That is what I have argued in earlier chapters. But it is not sufficient. In this chapter I will formulate an additional necessary condition that invokes the idea of a probable explanatory connection between hypothesis and evidence. A concept of explanation that can be used to understand this condition will then be explicated. (In what follows, unless otherwise indicated, when I speak of evidence I mean potential evidence; as in chapter 6, probability is being construed as objective epistemic probability.)

1. High Probability Is Not Sufficient for Evidence

To see why, recall the "irrelevant information counterexample" of chapter 4:

- e: Michael Jordan eats Wheaties
- b: Michael Jordan is a male basketball star
- h: Michael Jordan will not become pregnant

The probability of h given b is close to 1, and this probability is unchanged if we add information e to b. That is, $p(h/e\&b)$ is approximately 1. Yet it seems absurd to claim that, given that Michael Jordan is a male basketball star, the fact that he eats Wheaties is evidence that he will not become pregnant. In general, I have been claiming, for any e, b, and h, if e is evidence that h, given b, then, given b, e provides a good reason to believe h. The fact that Michael Jordan is male provides a good reason to believe he will not become pregnant. But given that he is a male basketball star, the fact that Michael Jordan eats Wheaties fails to provide such a reason.

More generally, suppose that the background information b itself is, or contains, evidence that h. (In our example, b contains the information that Michael Jordan is male.) And suppose we append to the background information b some item of information e that is, evidentially speaking, irrelevant for the hypothesis

h, and that does not change the probability of h on b. (In our example, such information is that Michael Jordan eats Wheaties.) Then although $p(h/e\&b) > \frac{1}{2}$, and b itself, or something contained in b, is evidence that h, e is not evidence that h, given b.

Nor is the *conjunction* of b (or a part of b) with e evidence that h. Although the fact that Michael Jordan is male is evidence that he won't become pregnant, there is something amiss with taking the conjoined fact that he is a male (or a male basketball star) *and* that he eats Wheaties to be evidence that he won't become pregnant. In this respect evidence is similar to causation and explanation, and unlike deductive implication. If the fact that some event x_1 occurred is what *caused y* to occur, or what *explains* why it occurred, then the conjunction of the former fact with the fact that some other event x_2 occurred is what caused or explains y's occurrence only if x_2's occurring is somehow a part of the cause or explanation of y. If x_2's occurrence is causally or explanatorily irrelevant for the occurrence of y, then it is false to claim that x_1 and x_2 caused or explains y. By contrast, if some proposition p deductively implies q, then the conjunction of p with any other proposition also deductively implies q.

In chapter 4, section 9, I spoke of the causal relation as being *selective:* If some fact or event E_1 caused an event E, it will be false to say that E_1 and E_2 caused E, even if E_2 is a fact or is an event that occurred, unless E_2 was causally involved in producing E. By contrast, probability is not selective. It can be the case that h's probability on e and b is r even though e makes no contribution to that probability (even though $p(h/e\&b) = p(h/b) = r$ and $p(h/e) = p(h)$). The high-probability definition of evidence has as a consequence that evidence is not selective. In the basketball example, the hypothesis h is highly probable given e and b, even though $p(h/e\&b) = p(h/b)$, which is approximately 1, and $p(h/e) = p(h)$. The reason we resist the conclusion that e is evidence that h, given b, or that $e\&b$ together constitute evidence that h, is that e makes no contribution to the evidence that h. Because evidence, by contrast with probability, is selective, the high-probability definition is not sufficient.

2. Deductive Connection

We seek a condition (or set of them) that together with high probability will be necessary and sufficient for e's being evidence that h. Such a condition must guarantee that e is a good reason to believe h. Since we are trying to define the concept of potential, and then veridical, evidence, both of which are objective, the condition must be objective. Its application must not depend on who knows or believes what.

Why is the fact that Michael Jordan eats Wheaties not evidence that he will not become pregnant, even under the assumption that he is a male basketball star? An immediate response is that his eating Wheaties has no connection with his not becoming pregnant. The obvious question, then, is what sort of connection is required. The answer I shall propose is usefully developed by looking at some connections between evidence and hypothesis that have been developed by others and may be thought of as first approximations to the one we seek.

The first derives from a central view in the philosophy of science concerning methodology, viz. hypothetico-deductivism. On this view, the scientist proposes a hypothesis from which, together with auxiliary assumptions as part of the background information, observational conclusions can be drawn that cannot be drawn from the auxiliary assumptions themselves. If these conclusions are true, they constitute evidence that the hypothesis is true. Accordingly, we have

Basic hypothetico-deductive (h-d) condition: If h together with b entails e, but b by itself does not, and if e is true, then e is evidence that h, given b.

This h-d view yields an objective concept of evidence, since whether the entailments in question obtain is an objective matter that does not depend on who knows or believes what. However, as a sufficient condition it will not do.

One of the most important reasons for its failure derives from the problem of competing hypotheses, recognized by opponents of hypothetico-deductivism such as Isaac Newton and John Stuart Mill.[1] Suppose there are numerous hypotheses that together with b entail e, and suppose that all of them conflict with h and with each other. Then e is evidence that each of them is true. For example, let b be the information that there is a lottery consisting of 1000 tickets, each marked with a different number from 1 to 1000. Let e be the information that Mike owns the ticket marked 1. Let h_1 be that Mike owns just one ticket, viz. the one marked 1. Let h_2 be that Mike owns exactly 2 tickets, including the one marked 1. Let h_3 be that Mike owns exactly 3 tickets, including the one marked 1. And so forth. In accordance with the h-d condition, the fact (e) that he owns the ticket marked 1 would be evidence that each of the hypotheses h_1, $h_2, \ldots, h_{1000}$ is true, which seems absurd. Although e is *consistent* with each of these hypotheses, it is not evidence that each is true; it is not a good reason for believing each.

My main purpose, however, is not to refute the h-d condition, but to see what happens if we add this condition to the high-probability requirement, as follows:

h-d plus high probability: If h together with b entails e, but b by itself does not, and if e is true, *and if $p(h/e\&b) > \frac{1}{2}$, then e is evidence that h, given b.*

On this h-d view, in addition to high probability, the "connection" we seek between evidence and hypothesis is deductive: h, together with b, must entail e.

Combining the high-probability condition with this deductive connection has advantages. It avoids the previous lottery counterexample since, we may suppose, the probability of each h_i, given just $e\&b$, is not greater than $\frac{1}{2}$. More generally, it avoids the competing-hypothesis objection. Even if each member of a set of competing hypotheses entails e, if one of these hypotheses has a probability on e greater than $\frac{1}{2}$, then each of the other members of the set must have a probability on e less than $\frac{1}{2}$. So e cannot be evidence that any of the other competing hypotheses is true.

Combining high probability with the deductive connection also does not permit the Wheaties counterexample. The hypothesis (h) that Michael Jordan will

---

1. Isaac Newton, *The Correspondence of Isaac Newton*, eds., H. W. Turnbull, J. F. Scott, A. Rupert Hall, and Laura Tilling (Cambridge: Cambridge University Press, 1959–1977), vol. 1, p. 169; John Stuart Mill, *A System of Logic* (London, 1959).

not become pregnant, together with the background information (b) that he is a male basketball star, does not entail (e) that he eats Wheaties. So even though $p(h/e\&b) > \frac{1}{2}$, we cannot conclude that e is evidence that h, given b.

Unfortunately, "h-d plus high probability" has serious problems. It fails to provide a set of sufficient conditions. This can be seen in the Wheaties case if we conjoin the hypothesis h with the evidence e to form a new hypothesis H:

- H: Michael Jordan eats Wheaties and will not become pregnant.
- e: Michael Jordan eats Wheaties.
- b: Michael Jordan is a male basketball star.

Our new hypothesis H (with or without b) entails e (which is not entailed by b itself). Furthermore, it should be obvious that the probability of H, given $e\&b$, is close to 1. Yet e is not evidence that H, given b. The fact that Michael Jordan eats Wheaties is not a good reason to believe that Michael Jordan eats Wheaties and will not become pregnant, even given that he is a male basketball star.

More importantly, "h-d plus high probability" attempts to provide only a set of sufficient conditions, not necessary ones. If we required that the hypothesis h deductively imply the evidence e, we would rule out cases in which it seems appropriate to claim that e is evidence that h. Here are two examples:

- e: This coin landed heads on each of the past 1000 tosses.
- h: This coin will land heads on the 1001st toss.

- e: In a large, varied sample in which 10,000 patients were given medicine M to relieve symptoms S, 9500 patients had their symptoms relieved.
- h: Medicine M is approximately 95% effective in relieving symptoms S.

In neither case is the purported evidence deductively entailed by the hypothesis h. Yet in each case e is (or could be) evidence that h.

3. Explanatory Connection

Another connection that is often suggested between evidence and hypothesis is explanatory. The basic idea is that the hypothesis h should explain the putative evidence e, that is, why e is true. Most frequently defenders of such a view do not define the concept of explanation involved.[2] However, they do assume that it is objective: whether h explains e does not depend on who knows or believes what. Moreover, the concept of explanation invoked is what might be called *potential* explanation: hypothesis h, if true, would correctly explain e. Finally, although some explanations are deductive, it is not assumed that all of them are, or that the fact that h entails e is sufficient for h to explain e.

The simplest version of the explanation view is this:

> *Basic explanation condition*: If h, if true, would correctly explain e, and if e is true, then e is evidence that h.

2. These include William Whewell, *Philosophy of the Inductive Sciences*, vol. 2 (New York: Johnson Reprint Corporation, 1967; reprinted from the 1847 edition); Gilbert Harman, "The Inference to the Best Explanation," *Philosophical Review*, 74 (1965); and Peter Lipton, *Inference to the Best Explanation* (London: Routledge, 1991).

This condition, like the basic h-d condition, is subject to the competing hypothesis objection. There may be numerous hypotheses, some of them quite crazy, which, *if true*, could correctly explain *e*. This is not enough to make *e* evidence that each is true.

Accordingly, as with the deductive condition, a natural idea is to combine this with "high probability":

Explanation plus high probability: If *h*, if true, would correctly explain *e*, and if *e* is true, then *e* is evidence that *h*, provided that $p(h/e) > \frac{1}{2}$.

This will avoid the competing hypothesis objection. Even if there are numerous competing hypotheses that, if true, would correctly explain *e*, if $p(h/e) > \frac{1}{2}$, then the probability of each of *h*'s competitors on *e* is less than $\frac{1}{2}$. Indeed, on this view, if *e* is evidence that *h*, then *h* is more probable than any competing explanatory hypothesis. Accordingly, this is one version of a view known as "inference to the best explanation."[3]

This view has two advantages over the corresponding "h-d plus high probability" view. First, it avoids the type of problem generated by concocting a hypothesis consisting of a conjunction of the evidence *e* with a claim that is completely irrelevant to *e* but has a high probability. Our previous example was

H: Michael Jordan eats Wheaties and will not become pregnant.
e: Michael Jordan eats Wheaties.
b: Michael Jordan is a male basketball star.

The problem with the h-d account arises because *H&b* entails *e*, but *b* alone does not, and $p(H/e\&b)$ is approximately 1. With an explanation condition, however, the situation is different. Although *H* entails *e* it does not explain why *e* is true. Explanation is a richer, more demanding idea than mere entailment.

A second advantage of the explanation view is that, unlike the deductive view, it permits us to speak of evidence in cases where the explanatory hypothesis does not deductively entail the putative evidence. A previous example was this:

e: In a large, varied sample in which 10,000 patients were given medicine M to relieve symptoms S, 9500 patients had their symptoms relieved.
h: Medicine M is approximately 95% effective in relieving S.

Although *h* does not deductively entail *e*, it seems very plausible to say that *h*, if true, correctly explains why *e* is true: 9500 of the 10,000 patents in the sample

---

3. "Inference to the best explanation" is a rather vague and general view, according to which scientific inferences are inferences to a hypothesis that provides the best explanation of the data. The question, then, is what counts as the "best" explanation. Lipton distinguishes two interpretations of "the best": the one that, if true, would provide a correct explanation and is the most probable on the data; and the one that is (what he calls) the "loveliest," which, although it may not be the most probable, is the simplest, most elegant, most unifying. The latter is Lipton's own version. The former is, I think, captured above in the "explanation plus high probability" condition. For a critique of Lipton's "loveliest" view, see my "Inference to the Best Explanation: Or, Who Won the Mill-Whewell Debate?," *Studies in History and Philosophy of Science*, 23 (1992), 349–364.

had their symptoms relieved because medicine M is approximately 95% effective in relieving those symptoms (assuming it is).

However, the "explanation plus high probability" view is too strong. This can be demonstrated by reference to a previous coin tossing example:

    e: This coin landed heads on each of the past 1000 tosses.
    h: This coin will land heads on the 1001st toss.

That e is true is (or could well be) evidence that h. In this case, we may suppose, $p(h/e) > \frac{1}{2}$, yet h if true does not correctly explain why e is true: the fact that the coin will land heads on the 1001st toss, assuming it is a fact, does not correctly explain why it landed heads on each of the past 1000 tosses.

Before considering how we might alter the explanation condition to account for this example, let me introduce another:

    e: John, who has symptoms S, took medicine M, which is 95% effective in relieving S.
    h: John's symptoms S were relieved.

Here again e is, or could well be, evidence that h, even though h, if true, does not correctly explain why e is true. In this case the explanation goes in the reverse direction: e if true correctly explains (or might) why h is true: the fact that John, who has symptoms S, took M, which is 95% effective in relieving S, correctly explains (or might) why John's symptoms were relieved (not vice versa).

In the coin tossing example, however, there seems to be no explanation in either direction. The reason the coin landed heads on the 1001st toss is not that it landed heads on each of the previous 1000 tosses; nor is the reason that the coin landed heads on tosses 1 through 1000 that it landed heads on toss 1001. Yet there is, or could be, an explanatory connection between these two facts. Some third fact could correctly explain why each of these facts obtains. For example, it could be that the coin is severely biased in favor of heads, or that it is a coin that shows heads on both sides.

Let us say that there exists an *explanatory connection* between h and e if h correctly explains why e is true, *or* if e correctly explains why h is true, *or* if some hypothesis correctly explains both why e is true and why h is true. We might now formulate the following conditions for evidence:

> *Explanatory connection plus high probability:* e is evidence that h only if
>   (i)   there is an explanatory connection between h and e;
>   (ii) $p(h/e) > \frac{1}{2}$.

This allows the previous two examples in which e is evidence that h even though h does not correctly explain e. In both of these examples although h does not correctly explain e, we may suppose that e correctly explains h, or that some hypothesis correctly explains both h and e.

Although an improvement, the explanation condition (i) above is still too strong. To revert to the medicine M case, suppose that (e) John, who has symptoms S, took medicine M, which is 95% effective; and that (h) John's symptoms were in fact relieved. But suppose e, although true, does not in fact correctly explain why h is true. Although he took medicine M, John's symptoms disappeared "spontaneously," not as a result of his taking M; this happens, let us say, 1% of the time. If so, the explanatory connection condition (i) is not satisfied. Yet in such a case we may still want to claim that the fact that John took M is (potential) evidence that his symptoms were relieved. (Whether it is veridical evidence, I will consider in chapter 8.)

What is needed for potential evidence is not that there is an explanatory connection between h and e, but that, assuming the truth of both e and h, there *probably* is such a connection. Given that (e) John, who has symptoms S, took medicine M, which is 95% effective in relieving S, and that (h) John's symptoms were relieved, there is probably an explanatory connection between h and e.

Accordingly, the following seems promising:

High probability of an explanatory connection plus high probability of h: e is (potential) evidence that h only if

(i) $p(\text{there is an explanatory connection between } h \text{ and } e/h\&e) > \frac{1}{2}$;
(ii) $p(h/e) > \frac{1}{2}$.

Will this work? Three issues need to be considered. First, does it avoid the counterexamples? Second, do these conditions suffice for making e a good reason to believe h, as is required for potential (and veridical) evidence? Third, what concept of explanation is being employed?

4. The Counterexamples

We need only consider those counterexamples that we introduced to show that high probability of h on e is not sufficient for e to be evidence that h. The question is whether the explanatory connection condition, as formulated above, precludes these counterexamples.

The first and perhaps most important type of counterexample involves one introducing irrelevant information:

e: Michael Jordan eats Wheaties.
b: Michael Jordan is a male basketball star.
h: Michael Jordan will not become pregnant.

Our latest conditions for evidence preclude saying that e is evidence that h, given b, because condition (i)—the explanation condition—is violated: Given that Michael Jordan will not become pregnant, and that Michael Jordan eats Wheaties, and that he is a male basketball star, the probability is not high that there is an explanatory connection between his not becoming pregnant and his eating Wheaties. It is not true that he probably won't become pregnant because he eats Wheaties, or that probably he eats Wheaties because he won't become pregnant, or that probably there is some true hypothesis that correctly explains both why he won't become pregnant and why he eats Wheaties. His eating Wheaties probably has "nothing to do" with his not becoming pregnant, even though the probability that he won't become pregnant, given that he eats Wheaties, is very high. (It is also very high without this assumption.)

The present definition also allows us to avoid counterexamples to explanatory connection conditions that require the hypothesis to correctly explain the evidence. Sometimes the reverse is true. Sometimes neither is true, but something explains both why the hypothesis and the evidence are true. In view of counterexamples we also saw the need to require not that there be an explanatory connection between h and e, but that given the truth of h and e the probability of such a connection is high.

Finally, since the present conditions require the high probability of h on e, they avoid the lottery and swimming counterexamples to the positive relevance account in chapter 4. Whether new counterexamples emerge is a question I will take up later.

5. Good Reason to Believe

Do these conditions guarantee that e is a good reason to believe h? That is a more difficult question to answer. To do so, I will develop a general principle that relates explanation and a good reason to believe.

I begin with a simple example. Let

- e: Arthur has a rash on his arm that itches.
- b: Arthur was weeding yesterday bare-armed in an area filled with poison ivy, to which he is allergic.
- h: Arthur's arm was in contact with poison ivy.

Given e and b, the probability is high that the reason (e) Arthur has a rash on his arm that itches is that (h) his arm was in contact with poison ivy. Under these conditions, given e and b, e is a good reason to believe h. Given that Arthur has a rash on his arm that itches and that he was weeding yesterday bare-armed in an area filled with poison ivy, to which he is allergic, the fact that he does have that rash is a good reason to believe that his arm was in contact with poison ivy.

More generally, suppose that given just the putative evidence e and the background b (not e and b and h), the probability is high that the reason that e is true is that h is. Then e is a good reason to believe h.

Here is a different case in which given e and b (not e and b and h) the probability is high that the reason that h is true is that e is (simply switch the e and h of the previous example):

- e: Arthur's arm was in contact with poison ivy.
- b: (same as above)
- h: Arthur has a rash on his arm that itches.

In this case, given that Arthur's arm was in contact with poison ivy and that he was weeding yesterday bare-armed in an area filled with poison ivy, to which he is allergic, the fact that his arm was in contact with poison ivy is a good reason to believe that he has a rash on his arm that itches. More generally, if given e and b, the probability is high that the reason that h is true is that e is, then, given e and b, e is a good reason to believe h.

Finally, let

- e: Arthur has a rash on his right arm that itches.
- b: Arthur was weeding yesterday using both hands with both arms bare in an area filled with poison ivy, to which he is allergic.
- h: Arthur has a rash on his left arm that itches.

In this case, given e and b, the probability is high that some hypothesis (viz. that his arms were in contact with poison ivy) explains why both e and h are true. In this case also, given b, e is a good reason to believe h.

The general principle I am proposing is this:

General Principle: If, given e and b, the probability is high that there is an explanatory connection between h and e, then, given e and b, e is a good reason to believe h.[4]

Let us suppose that this principle is true (as I think it is), and, as we have been doing, let us construe "high probability" as probability greater than $\frac{1}{2}$. Then we can demonstrate the following: If the conditions for evidence given at the end of section 3 are strengthened in a certain way, they will entail that if e is evidence that h, given b, then, given b, e is a good reason to believe h.[5]

The demonstration makes use of the following

Theorem: If A entails h, then for any e,
$$p(A/e) = p(A/h\&e) \times p(h/e).[6]$$

Now from the definition of *explanatory connection*, "there is an explanatory connection between h and e" entails h. Accordingly, from the theorem it follows that

p(there is an explanatory connection between h and e/e) = p(there is an explanatory connection between h and $e/h\&e$) x $p(h/e)$.

So, letting A = there is an explanatory connection between h and e, it follows that $p(A/e) > \frac{1}{2}$ if and only if $p(A/h\&e)xp(h/e) > \frac{1}{2}$. This is possible only if both $p(A/e) > \frac{1}{2}$ and $p(h/e) > \frac{1}{2}$. This means that if, given e, the probability is greater than $\frac{1}{2}$ that there is an explanatory connection between h and e (and hence that, given e, e is a good reason to believe h), then both of the conditions for evidence given at the end of the previous section are satisfied:

4. As can be seen from the above examples, it is also the case that if, given e and b, the probability is high that there is an explanatory connection between h and e, then given h and e, h is a good reason to believe e. This "reciprocity" and its implications for evidence will be discussed in chapter 8, section 5.
5. Since e is evidence that h only if e is true (see chapter 2, section 4), from the fact that e is evidence that h we may conclude that e *is* a good reason to believe h (and not just that e *if true* is such a reason).
6. Proof:
   1. $p(A/h\&e) = p(A\&h\&e)/p(h\&e)$
   2. Since A entails h, $p(A\&h\&e) = p(A\&e)$
   3. So from 1, $p(A/h\&e) = p(A\&e)/p(h\&e)$
   4. From 1 and 3, $p(A/h\&e) = p(A\&e)/p(h\&e) = p(e)xp(A/e)/p(e)xp(h/e)$
   5. From 4, $p(A/e) = p(A/h\&e)xp(h/e)$. QED

(i) p(there is an explanatory connection between h and $e/h\&e) > \frac{1}{2}$, and
(ii) $p(h/e) > \frac{1}{2}$.

The converse, however, does not obtain. That is, conditions (i) and (ii) might be satisfied without its being the case that p(there is an explanatory connection between h and $e/e) > \frac{1}{2}$. For example, if p(there is an explanatory connection between h and $e/h\&e) = .6$ and $p(h/e) = .7$, then, by the theorem above, p(there is an explanatory connection between h and $e/e) = .42 < \frac{1}{2}$. Accordingly, the General Principle will not permit us to conclude that if e is evidence that h (by the conditions of evidence at the end of section 3), then e if true is a good reason to believe h. Should these conditions be strengthened, and if so, how? Let us consider a case.

I will use the example of section 3 involving symptoms S and medicine M, but change the figures a bit. Suppose that 70% of those with symptoms S have relief of those symptoms in a week. Writing R for getting relief in a week, and S for having symptoms S, we have

$p(R/S) = .7.$[7]

Suppose that 70% of those with S who take medicine M get relief in a week, so

$p(R/S\&M) = .7.$

Suppose finally that among those with S who take M and get relief in a week, 60% get relief *because* they took M (while 40% do so for other reasons), so

$p(R$ because of $M/R\&S\&M) = .6.$

From the previous theorem

$p(R$ because of $M/S\&M) = p(R$ because of $M/R\&S\&M) \times p(R/S\&M)$
$= .6 \times .7 = .42.$

That is, 42% of those with symptoms S who take medicine M get relief because of M. M is effective in relieving S only 42% of the time. (And, in fact, taking M doesn't increase the chances of relief in a week, since $p(R/S\&M) = p(R/S) = .7$.
Now let

e: John, who has symptoms S, is taking M.
h: John will get relief in a week.

And let the background information b include general information supplied above about the effectiveness of M. Is e if true a good reason to believe h? I don't believe it is. Given e, and the background information b which includes the fact that 70% of those with S get relief in a week, *there is* quite a good reason to believe (h) that John will get relief in a week. But the fact that he is taking medicine M is not a particularly good reason, since M is effective only 42% of the time. If

---

7. This probability can be construed as an objective epistemic one by taking R and S to be "open sentences" of the form "x gets relief in a week" and "x has symptoms S." Alternatively, with R and S as types of events, this probability statement can be interpreted in a relative frequency or propensity sense. So construed, it can then be taken as a basis for objective epistemic probability claims involving propositions about particular persons with S getting relief.

taking medicine M is to count as being a good reason to believe h, then M should be effective more than half the time. (A fair coin's being tossed is not a good reason to believe it will land heads.)

If so then the fact that p(there is an explanatory connection between h and $e/h\&e\&b) > \frac{1}{2}$ and that $p(h/e\&b) > \frac{1}{2}$ does not suffice to make e if true a good reason to believe h, assuming the truth of b. If not then the two conditions for evidence given at the end of section 3 will not guarantee that if e is evidence that h then e if true is a good reason to believe h. In the present case both conditions are satisfied, since p(there is an explanatory connection between h and $e/h\&e\&b) > \frac{1}{2}$ and $p(h/e) > \frac{1}{2}$.

We need something stronger, viz.

Stronger condition: e is evidence that h, given b, only if p(there is an explanatory connection between h and $e/e\&b) > \frac{1}{2}$.

By the previous theorem, this entails that e is evidence that h only if *the product of the two probabilities in the previous conditions for evidence is greater than $\frac{1}{2}$,* that is,

e is evidence that h, given b, only if p(there is an explanatory connection between h and $e/h\&e\&b) \times p(h/e) > \frac{1}{2}$.

In accordance with the General Principle, the stronger condition entails that if e is evidence that h, then e if true is a good reason to believe h. The stronger condition also entails the two conditions for evidence given at the end of section 3, but not conversely: If p(there is an explanatory connection between h and e/e) $> \frac{1}{2}$, then p(there is an explanatory connection between h and $e/h\&e) > \frac{1}{2}$ and $p(h/e) > \frac{1}{2}$, since by the previous theorem, p(there is an explanatory connection between h and e/e) $= p$(there is an explanatory connection between h and $e/h\&e$) $\times p(h/e)$. But the converse is not true.

Let us assume that if the condition $p(h/e) > \frac{1}{2}$ is satisfied, then if e is true there is a good reason to believe h.[8] *But e itself may not be that reason.* That is what the Michael Jordan ("irrelevant information") counterexample demonstrates. The question then for evidence is what is necessary to make e itself that reason. The explanation condition p(there is an explanatory connection between h and $e/h\&e) > \frac{1}{2}$ won't suffice if added to $p(h/e) > \frac{1}{2}$. This is shown by the previous "relief of symptoms" example in which

p(John will get relief in a week/John, who has symptoms S, is taking M) = .7

From this we can conclude that if it is true that John, who has S, is taking M, then *there is* a good reason to believe he will get relief in a week. But we cannot conclude that *his taking M* is that reason. In the example, among those who have S and take M, although 70% get relief in a week, less than half get the relief *from* M. What I am suggesting is that if his taking M is to be a good reason to believe John will get relief in a week, then taking M needs to be more effective than this. Given that he takes M, the probability that John gets relief because he takes M, or more generally, the probability of an explanatory connection between his get-

---

8. This assumption will be discussed in the next section.

ting relief and his taking M, needs to be higher. This would be achieved, I suggest, only if the following were true:

> p(there is an explanatory connection between his getting relief in a week and his taking M/John takes M) $> \frac{1}{2}$.

Or, what comes to the same thing, it would be achieved only if

> p(there is an explanatory connection between his getting relief in a week and his taking M/John takes M and gets relief in a week) $\times p$(John gets relief in a week/John takes M) $> \frac{1}{2}$.

6. Is the Probability Threshold High Enough?

In chapter 6 I argued that a necessary condition for e's being a good reason to believe h is that $p(h/e) > \frac{1}{2}$. Since (potential and veridical) evidence must provide a good reason to believe, this is a probability requirement I impose on evidence. Indeed, I have argued in the previous section, evidence requires not just that $p(h/e)$ be greater than $\frac{1}{2}$ but that the product of this probability and p(there is an explanatory connection between h and $e/h\&e$) be greater than $\frac{1}{2}$. Since both of these probabilities are usually less than 1, and since their product must exceed $\frac{1}{2}$, both probabilities will usually be significantly greater than $\frac{1}{2}$.

At this point we might ask whether a probability threshold for evidence higher than $\frac{1}{2}$ should be imposed. Two answers will be considered.

1. *The probability threshold should remain $\frac{1}{2}$.* Cases in which the product of the two requisite probabilities is slightly larger than $\frac{1}{2}$ will be understood as implying that although e is evidence that h, it is not particularly strong evidence. Let us change the percentages in the previous medical example once again. Suppose that 80% of those with symptoms S who take medicine M get relief in a week, so $p(R/S\&M) = .8$. And let us suppose that among those with S who take M and get relief in a week, 70% get relief because they took M, so $p(R$ because of $M/R\&S\&M) = .7$. From the theorem of section 5,

> $p(R$ because of $M/S\&M) = .56$.

Again, letting

  e: John, who has symptoms S, is taking M
  h: John will get relief in a week
  b: the probability information given above,

we can conclude that

> $p(h$ because of $e/e\&b) = .56$.

On the current viewpoint, since this probability is greater than $\frac{1}{2}$, enough probability is generated for e to be evidence that h, given b. However, that evidence is not particularly strong, since only a little more than half of those with symptoms S who take M get relief from M.

2. *The probability threshold should be significantly greater than $\frac{1}{2}$.* But, on this view, there is no precise threshold value that is both necessary and sufficient. $\frac{1}{2}$

is a necessary threshold but it is not sufficient; .95 is a sufficient threshold but it is not necessary. No number is both. If Sam owns 5 of the 1000 lottery tickets, then this fact is not evidence that he will win. If he owns 999 of the 1000 tickets, that is evidence that he will win. What if he owns 550? There is a gray area with no precise boundaries in which there is no definitive answer. On this viewpoint, evidence is a threshold concept with respect to probability in the following sense: although a necessary condition for e's being evidence that h is that the probability product in question exceed a threshold value of $\frac{1}{2}$, there is no probability value which is such that all probabilities higher than it satisfy the probability requirement for evidence and all lower probabilities fail to satisfy it.

This does not make evidence a subjective or relativized concept. It is not being claimed that a probability threshold value can vary from one person or community to another. (There may be variations in what probabilities different individuals or communities *take* to suffice, but that is something else.) Nor is the claim that the probability threshold can vary from one context to another, depending on the interests of those in the context. Context can affect what actions one should take on the basis of a hypothesis for which there is (or is not) evidence, not whether some piece of information is evidence for that hypothesis. The latter, in the case of potential and veridical evidence, is objective; it is independent of interests as well as knowledge and beliefs. The present viewpoint is that, objectively speaking, there exists no precise threshold probability exceeding which is both necessary and sufficient for the probability required by evidence. What can be said is imprecise, viz. that a probability significantly greater than $\frac{1}{2}$ is both necessary and sufficient for this.

These represent two different answers to the question of whether a probability threshold for evidence higher than $\frac{1}{2}$ should be imposed. An advantage of the first answer (keep $\frac{1}{2}$) over the second is precision. We have a concept that imposes a threshold value of probability that is both necessary and sufficient. Another advantage of the first answer over the second is that a more liberal concept of evidence is sanctioned. However, for the same reason defenders of the second answer may reject the first answer. What is wanted, they may assert, is a more stringent concept in accordance with which evidence provides a strong reason to believe. In what follows I will retain the more liberal and more precise "$>\frac{1}{2}$." Those with less tolerance for liberality in a concept of evidence and more tolerance for vagueness can replace this with "$\gg\frac{1}{2}$."

7. Is the Explanatory Connection Condition Too Strong?

How, if at all, does the explanatory connection condition

(i) e is evidence that h, given b, only if p(there is an explanatory connection between h and $e/e\&b) > \frac{1}{2}$

square with the fact that scientists frequently defend a hypothesis h by showing that it explains some phenomenon e? If it does not square with this fact, shouldn't we conclude that (i) is too strong a requirement for evidence?

In section 3 I rejected the "Basic Explanation Condition":

If h, if true, would correctly explain e, and if e is true, then e is evidence that h.

I rejected this because numerous conflicting (and crazy) hypotheses could exist which, if true, would correctly explain e. That e is true is not evidence that each of these hypotheses is true. What is usually involved in a scientist's claim that e is evidence that h because h explains e is, I believe, more complex and more interesting than that suggested by the Basic Explanation Condition. It will be useful to invoke an example.

During the first four decades of the nineteenth century the wave theory of light was defended by Young, Fresnel, Herschel, Lloyd, and others, in part at least, by showing how this theory could explain a host of optical phenomena including the rectilinear propagation of light, reflection, and refraction.[9] Let e be that light travels in straight lines, that light is reflected in such a way that the angle of incidence equals the angle of reflection, and that light is refracted in such a way that Snell's law is satisfied. Wave theorists showed how their theory explains e. The problem is that the rival particle theory of light (espoused by Newton in the eighteenth century, and by various physicists in the nineteenth century, including Brougham and Brewster) explained these same phenomena on the conflicting assumption that light is a stream of particles, not a wave motion in a medium. Simply on the basis of e it could not be claimed that the probability is high that there is an explanatory connection (that is, a correct one) between the wave theory and e.

Did the claim of the wave theorists violate (i) above? It did not. The argument strategy of the wave theorists was more complex than simply showing how the wave theory can explain various optical phenomena. Although it contained important explanatory components, it was an eliminative one, as follows.

From the observed fact that light travels from one point to another in a finite time, and the fact that the only known modes of propagation are via waves and particles, wave theorists assumed that the probability is very high that either light is a wave motion spread with a finite velocity in some medium, or that it consists of discrete particles moving with a finite velocity. Using b for the observed facts from which this inference was drawn, W for the wave theory, and P for the particle theory, we have

(1) $p(W \text{ or } P/b)$ is close to 1.

Wave theorists now appeal to another observed phenomenon of light, viz. diffraction patterns, and argue that to explain these, the rival particle theory introduces attractive and repulsive forces that are extremely improbable because, unlike other known forces, they do not vary with the mass or shape of the object exerting the force. Using e_d for a description of the observed diffraction phenomena and h for the hypothesis postulating the forces mentioned, and letting (the background information) b also contain information about known forces in nature, wave theorists claimed that

(2) $p(h/e_d \& b)$ is close to 0.

9. See my *Particles and Waves* (New York: Oxford University Press, 1991), chapters 3 and 4.

However, they also supposed that, *on the assumption that the particle theory P is true*, the probability that the forces in question exist (h) is very high, that is,

(3) $p(h/P\&e_d\&b)$ is close to 1.

From (2) and (3), it follows that

(4) $p(P/e_d\&b)$ is close to 0.[10]

Now wave theorists also assumed that the fact that e_d (diffraction patterns exist) did not change the probability that either the particle *or* the wave theory is true. So from (1),

(5) $p(W \text{ or } P/e_d\&b)$ is close to 1.

But from (4) and (5) we get

(6) $p(W/e_d\&b)$ is close to 1.

So far we have an eliminative argument to the wave theory based on diffraction and the finite motion of light. It is at this point that the explanation of other optical phenomena, including rectilinear propagation, reflection, and refraction, can be shown to enhance the argument. Wave theorists explain these phenomena (which we previously called e) by deriving them from the assumptions of their theory. Since e is derivable from W, it follows that

(7) $p(W/e\&e_d\&b) \geq p(W/e_d\&b)$.

Therefore, from (6) and (7),

(8) $p(W/e\&e_d\&b)$ is close to 1.

Accordingly, even if the optical phenomena e, including rectilinear propagation, reflection, and refraction, do not by themselves give the wave theory high probability, at least they sustain the high probability afforded that theory by diffraction and finite motion. What the wave theorist seeks to show is that his theory is highly probable *given all the (known) optical phenomena*. This he attempts to do by establishing high probability on the basis of some phenomena and explaining the others by derivation.

Finally, returning to the question of evidence, do the optical phenomena reported in e constitute potential evidence that the wave theory is true? Not by themselves, and not by virtue of the fact that they are explainable by derivation from the theory (since they are also derivable from the particle theory). However, the wave theorist argued, in conjunction with diffraction and finite motion, they do constitute evidence. That is, $e\&e_d\&b$ together provide evidence that W. Given this set of phenomena, the wave theorist is arguing, the probability is very high that the reason these phenomena obtain is that light is a wave motion in a medium. This comports with the explanatory connection condition (i).[11]

---

10. For a proof, see *Particles and Waves*, p. 86.
11. My claim is that this is the argument proposed by wave theorists, not that its conclusion is correct. The "logic" is correct, but there is a fundamental flaw in premise (1), to which I will return in chapter 10.

More generally, even if some fact explained by a theory does not itself constitute evidence that the theory is true, it may when conjoined with other facts. What needs to be established is that given the set of facts it is very probable that there is an explanatory connection between the theory and this set of facts. Merely showing that the theory can explain these facts is not sufficient to demonstrate that these facts constitute evidence that the theory is true.

8. Correct Explanation

The concept of an explanatory connection is defined by reference to that of a correct explanation: There is an explanatory connection between h and e if and only if either h correctly explains why e is true, or e correctly explains why h is true, or some true hypothesis correctly explains why h is true and why e is true. What is this concept of correct explanation?

First, whatever else it is, it is (or, for potential evidence we need) a concept of correct explanation that is objective, rather than subjective. Whether some h correctly explains some e does not depend upon what anyone knows or believes. In this respect the concept needed is like that which Hempel attempts to explicate by means of his deductive-nomological (D-N) model of explanation.[12] A D-N explanation of a particular event is a deductive argument in which the premises contain laws and statements describing "initial conditions," and the conclusion is a sentence describing the event to be explained. Whether some argument is a correct D-N explanation does not depend upon whether anyone knows or believes the truth of the premises or conclusion or knows or believes that the conclusion follows deductively from the premises. Without accepting Hempel's D-N model, I am accepting his idea that there is an objective concept of correct explanation.

Second, the required concept of explanation is that of a *correct* one, not necessarily one that is good or appropriate for one context of inquiry but not another. In this respect the concept of correct explanation is like that of causation. If John's taking medicine M caused his symptoms S to be relieved, then it did so whether or not the context of inquiry calls for a deeper or more detailed causal story (such as one that indicates how M causes relief, or what caused John to select M). In this respect also the concept required is similar to one that Hempel's D-N model attempts to explicate: Whether a D-N explanation is correct does not depend upon the context of inquiry. Its correctness is not determined by standards that can vary with the knowledge and interests of different inquirers.

Third, on pain of circularity, the required concept of correct explanation must not itself be understood or explicated in terms of evidence. In this respect also the concept of correct explanation needed is like that of Hempel's D-N model. Whether an argument satisfies the D-N conditions for being a correct explanation does not depend on the satisfaction of any requirement of evidence. For example, it is not required that there be evidence that supports the premises or conclusion of a D-N argument.

12. Carl G. Hempel, *Aspects of Scientific Explanation* (New York: Free Press, 1965), Part IV.

Unfortunately, Hempel's D-N model of explanation is subject to numerous counterexamples and other objections that make it impossible to employ for purposes of defining evidence.[13] In any case the model requires a deductive connection between the explanatory sentences and the sentence describing what is to be explained. As noted in section 2, we want to consider cases where e is evidence that h, yet there is no such deductive connection between h and e, or between some other hypothesis H and both h and e, where H correctly explains h and e.[14]

What can we say further, then, about the concept of correct explanation needed for potential evidence? One option is to say nothing further, but to treat "correct explanation" as a more basic concept than evidence, simply asserting that it has the three features noted above: it is objective, noncontextual, and not to be understood in terms of evidence. This is the procedure of those such as Whewell, Peirce, Hanson, Harman, and Lipton, who champion retroductive (explanatory) reasoning, or "inference to the best explanation."[15] Their idea is to say that reasoning from evidence to hypothesis involves reasoning that the hypothesis supplies an explanation, or the best explanation, of something (which may include the evidence). But they provide little if any clarification of the concept of explanation they invoke. Their view does seem to require a concept of explanation that is objective, noncontextual, and not to be understood in terms of a nonexplanatory concept of evidence.

In other writings I have proposed an account of explanation that can be used for present purposes. The account is complex, but I will attempt to simplify appropriate parts of it for presentation here.[16] I will then show that it generates a concept of explanation of a type needed for evidence, and I will contrast this concept with Hempel's D-N explanation and indicate advantages. It is not my claim that my particular account of explanation is required to understand evidence. Everything I have said so far about evidence could be accepted without endorsing this account of explanation. My claim is only that the account supplies a concept that will suffice for evidence.

9. Explanation and Content

I begin with a category of *content-giving sentences for a content-noun*. Here are some examples:

(1)   The *reason* John's symptoms were relieved is that he took medicine M.
      The *excuse* was that Sam got sick.

---

13. For a review of these problems, see my *The Nature of Explanation* (New York: Oxford University Press, 1983), chapter 5.
14. Hempel's alternative model for nondeductive cases, the inductive-statistical (I-S) model, will not suit our purposes, since it does not provide a notion of *correct* explanation. An I-S explanation may satisfy Hempel's criteria—it may be a good inductive-statistical explanation—without correctly explaining the event it purports to explain.
15. Charles Peirce, *Collected Papers* (Cambridge, MA: Harvard University Press, 1960), Charles Hartshorne and Paul Weiss, eds., vol. 5, 5.189; N. R. Hanson, *Patterns of Discovery* (Cambridge: Cambridge University Press, 1958); for other references, see note 2.
16. For details see *The Nature of Explanation*, chapters 1–3.

The *danger* in climbing the Matterhorn in the afternoon is that the ice will melt.

The *penalty* for trespassing is that the person convicted will be subject to a $1000 fine.

These sentences contain content-nouns (ones in italics) together with that-clauses, or, more generally, nominals, that give content to the noun. In sentence (1) "reason" is a content-noun and "that he took medicine M" is a nominal that gives the content of the reason. Content-nouns are abstract nouns ("reason," "explanation," "excuse," "danger") rather than nouns for physical objects, properties, or events. They can be used to form content-giving sentences, such as ones above, with this form:

The + content-noun + {prepositional phrase} + form of *to be* + nominal
{that-phrase}

The nominals are noun phrases with a verb or verb derivative. They include that-clauses, infinitive phrases ("the purpose of the flag is *to warn drivers of danger*"), and many others.

Content-giving sentences are to be contrasted with ones that contain a content-noun but do not give a content to that noun. For example,

The reason John's symptoms were relieved is difficult to grasp.
The excuse was unacceptable.
The penalty for trespassing is severe.

These sentences do not say what the reason, excuse, or penalty is.

A content-giving sentence such as (1) above may be equivalent in meaning to another, such as,

(2)   The reason John's symptoms were relieved is his taking medicine M.

I shall say that sentences (1) and (2) express the same proposition, and that this proposition is a content-giving proposition for a concept expressed by the noun "reason."

Sentences (1) and (2) contain an answer to the question "Why were John's symptoms relieved?" That answer is given by the content of the reason ("that he took medicine M"). The proposition expressed by sentences (1) and (2) above will be said to be a content-giving proposition *with respect to that question*.

A question such as "Why were John's symptoms relieved?" presupposes various propositions, for example,

John had symptoms.
His symptoms were relieved.
(3)   John's symptoms were relieved for some reason.

A *complete presupposition* of a question is a proposition that entails all and only the presuppositions of that question. Of the three propositions just given, only (3) is a complete presupposition of the question "Why were John's symptoms relieved?" Sentence (3) can be transformed into a *complete answer form* for the question "Why were John's symptoms relieved?," viz.

(4) The reason that John's symptoms were relieved is _____

by dropping "for some reason" in (3) and putting the expression "the reason that" at the beginning and "is" followed by a blank at the end to yield (4). (For more details and a generalization of this to various types of questions, see my *The Nature of Explanation*, 28ff.)

We can now say that

p is a *complete* content-giving proposition with respect to question *Q* if and only if (a) *p* is a content-giving proposition (for a concept expressed by some noun *N*); (b) *p* is expressible by a sentence obtained from a complete answer form for *Q* (containing *N*) by filling in the blank; (c) *p* is not a presupposition of *Q*.

Finally, *Q* will be said to be a *content-question* if and only if there is a proposition that is a complete content-giving proposition with respect to *Q*. By these definitions, (1) and (2) express complete content-giving propositions with respect to the content-question "Why were John's symptoms relieved?"

If *Q* is a content-question (whose indirect form is *q*), and *p* is a complete content-giving proposition with respect to *Q*, then *p provides an explanation of q*. Thus, the proposition expressed by

(1) The reason John's symptoms were relieved is that he took medicine *M*

provides an explanation of why John's symptoms were relieved. Any explanation that is offered is, or can be transformed into, one that answers some content-question by supplying a complete content-giving proposition with respect to that question. Thus, an explanation of the relief of John's symptoms that appeals to his taking medicine *M* can be construed as answering the content-question "Why were John's symptoms relieved?" by furnishing a complete content-giving proposition with respect to that question, viz. (1). The latter provides an explanation of the relief of John's symptoms.

The following simple condition holds for *correctness* of such explanations:

(5) If *p* is a complete content-giving proposition with respect to *Q*, then *p* provides a correct explanation of *q* if and only if *p* is true.

So, for example, since (1) expresses a complete content-giving proposition with respect to the content-question

Q: Why were John's symptoms relieved?

Proposition (1), assuming it is true, provides a correct explanation of why John's symptoms were relieved. Condition (5) for correctness of explanations has considerably more bite than it might seem. For example, compare (1) with

(6) John took medicine *M*.

Even if (6) is true, that will not suffice to guarantee that (6) provides a correct explanation of why John's symptoms were relieved. (His symptoms might not have been relieved *because* he took *M*.) (6) is not a complete content-giving proposition with respect to *Q*. By contrast, if (1) is true, then, since it does express a complete content-giving proposition with respect to *Q*, it provides a correct explanation of *q*.

A correct explanation may or may not be a (particularly) good one. Nor must a good explanation be correct. Goodness in explanations is a broader concept than correctness, and unlike the latter is context-dependent.[17] Whether (1) provides a good explanation of why John's symptoms were relieved depends on the "appropriateness" of the answer it provides, which is determined by the knowledge and interests of those for whom the answer is provided. If the context is one in which the intended audience already knows that medicine M is the agent that produced relief but not how it did so, then although (1) provides a correct explanation of why John's symptoms were relieved, it is not a good one. Moreover, there are contexts of evaluation in which correctness is not required. Explanations such as the Ptolemaic or Newtonian ones of the observed motions of the planets may be good ones in virtue of their comprehensiveness, precision, or predictive qualities, even though they are incorrect. What is needed for the concept of (potential and veridical) evidence is not a contextual concept of good explanation but a noncontextual one of correct explanation.

In section 8 I indicated three criteria that need to be satisfied by any concept of correct explanation employed for potential evidence. First, it must be objective rather than subjective. Whether h correctly explains e cannot depend upon what anyone knows or believes. Second, it must be noncontextual. It cannot depend upon standards of goodness appropriate in one context of evaluation but not another. Third, the concept must not itself be understood in terms of evidence. I maintain that the concept of correct explanation given in (5) satisfies these conditions.

10. But Is It Circular?

A pair of circularity charges may be levelled against this account. One involves circularity in the account of explanation itself. Even if "explanation" is not defined by reference to evidence, it is defined by reference to concepts dangerously close to that of explanation itself, viz. by reference to the concept of a content-noun, where the latter category includes the noun "explanation." This charge of circularity I reject, since the category of content-nouns, and more generally of content-giving propositions, is characterized, as I have done in the previous section, without appeal to any concept of explanation.

The second and more serious charge of circularity, or at least triviality, concerns condition (5) for being a *correct* explanation. On this condition, whether

(1) The reason that John's symptoms were relieved is that he took medicine M

is a correct explanation of why John's symptoms were relieved depends not only on whether (1) is a complete content-giving proposition with respect to the question, but also on whether (1) is true. But whether (1) is true is just what we want a definition of "correct explanation" to tell us how to determine. And condition (5) in the previous section doesn't do that at all!

---

17. See *The Nature of Explanation*, chapter 4.

My response to this is to agree that (5) does not provide a way of determining whether a proposition such as (1) is true. What (5) does, which is not trivial, is to say that if some proposition p is a complete content-giving proposition with respect to Q, then it provides an explanation of q that is correct if and only if p is true. This cannot be said for propositions generally, even if they are true and provide answers to Q. It is possible for the proposition "John took medicine M" to be both true and to provide an answer to "Why were John's symptoms relieved?" without that proposition's being a correct explanation of why his symptoms were relieved. But it is impossible for the complete content-giving proposition (1) to be true without its also being a correct explanation of why John's symptoms were relieved. (5) does not, however, provide conditions for determining whether (1) is true. In this respect it is not, and is not intended to be, a definition of "correct explanation."

There is an analogy between my procedure here and what Hempel does in his account of D-N explanation. He offers conditions for being a *potential* explanation (which may be a correct or an incorrect explanation). The conditions—such as the requirement of lawlike sentences in the explanans and a deductive relationship between explanans and explanandum—do not invoke any notion of explanation itself. To give the further condition for a *correct* D-N explanation he simply adds that the sentences in the explanans must be *true*. In an analogous fashion I define the notion of explanation in terms of content-giving propositions, etc., without invoking any notion of explanation itself. The further condition I give for a *correct* explanation is simply that the relevant content-giving proposition must be *true*.

The difference between what Hempel does and what I do is this. Hempel's explanans sentences do not include ones such as (1) above. He banishes from an explanans any terms such as "reason," "causes," and "explanation" (itself). This idea is required by what (in *The Nature of Explanation*) I call the NES (No Entailment by Singular sentences) requirement. NES, to which Hempel is committed, precludes from an explanans any singular sentences (ones describing particular events), including (1), that by themselves, without the necessary laws, entail the explanandum.

Hempel's NES requirement, however, leads to serious counterexamples in which, although the explanans is true (and satisfies all of Hempel's conditions, including NES), it does not *correctly* explain the explanandum. Reverting to a previous example, suppose it is a law of nature that anyone who eats a pound of arsenic dies within 24 hours. Suppose Ann eats a pound of arsenic at time T and dies within 24 hours. The following explanation of this event satisfies all of Hempel's conditions for being a correct D-N explanation:

At time T, Ann ate a pound of arsenic
Anyone who eats a pound of arsenic dies within 24 hours
    Therefore,
Ann died within 24 hours of T.

Suppose, however, Ann was killed not by the arsenic but by being hit by a truck, which had nothing to do with the arsenic. Then the above explanation is incorrect—it does not give a correct reason that Ann died within 24 hours of T—despite the fact that the sentences in the explanans are true, the explanans contains a true law, the explanans deductively implies the explanandum, and NES is satisfied. (The only singular sentence in the explanans, the first sentence in the argument, does not by itself deductively imply the explanandum.)

This problem arises because of the occurrence of an intervening cause. The only way to avoid it is, I suggest, to include in the explanans a sentence such as

> The reason that Ann died within 24 hours of T is that she ate a pound of arsenic at T
> The cause of Ann's death within 24 hours of T is her eating a pound of arsenic at T.[18]

These sentences, if true, would correctly explain why Ann died within 24 hours of T. They express complete content-giving propositions with respect to the content-question "Why did Ann die within 24 hours of T?" So, in accordance with my condition (5) of the previous section, if they are true they provide correct explanations. But these sentences violate Hempel's NES requirement. They are singular sentences that by themselves, without the need of laws, entail the explanandum sentence "Ann died within 24 hours of T."

I have described a concept of explanation that meets three criteria required for potential evidence. It is objective, noncontextual, and not itself defined in terms of evidence. Nor is the account circular in the sense of defining explanation by reference to itself. The general notion of ("potential") explanation is defined by reference to complete content-giving propositions, without appeal to any notion of explanation. But conditions for correctness in such explanations will need to include the requirement of the truth of propositions that violate NES: the truth of complete content-giving propositions that contain terms such as "reason" and "cause."

11. Conclusions

I have argued that high probability, although necessary for potential evidence, is not sufficient. The reason it is not sufficient is that evidence, unlike probability, is selective. It can be the case that h's probability on $e\&b$ is r, even though e is probabilistically irrelevant for h. But e is evidence that h, given b, or $e\&b$ is evidence that h, only if e makes an evidential contribution beyond simply not preventing b from being evidence that h. Because probability is not selective, the hypothesis (h) that Michael Jordan will not become pregnant has high probability, given both (e) Michael Jordon eats Wheaties and (b) Michael Jordan is a male basketball star, even though $p(h/e\&b) = p(h/b)$ and $p(h/e) = p(h)$. But because evidence is selective, from the fact that (part of) b is evidence that h, we cannot conclude that $e\&b$ is evidence that h or that e is evidence that h, given b, even though e does not prevent b from being evidence that h.

The condition on potential evidence that needs to supplement high probabil-

---

18. Other attempts to avoid the problem are unsuccessful; see my *The Nature of Explanation*, chapter 5.

ity is one involving an explanatory connection between h and e. It is that the probability be greater than $\frac{1}{2}$ that there exists an explanatory connection between h and e, given $h\&e$. In fact I argue that a stronger condition is required, viz. that given just e, the probability is greater than $\frac{1}{2}$ that there is an explanatory connection between h and e. This condition is shown to be equivalent to one that requires that the product of two probabilities be greater than $\frac{1}{2}$, viz. the probability of h on e and the probability that there is an explanatory connection between h and e, given both h and e. I consider whether the explanatory connection condition is too weak (should the threshold probability be greater than $\frac{1}{2}$?) and also whether this condition is too strong (would it preclude explanatory arguments to hypotheses frequently given in science?). Reasons are offered for retaining the present condition.

The remainder of the chapter develops an account of "correct explanation" that can be used to explicate the idea of an explanatory connection. That account, I argue, is objective, and noncontextual, and it does not define "correct explanation" by invoking other explanatory or evidential concepts.

8

FINAL DEFINITIONS AND REALISM

In this chapter I complete the definition of potential evidence and show how the other concepts of evidence distinguished in chapter 2 (ES, subjective, and veridical) can be defined by reference to potential evidence. I also address the question of whether the concepts of evidence defined require a realist, or permit an instrumentalist, stance toward hypotheses.

1. Truth and Entailment

I have argued that e is potential evidence that h only if

p(there is an explanatory connection between h and e/e) $> \frac{1}{2}$.

This, however, is not sufficient. One condition also necessary is that e be true. That John took medicine M is evidence that his symptoms S will be relieved only if it is in fact true that he took medicine M. If it is not true, then we can say only that *if* it were true it *would be* evidence that this is so.

Should we also require that any background assumptions to which we are relativizing the evidence claim be true? Suppose we say that, given that medicine M is 95% effective in relieving symptoms S, the fact that John took M is evidence that his symptoms will be relieved. Should this relativized evidential claim be counted as false if in fact medicine M is not 95% effective?

There is an ambiguity in such relativized claims, just as there is with probabilistic claims (as was noted in chapter 5, section 5). The "given that" clause can mean the same as "since it is the case that." If so, what follows the "given that" is (supposed to be) true. It can also mean "supposing that," where what follows need not be true. So construed, a relativized evidential statement can be understood as a conditional: if b is (or were) true, e is (or would be) evidence that h. A *non*conditional evidential claim that is relativized to background information b will then be understood as requiring the truth not only of e but of b as well. In what follows, evidential claims relativized to background information will be understood in the nonconditional way.

Two objections might be made to the requirement of truth of e (and b). One is that what is required is only the *belief* that e is true. Suppose John's doctor, who prescribed medicine M, believes John has taken it, although in fact he has not. Can't the doctor truly claim that John's taking M is evidence that his symptoms will be relieved? Yes, he can, but this is what, in chapter 2, was called *subjective* evidence: it is the doctor's evidence, what the doctor takes to be evidence. In fact it is not (potential or veridical) evidence at all, since John never took M. An alleged fact cannot be a good reason to believe a hypothesis (in the sense of "good reason" required for potential and veridical evidence) if there is no such fact.

The second objection is that more than truth is required: e must be *known* to be true.[1] Facts are not evidence if nobody knows about them. Evidence is important and useful to us only if we have it. Who cares if some fact no one knows points to some conclusion?

To be sure, if no one knows that e is true, then no one knows that e is evidence that h. But the claim that if the latter is unknown, then the fact that e is evidence that h is of no interest cuts no ice. If discovered, this fact could be of considerable interest.

A final condition I suggest for (potential and veridical) evidence is that e does not logically entail h. The fact (e) that I am wearing a blue suit today is not evidence that (h) I am wearing a suit. It is too good to be evidence. Yet given appropriate background information b which includes the fact that the only piece of clothing I own for work is a blue suit, and given e, the probability is very high that some hypothesis (viz. that the only piece of clothing I own for work is a blue suit) correctly explains both h and e.

It might be agreed that where the entailment is immediate and obvious, as it is with the e and h above, we do not speak of evidence. But what if e entails h but showing that it does requires a complex proof? Aren't there cases of this sort where we do want to claim that e is evidence that h?

There are, but, I believe, they involve an entailment not from e to h but from e *plus background assumptions* to h. For example, let

e: This projectile has been launched from the earth's surface with a velocity greater than 11.2 km/sec.
h: This projectile will escape from the earth's gravitational pull.

The truth of e is evidence that h. Now h is derivable from e, but not from e alone. Several crucial assumptions are involved in the derivation, viz. Newton's law of gravity, plus specific values for the radius and mass of the earth and for Newton's gravitational constant. If we combine all of these facts into one "big fact," the latter provides, or allows us to provide, a proof or an argument demonstrating that the projectile will escape. But I am disinclined to classify this "big fact" as *evidence* that the projectile will escape; it is too strong for that. In general, evidence is to be distinguished from proof or demonstration. The entailment requirement, then, is simply that for e to be evidence that h, e (by itself) does not entail h.

---

1. We might recall that Patrick Maher has such a requirement. See above, chapter 4, section 11.

2. Final Definitions

Here, then, are my final necessary conditions for potential evidence:

(PE) e is potential evidence that h, given b, only if
1. p(there is an explanatory connection between h and e/$e\&b$) $> \frac{1}{2}$
2. e and b are true
3. e does not entail h.

Since e and b must both be true, we can rewrite these conditions more simply by retaining 3 and combining 1 and 2 as follows:

e is potential evidence that h, given b, only if
1. $p_{e\&b}$(there is an explanatory connection between h and e) $> \frac{1}{2}$
2. e does not entail h.

The first condition in (PE) above is equivalent to

$p(h/e\&b) \times p$(there is an explanatory connection between h and e/$h\&e\&b$) $> \frac{1}{2}$,

which requires that both of the probabilities involved be greater than $\frac{1}{2}$.

The concept of probability in the definitions above is to be interpreted as "the degree to which it is reasonable to believe," where this is construed as an objective epistemic probability of the sort developed in chapter 5. With this concept of probability, potential evidence is an objective epistemic concept. It is epistemic because it is tied to the degree of reasonableness of belief. It is objective because whether e is evidence that h does not depend on whether anyone knows or believes e or h or that e is evidence that h. It is not evidence for some particular person or group. Nor is it evidence for someone in a particular type of epistemic situation.

Are the conditions in the definition (PE) sufficient for potential evidence? In chapter 7 the question was raised as to whether the probability threshold for evidence should be greater than $\frac{1}{2}$. Two strategies were proposed. One, more liberal but more precise, retains the "$> \frac{1}{2}$" condition. The other, more stringent but less precise, requires the probability value to be significantly greater than $\frac{1}{2}$, without indicating what is to count as "significantly greater." With the first strategy, the one I adopt, the conditions in (PE) above remain as indicated. With the second strategy, "$> \frac{1}{2}$" is changed to "$>> \frac{1}{2}$." On either strategy, the conditions in (PE) are proposed as both necessary and sufficient.

These conditions suffice not only for evidence that h is true but also for evidence that h is false. The latter is simply evidence that the negation of h is true, which, of course, is not the same as saying that e is not evidence that h. For example, let

- e: Al owned at least one ticket in the lottery
- e': Al owned *only* one of the tickets in the lottery
- b: The lottery was a fair one involving 1 million tickets, in which one ticket was chosen to win
- h: Al won.

Neither e nor e' is evidence that h is true, given b, since the probability of h on either is not sufficiently high. But e' (and not e) is evidence that h is false. Given that the lottery was a fair one involving 1 million tickets and that Al owned only one of the tickets, the probability is high that the reason he did not win is that he owned only one ticket. Here the "only" is important. In conjunction with *b*, it indicates not only that Al owned one ticket *and no more*, but also that there were in existence many more tickets that he did not own. It is very probable that because of this he did not win. In this situation it seems quite plausible to say that, given that the lottery contained 1 million tickets, the fact that Al owned *only* one of the tickets is evidence that he did not win. By contrast, it seems much less plausible to say that, given that the lottery contained 1 million tickets, the fact that Al owned *at least* one ticket is evidence that he did not win.

3. Special Relativizations

As noted in chapter 5, objective epistemic probability statements are frequently relativized to "no interference" and "disregarding" conditions. How do such conditions affect evidential claims? Let us consider two examples, the first a gambling case in which both relativizations occur. Let *e* and *h* be as follows:

- *e*: John owns 95% of the tickets in a lottery, one ticket of which will be selected.
- *h*: John will win.

Suppose we claim that *e* is (potential) evidence that *h*, and that when we do so we are assuming that there are no interference conditions (such as cheating or invalidation of tickets), and that microconditions of the selection device are being disregarded. One of the probability claims implied by such a claim that *e* is potential evidence that *h* is

$p_{e\&n.i.\&d(mi)}(h) > .5$,

or else

$p(h/e\&n.i.\&d(mi)) > .5$.

Using the first of these, our evidential claim is that since there is no interference in the gambling situation and since microconditions are disregarded, the fact that John owns 95% of the tickets is potential evidence that he will win. If microconditions are not disregarded, and the ones under which the selection is made preclude John's winning, or if cheating is introduced which precludes this, then the fact that John owns 95% of the tickets is not potential evidence that he will win.

Second, consider Hertz's case in which *e* reports that in Hertz's experiments no electrical deflection of cathode rays was detected, and *h* is that cathode rays are electrically neutral. Since Hertz was assuming that there was no interference (of a sort later shown by Thomson), his evidential claim, relativized to this assumption, can be understood as stating that *e* is potential evidence that *h*, which implies

$p_{e\&n.i.}$(there is an explanatory connection between *h* and *e*) $> .5$.

Thomson refuted the latter by showing that there is interference (*n.i.* is false).
Could Hertz, then, have replied making the claim that

e is potential evidence that *h*, given *n.i.*,

in which the "given n.i." means "on the assumption of n.i." and not "since n.i. is the case"? Yes, if (as seems very unlikely) he intended to make only a conditional claim of the form "if n.i. is true, then e is potential evidence that h." If (as seems much more plausible given what he says), his evidential claim was nonconditional, then its truth requires the truth of both e and n.i. But n.i. is false, so, in accordance with (PE), e is not potential evidence that h, given n.i.

Finally, desperate to salvage something, the claim might be that Hertz was invoking a disregarding condition, saying in effect that electrical interference, if any, is being disregarded. Denoting the latter by d(e.i.), Hertz might have been claiming that

e is potential evidence that h, given d(e.i.).

On (PE) this would be true if

(i) $p_{e \& d(e.i.)}$(there is an explanatory connection between h and e) > .5

and

(ii)   e does not entail h.

But in this case, although (ii) is true, (i) is not. How reasonable it is to believe that there is an explanatory connection between h (cathode rays are electrically neutral) and e (no electrical effects were detected in Hertz's experiments) depends crucially on whether there was electrical interference. If electrical interference, if any, is being disregarded, the question of how reasonable it is to believe that e is true because h is has no answer. The probability in (i) is undefined.

How is this different from the earlier lottery case in which microconditions are being disregarded? Again, let e be that John owns 95% of the tickets in a lottery, one ticket of which will be selected; and let h be that John will win. Suppose microconditions of the selection device are disregarded, that is, d(mi). The claim was that e is potential evidence that h, given d(mi), which requires the truth of

$p_{e \& d(mi)}$(there is an explanatory connection between h and e) > .5.

Now suppose that microconditions M, if they obtain, determine that John will win, and that in all probability M will obtain. Then, it might be objected, it cannot be the case that John will probably win because he owns 95% of the tickets, since he will probably win because microconditions M will (probably) obtain. If one explanation is (probably) correct, the other cannot be. I reject this claim. Both explanations can be correct, one at the "microlevel" and one at the "macrolevel." It can be true that he will win because he owns almost all the tickets; it can also be true that he will win because of the specific microconditions that will obtain in the selection process. Accordingly, given that microconditions are being disregarded but the percentage of tickets owned by John is not, the fact that he owns 95% of the tickets makes it very reasonable to believe that he will win because he owns 95% of the tickets.

Nevertheless, it might be objected, an evidential claim can be trivialized by introducing a disregarding condition. Suppose that we are tossing a coin. Let information e, which contains the time at which each toss was made and the outcome, report that in 1000 tosses the coin landed heads 500 times. From e construct e' which contains only the time at which an outcome was *heads* and which reports that in 500 tosses there were 500 heads. Hypothesis h is that the next toss will yield heads. Now we introduce a disregarding condition $d(T)$, which means that tails outcomes are being disregarded. Is e' potential evidence that h, given $d(T)$? If it is, then we can always cook results with respect to a hypothesis by disregarding parts of the results, making the remainder potential evidence for that hypothesis.

The crucial question here is whether

$$p_{e'\&d(T)}(\text{there is an explanatory connection between } h \text{ and } e') > .5,$$

which is true only if

$$p_{e'\&d(T)}(h) > .5.$$

But how reasonable it is to believe that the next toss will yield heads, given the 500 heads reported, depends crucially on how many tosses also yielded tails. With tails outcomes disregarded, the question of how reasonable it is to believe the next toss will yield heads, given e', has no answer. The required probability does not exist.

4. Veridical, ES-, and Subjective Evidence

The conditions for e's being veridical evidence that h, given b, will include at least that e is potential evidence that h, given b, and that h is true. Does veridical evidence require more? Suppose that (e) this coin has landed heads each time on the first 1000 tosses. Let the hypothesis (h) be that it will land heads on the 1001st toss, and let the background information (b) contain the fact that this coin has a very strong physical bias toward heads. Suppose that the coin does in fact land heads on the 1001st toss; but this is due not to its physical bias, which is responsible for the previous results, but to the fact that, on this toss only, an extremely unlikely external force intervened, which has never before occurred. This force, not the coin's physical bias, caused it to land heads. If, as seems reasonable to suppose, e is potential evidence that h, given b, then, since h is true, shall we conclude that e is veridical evidence that h, given b?

In chapter 2 veridical evidence was described as providing a good reason to believe a hypothesis in a sense of "good reason" that requires the hypothesis to be true. In view of the fact that the coin has a very strong physical bias for heads, isn't the fact that the coin landed heads each time in the first 1000 tosses a good reason (in this sense) to believe that it will land heads on the 1001st toss, given that the latter is indeed true? Or does it fail to be a good reason, since the physical bias, which does explain the results obtained in the first 1000 tosses, had nothing to do with causing the result of the 1001st toss?

I suggest that both options are possible. There is a sense or type of good (veridical) reason for belief that requires (a) that what is believed is true, and (b) that there *probably* is an explanatory connection between what is believed and the reason for it, but does not require (c) that such an explanatory connection actually exist. And there is a sense or type of good (veridical) reason for belief that requires (a), (b), and (c). In this sense, we might say that the fact the coin landed heads on the first 1000 tosses is a misleading reason (even if veridical in the former sense) to think it will land heads on the 1001st toss, because it falsely suggests that there is an explanatory connection between these facts.

Accordingly, we can write

e is veridical evidence that h, given b, if and only if
1. e is potential evidence that h, given b
2. h is true
(3. There is an explanatory connection between e's being true and h's being true.)

With the third condition we have a strong concept of veridical evidence, without it we have a weaker one. Employing the weaker concept, this coin's landing heads on the first 1000 tosses is veridical, albeit misleading, evidence that it will land heads on the 1001st toss; on the strong concept, it is not veridical evidence at all, since there is no explanatory connection between the two.[2]

In chapter 2, section 10, I claimed that scientists seek veridical evidence, not just potential or ES-evidence. They want their hypotheses to be true, and they want to provide a good reason for believing them in a sense of "good reason" that requires truth. But they want something in addition. They want evidence to provide a reason for belief that is not misleading. Accordingly, my claim is that scientists seek evidence that is veridical in the strong sense requiring the satisfaction of all three conditions above. It is this concept of veridical evidence that will be employed in what follows.

So far I have proposed definitions for potential and veridical evidence. The two remaining concepts of evidence—subjective and ES—can be defined by reference to veridical (and hence potential) evidence, as follows:

e is ES-evidence that h (with respect to an epistemic situation ES) if and only if e is true and anyone in ES is justified in believing that e is (probably) veridical evidence that h.

e is X's subjective evidence that h at time t if and only if at t, X believes that e is (probably) veridical evidence that h, and X's reason for believing h true (or probable) is that e is true.

These definitions are in accord with the preliminary characterizations of ES- and subjective evidence given in chapter 2. In section 4 of that chapter, ES-evidence that h was characterized as providing a justification for believing h for anyone in epistemic situation ES. This follows from the definition above, since anyone justified in believing that e is (probably) veridical evidence that h is justi-

---

2. The strong concept is relevant for certain "Gettier cases" purporting to show that justified, true, belief is not sufficient for knowledge. Suppose I am justified in believing that (h) this coin will land heads on the 1001st toss, since I know that (e) it has landed heads each time on the first 1000 tosses, and (b) that it has a strong physical bias toward heads; but I have no way of knowing that an extremely unlikely external force, not its physical bias, will cause it to land heads on the 1001st toss. The usual view is that under these circumstances, although I have a justified, true, belief that h, I do not know that h is true. Requiring that my justification contain veridical evidence in the strong sense precludes this sort of case.

fied in believing that h is (probably) true. Moreover, ES-evidence, so defined, will satisfy the other criteria noted in chapter 2. For example, it will be objective (it will not depend on who in fact believes what), and it will allow h to be false.

In section 5 of chapter 2, I said that e is X's subjective evidence that h at time t if and only if at time t, X believes that e is evidence that h, X believes that h is true or probable, and X's reason for believing this is that e is true. This is entailed by the definition of subjective evidence above. For a defense of the claim that "evidence" in this definition (as well as in that for ES-evidence) requires veridical, and not simply potential, evidence, see chapter 2, section 9.

One point needs emphasizing regarding evidence of all four types. I begin with subjective evidence. My reason for believing that h is true may be that e is true, even though e is not my subjective evidence that h. For example, let the hypothesis be

h: There is a bear in that tree.

My reason for believing h may be the fact that

e: I see a bear in that tree.

But e is not my (subjective) evidence that h is true. It is too close to be evidence, and indeed, in the case of potential and veridical evidence, it is precluded by the requirement that e does not entail h. Since I know that e entails h, I do not believe that e is veridical evidence that h. So by the definition above, e is not my subjective evidence that h. If I do have subjective evidence that h, it would something like

e': There is a large heavy animal in that tree with long shaggy hair, eating berries, etc.,

which does not entail h. Your subjective evidence that there is a bear in that tree may be the fact that I claim to see a bear in that tree. This is possible, since the latter fact does not entail h.

Analogous points can be made for the remaining types of evidence. In the case of potential and veridical evidence, this means that although e's being a good reason to believe h is a necessary condition for evidence it is not sufficient. In the case of ES-evidence, it means that although the fact that e justifies a belief that h is necessary for ES-evidence it is not sufficient.

5. Reciprocity

It is a consequence of the definitions of potential, veridical, and ES-evidence that e can be evidence that h while at the same time h is evidence that e. In chapter 7, section 5, a case of this sort was introduced in which

e: Arthur has a rash on his arm that itches
h: Arthur's arm was in contact with poison ivy
b: Arthur was weeding yesterday bare-armed in an area filled with poison ivy, to which he is allergic.

Suppose that e and b are both true. Since e does not entail h, and since p(there is an explanatory connection between h and $e/e\&b) > \frac{1}{2}$, it follows that e is potential evidence that h, given b. Suppose that h is also true. Since h does not entail e, and since p(there is an explanatory connection between e and $h/h\&b) > \frac{1}{2}$, it follows that h is potential evidence that e. If h is true and there is an explanatory connection between h and e, then e is veridical evidence that h, given b, and h is also veridical evidence that e, given b. It can also be shown that it is possible for an epistemic situation ES to be such that e is ES-evidence that h, given b, and that h is ES-evidence that e, given b.

Is this consequence of these definitions undesirable? I suggest not. Concentrating just on the notion of a good reason for believing, which is provided by potential and veridical evidence that h, in the case above the fact that Arthur has a rash on his arm that itches is a good reason to believe that his arm was in contact with poison ivy, since he was weeding yesterday bare-armed in an area filled with poison ivy, to which he is allergic. But also, in view of the same background information, the fact that Arthur's arm was in contact with poison ivy is a good reason to believe that he has a rash on his arm that itches.

For such reciprocity to hold for subjective evidence, the definition requires the truth of both of the following: the fact that e is true is X's reason for believing h, and the fact that h is true is X's reason for believing e. Could this happen? And if so, is it reasonable? My answer is that it could happen, but if it did, there would be an undesirable circularity here that is not present with the other (objective) concepts of evidence. Why is this so?

Suppose that (A) Mary saw Arthur going out to weed bare-armed in an area full of poison ivy. And suppose that (B) she later saw Arthur scratching the rash on his arm. Then, we might conclude, her reason for believing (h) that Arthur's arm was in contact with poison ivy is (A) above; and her reason for believing (e) that Arthur has a rash on his arm that itches is (B) above. In this case we would not conclude that her reason for believing h is that e and her reason for believing e is that h.

By contrast, suppose that Mary did not see Arthur at all, so that both (A) and (B) are false. And suppose that, when asked for her reasons for believing h and e, she claims simply that her reason for believing h is that e is true and her reason for believing e is that h is true. I do not claim that this is conceptually impossible, only that it is circular. A person's reason for believing is an asymmetrical concept. There is a "starting point" and a "direction." Abstract reasons for believing have neither.

A corresponding situation arises with the concept of a logical or mathematical proof. Suppose that there is a valid proof showing that proposition $P1$ is derivable from proposition $P2$, and one showing that proposition $P2$ is derivable from $P1$. If the derivations are not too "immediate," it might be the case that $P2$ provides a good reason for believing $P1$ and also that $P1$ provides a good reason for believing $P2$. In the abstract the starting point and direction are irrelevant. For example, from the probability axiom

(P1) $p(h_1 \& h_2) = p(h_1) \times p(h_2/h_1)$

one can derive the following generalization as a theorem

(P2) $p(h_1 \& h_2 \& \ldots h_n) = p(h_1) \times p(h_2/h_1) \times \ldots p(h_n/h_1 \& h_2 \& \ldots h_{n-1})$

But equally one could take the generalization (P2) as an axiom and derive (P1) as a theorem. The fact that (P1) is true is a good reason for believing (P2). And the fact that (P2) is true is a good reason for believing (P1). But if *your* reason for believing (P1) is that (P2) is true, and *your* reason for believing (P2) is that (P1) is true, you are guilty of circularity. Either you start with (P1) as an axiom and use it to prove (P2); or you start with (P2) as an axiom and use it to prove (P1). Doing both is prohibited, despite the fact that both (P1) and (P2) are true and are mutually entailing. There is a similar situation with subjective evidence.

In the case of ES-evidence, I claimed above, reciprocity can obtain. In the poison ivy case someone in a certain epistemic situation may be justified in believing that e is veridical evidence that h and also in believing that h is veridical evidence that e. Nothing follows from this about what such a person's reasons for believing h and e are or should be. Indeed, I may realize that e is (potential and veridical) evidence that h, and be justified in this realization, without its being the case that I have, or am justified in having, e as my reason for believing h and h as my reason for believing e.

6. Hempel's "Conditions of Adequacy"

In a classic article on confirmation Hempel discusses various broad conditions an adequate definition of confirming evidence might be claimed to satisfy.[3] He notes the following, the first three of which he accepts.

1. *Consequence condition:* If e is evidence that each member of a set K of hypotheses is true, and K entails h, then e is evidence that h.

This has two important corollaries:

   (a) *Special consequence condition:* If e is evidence that h, and if h entails h', then e is evidence that h'.
   (b) *Equivalence condition:* If e is evidence that h, and if h is equivalent to h', then e is evidence that h'.
2. *Entailment condition:* If e entails h, then e is evidence that h.
3. *Consistency condition:* e is logically compatible with the set of all hypotheses for which it is evidence.
4. *Converse consequence condition:* If e is evidence that h, and if h' entails h, then e is evidence that h'.

To understand some of the implications of my proposed definitions of evidence, it will be useful to see which, if any, of these conditions is satisfied. Since the basic concept is potential evidence, I will focus on this.

1. *Consequence condition.* Here I will consider just the two corollaries. I begin with an example for the special consequence condition. Let

   e: John bought 950 tickets in the lottery
   b: There were 1000 tickets sold in the lottery, one of which was selected at random as the winning ticket

---

3. Carl G. Hempel, "Studies in the Logic of Confirmation," reprinted in his *Aspects of Scientific Explanation* (New York: Free Press, 1965), 3–46.

> h: John bought a winning ticket
> h': Someone or other bought at least one ticket in the lottery.

According to the special consequence condition, if e is (potential) evidence that h, given b, then e is evidence that h', since h entails h'. But it seems incorrect to claim that the fact that John bought 950 tickets in the lottery is evidence that someone bought at least one ticket in the lottery. On the definition of potential evidence, although e is evidence that h, given b, it is not evidence that h', given b, for two reasons. First, e by itself entails h', which is precluded by potential evidence. Second, the explanatory connection condition is not satisfied. It is not the case that, given that John bought 950 tickets in the lottery, the probability is high that the reason John did so is that someone or other bought at least one ticket, or that the reason that someone or other bought at least one ticket is that John bought 950 tickets in the lottery, or that something correctly explains both why someone or other bought at least one ticket and why John bought 950 tickets.

An even more striking case would be one in which h' is a tautology (which is entailed by any sentence). Here, again, both the no-entailment and explanatory connection condition for potential evidence are violated.

Accordingly, the special consequence condition fails to be satisfied by potential evidence.

The equivalence condition is also violated by cases involving tautologies. Let us add a tautology to the previous hypothesis h, obtaining

> h_1: John bought a winning ticket, and either February is the coldest month or it isn't.

Hypotheses h and h_1 are equivalent (each entails the other). Yet, although e is evidence that h, given b, there is something amiss with the claim that e is evidence that h_1, given b. Here the explanation condition fails to be satisfied.

Another type of case violating the equivalence condition is one involving equivalent hypotheses with different word order indicating a difference in emphasis or focus.[4] Consider:

> h: It was on Monday that John went to the bank.
> h': It was John who went to the bank on Monday.
> e: The bank is open only on Monday.
> b: John is the only one who needed to go to the bank.

Given b, e is (potential) evidence that h. But, although h and h' are equivalent since they are mutually entailing, e is not evidence that h', given b. The explanation condition is satisfied for e and h, but not for e and h'. The probability is high that, given $e \& b$, the reason that (h) it was on Monday that John went to the bank is that (e) the bank is open only on Monday. But the probability is not high that, given $e \& h$, the reason that (h') it was John who went to the bank on Monday is that (e) the bank is open only on Monday.

Evidence, as well as explanation, is emphasis-sensitive. In

---

4. For a more detailed discussion of emphasis and its effect on explanation and evidence, see my *The Nature of Explanation* (New York: Oxford University Press, 1983), chapter 6.

Given b, the fact that (e) the bank is open only on Monday is evidence that
(h) it was on Monday that John went to the bank,

the term "evidence" selects, or focuses upon, the day on which the banking took place, not the person banking. It does so in virtue of the emphasis in (h) on the day, as given by the word order. In

Given e, the fact that (b) John is the only one who needed to go to the bank is evidence that (h') it was John who went to the bank on Monday,

the term "evidence" selects, or focuses upon, the person banking, not the day. It does so in virtue of the emphasis in (h') on the person, as given by the word-order.

2. *Entailment condition.* This too is violated by potential evidence, according to which if e entails h then e is not evidence that h. (See the lottery example for the special consequence condition.)

3. *Consistency condition.* This is satisfied by potential evidence. If e is evidence that h then e cannot be evidence that h' where h' is incompatible with h. Otherwise the high-probability requirement would be violated. That is, if e is evidence that h, then $p(h/e) > \frac{1}{2}$, so that $p(h'/e) < \frac{1}{2}$ for any h' incompatible with h.

4. *Converse consequence condition.* This is violated by potential evidence, since it fails to satisfy the high-probability condition. That is, it is possible for e to be potential evidence that h, and for h' to entail h, without e's being potential evidence that h', since $p(h'/e) < \frac{1}{2}$. For example, let

e: Bill owned 950 tickets in the lottery.
b: The lottery contains 1000 tickets, two of which were selected at random as winners; Sam owned one ticket.
h: Bill was a winner.
h': Bill was a winner, and Sam was a winner.

e is potential evidence that h, given b. Hypothesis h' entails h, but e is not potential evidence that h', given b, since $p(h'/e\&b) = .00095$.

Finally, I will consider a condition Hempel does not formulate as such, although it follows from his consequence condition, viz.

Conjunction condition: e is evidence that $h_1\&h_2\ldots\&h_n$ if and only if e is evidence that h_1, and e is evidence that $h_2, \ldots$, and e is evidence that h_n.

Let us separate the "if" and the "only if" as follows:

Conjunction condition 1: If e is evidence that h_1, and if e is evidence that $h_2, \ldots$, and if e is evidence that h_n, then e is evidence that $h_1\&h_2\&\ldots\&h_n$.

Conjunction condition 2: If e is evidence that $h_1\&h_2\&\ldots\&h_n$, then e is evidence that h_1, and e is evidence that $h_2, \ldots$, and e is evidence that h_n.

The first condition is violated by potential evidence, since it fails to satisfy the requirement of high probability. Let

e: This coin has a bias of .9 for heads
h_i: This coin will land heads on the i-th toss

Although the conditions for potential evidence are satisfied for e and any particular h_i, they are not in general satisfied for e and a conjunction of h_i's. For example, suppose the hypothesis is that the first 7 tosses will result in heads. Since $p(h_1 \& h_2 \& \ldots \& h_7) < \frac{1}{2}$, e is not evidence that $h_1 \& h_2 \ldots \& h_7$, although it is evidence for each member of the conjunction.

The situation is similar in the lottery paradox. The fact that there are 1 million tickets, one of which will be selected at random, is evidence that $not\text{-}h_i$, where h_i is that ticket i will win. But it is not evidence that $not\text{-}h_1$ & $not\text{-}h_2$ & $\ldots$ & $not\text{-}h_{1\,million}$.

Conjunction condition 2 is more plausible. For example, let

    e: Al's symptoms S disappeared at 4 P.M.
    h_1: Al took medicine M_1 at 3 P.M.
    h_2: Al took medicine M_2 at 3 P.M.
    b: M_1 and M_2 taken together, but not separately or at separate times, cause S to disappear in an hour in 95% of the cases; they are the only things that can cause S to disappear; S will not disappear "spontaneously."

Intuitively, given b, e is evidence that $h_1 \& h_2$. But also, given b, e is evidence that h_1 and e is evidence that h_2. This is in accord with the concept of potential evidence, if we understand an "explanatory connection" between h and e to include cases in which h is part of the explanation of e. In the example above p(there is an explanatory connection between h_1 and $e/e\&b$) is high, since, given $e\&b$, the probability is high that part of the reason that (e) Al's symptoms S disappeared at 4 P.M. is that (h_1) he took M_1 at 3 P.M. (the other part being that he also took M_2).

7. Realism versus Instrumentalism

My definitions of potential and veridical evidence reek of realism. First, the claim that e is evidence that h I have taken to mean that e is evidence that h is true. Second, both potential and veridical evidence require the high probability that h is true, and veridical evidence goes further and requires that h is true. Can an instrumentalist who rejects, or does not care about, the truth or high probability of truth of ("theoretical") hypotheses use, or at least modify, my definitions of evidence to suit instrumentalist purposes? How might this be done?

Instead of asking whether e is evidence that h is true, we can ask whether e is evidence that h saves the phenomena. Here is a definition an instrumentalist might offer:

    I. e is potential evidence that h saves the phenomena, given b, if and only if
        1. e and b are true (truth is possible for "observational" claims)
        2. e does not entail that h saves the phenomena
        3. $p_b(h$ saves the phenomena$/e) \times p_b$(there is an explanatory connection between the fact that h saves the phenomena and e/h saves the phenomena $\& e) > \frac{1}{2}$.

Veridical evidence that h saves the phenomena would require the truth of the claim that h saves the phenomena.

Two questions can be raised. First, is (potential) evidence that h saves the phenomena necessarily potential evidence that h is true? (That is, is the instrumentalist idea of evidence really different from the realist's?) Second, assuming it is, could an instrumentalist accept the sort of definition I have suggested for him?

My answer to the first question is no. The instrumentalist and realist ideas are different. To show this we need some notion of "saving the phenomena." Let us speak of a hypothesis or theory h as "potentially saving" some putative phenomenon described by e, and think of this quite broadly in terms of entailment: h potentially saves e if it entails e.[5] With this notion as basic, let us adopt the following

Definition 1: h saves e if and only if (a) h potentially saves e, and (b) e is true.

Introducing probability, from definition 1 we can say that if h potentially saves e, then it is probable to degree r that h saves e if and only if $p(e) = r$. More generally,

Definition 2: If h potentially saves e, then, given $e_1, \ldots, e_n$, it is probable to degree r that h saves e if and only if $p(e/e_1 \ldots e_n) = r$.

Now the following is provable from probability theory:

Theorem 1: If h together with b entails $e_1, e_2, \ldots,$ and if $p(h/b) > 0$, then $\lim_{m,n \to \infty} p(e_{n+1} \ldots e_{n+m}/e_1 \ldots e_n \& b) = 1$.[6]

From theorem 1 and definition 2, we get

Theorem 2: If h potentially saves $e_1, e_2, \ldots$ in such a way that h together with b entails these e's (e.g., h explains the e's by derivation), and if $p(h/b) > 0$, then $\lim_{m,n \to \infty} p(h \text{ saves } e_{n+1} \ldots e_{n+m}/e_1 \ldots e_n \& b) = 1$.

From theorem 2 it follows that it can be the case that

(1) $\lim_{m,n \to \infty} p(h \text{ saves } e_{n+1} \ldots e_{n+m}/e_1 \ldots e_n) = 1$

even though $p(h/b)$ is very small, so long as it is not zero. Could (1) obtain even if, for any n, $p(h/e_1 \ldots e_n \& b)$ is very small? Yes, it could. Suppose that some incompatible hypothesis h' also potentially saves the e's in such a way as to entail them, but that the probability of h' on b is very high. Then no matter how many e's h entails, the probability of h on the e's will be and remain very low.

More precisely, suppose that the rival hypothesis h' is such that $p(h'/b) = r$. If h' together with b entails the e's, then for any n,

$p(h/e_1 \ldots e_n \& b) < 1 - r$.

So if the rival hypothesis h' is initially very probable, say $p(h'/b) = .95$, then no matter how many e's the original hypothesis h saves its probability $p(h/e_1 \ldots e_n)$

---

5. This is stronger than van Fraassen's account of "saving" which requires only that e be consistent with h (Bas van Fraassen, *The Scientific Image* [Oxford: Oxford University Press, 1980]). My point here is not to attack but rather to defend one possible antirealist claim, viz. that it is possible for e to be evidence that h saves the phenomena without e's being evidence that h is true.
6. For a proof, see John Earman, "Concepts of Projectibility and Problems of Induction," *Nous*, 19 (1985), 521–535.

will be and remain less than or equal to .05 for any n. This can be true even if (1) obtains, that is, even if the probability that h saves the e's gets larger and larger, approaching 1 as a limit.

Here is a simple example. Suppose the background information b tells us that one side of a certain coin marked heads is magnetized but the other is not, and that a powerful magnetic field exists under the spot on which the coin will land. The coin is tossed and lands heads each time. Let $e_1,\ldots,e_n$ describe the results of the first n tosses. Now consider two incompatible hypotheses designed to save the phenomena:

h: When the coin is tossed God intervenes (not the magnetic field) and causes the coin to land heads.

h': When the coin is tossed the magnetic field (not God) causes the coin to land heads.

Let us suppose that the initial probability of each hypothesis is not zero. Since both hypotheses entail $e_1,e_2,\ldots$, it follows from theorem 2 that the probability that each hypothesis saves the phenomena approaches 1 as a limit. That is, (1) obtains for both hypotheses.

Let us suppose that the probability of h' on b alone is very high, say .95. It follows that for any n,

$$p(h'/e_1\ldots e_n \& b) \geq .95,$$

and so, for any n,

$$p(h/e_1\ldots e_n \& b) \leq .05.$$

That is, the probability that the God-intervention hypothesis is true is and remains low, no matter how many coin tossing phenomena it saves. Yet the probability that the God-intervention hypothesis saves the phenomena gets higher and higher, approaching 1 as a limit.

Accordingly, it is indeed possible for it to be highly probable that a hypothesis saves the phenomena while it is highly improbable that the hypothesis is true. One of the central conditions proposed above for e's being potential evidence that h saves the phenomena requires that

(i) $p_b(h$ saves the phenomena$/e) > \frac{1}{2}$.

One of the central conditions proposed for e's being potential evidence that h is *true* requires that

(ii) $p_b(h/e) > \frac{1}{2}$.

What I have been arguing is that (i) can be true even if (ii) is false.

Returning to the God-intervention hypothesis h above, let us consider a second central condition for e's being potential evidence that h saves the phenomena, viz.

(iii) $p_b($there is an explanatory connection between h's saving the phenomena and e/h saves the phenomena and $e) > \frac{1}{2}$.

Is this condition satisfied in the God-intervention case? Yes, it is. In light of the background information, it is very probable that the following correctly explains both why h saves the phenomena and why e is true (where e is the fact that the first n tosses were heads):

(iv) The reason that h saves the phenomena and that e obtains is that there is a magnetic field which causes heads on each toss and this allows h to save the phenomena, since h entails that each toss will result in heads.

Since (i) and (iii) obtain, since e and b are true, and since e does not entail that h saves the phenomena, it follows from the proposed definition **I** that, given b, e is potential evidence that h saves the phenomena. (I will assume that the product of the probabilities in (i) and (iii) is greater than $\frac{1}{2}$.) Yet since (ii) is false—since $p(h/e) << \frac{1}{2}$—it is not the case that, given b, e is potential evidence that h is true.

So far, then, an instrumentalist should be happy. The instrumentalist should want it to be the case that some e is potential evidence that h saves the phenomena without its being potential evidence that h is true.

What a committed instrumentalist will need to add is that, where h is a hypothesis about "unobservables," my conditions for potential evidence that h is true will never be satisfied. The most obvious such condition the instrumentalist will focus on is that

$$p_b(h/e) > \tfrac{1}{2}.$$

This condition he will (probably) argue is never satisfied for theoretical hypotheses, even though

$$p_b(h \text{ saves the phenomena}/e) > \tfrac{1}{2}$$

is satisfied for certain h's and e's. To be sure, if $p_b(h/e) < \frac{1}{2}$, then $p_b(-h/e) > \frac{1}{2}$. But then, where h is a theoretical hypothesis, all you get is that the probability such a theoretical hypothesis is *not* true is greater than $\frac{1}{2}$. Moreover, even in such cases, it does not follow that e is evidence that $-h$, since it does not follow that the explanation-condition (p(there is an explanatory connection between $-h$ and $e/-h$ & $e) > \frac{1}{2}$) is satisfied.

There is, however, a serious and I think insurmountable problem that remains for the instrumentalist who accepts my definition **I** for e's being potential evidence that h saves the phenomena. That definition requires that

$$p_b(\text{there is an explanatory connection between } h\text{'s saving the phenomena and } e/h \text{ saves the phenomena & } e) > \tfrac{1}{2}.$$

The problem is readily seen by focusing on (iv), the probable correct explanation of both why the God-intervening hypothesis h saves the phenomena and why the coin has landed heads each time. The explanation appeals to the truth of the competing hypothesis h', that there is a magnetic field which causes heads on each toss. An appeal to the probable truth of this ("theoretical") hypothesis is anathema to an instrumentalist. Moreover, the instrumentalist cannot offer as a probable explanation for both why h saves the phenomena and why e is true the fact that h' saves the phenomena. The fact that h' (the magnet hypothesis) *saves the phenomena,* even if it does, does not correctly explain why each toss so far has been heads (that is, *e*). By contrast, the *truth* of h' does correctly explain this.

Accordingly, an instrumentalist who seeks to define "*e* is potential evidence that *h* saves the phenomena" as I have done in **I** is confronted with an unhappy choice of alternatives. One is to admit that a claim that *e* is evidence that (theoretical hypothesis) *h* saves the phenomena commits one to the claim that *h* is probably true. No instrumentalist should want that. Another course of action is to drop my explanation condition in the definition **I** of potential evidence in favor of some other. What this alternative condition might be I have no idea. Without it, however, I doubt that the instrumentalist can formulate a plausible objective notion of evidence. Another possibility is to argue that since evidence that a theoretical hypothesis *h* saves the phenomena involves the idea that *h* is probably true, such evidence is impossible for a scientist to obtain; the result is scepticism not only for truth but for saving the phenomena as well. This instrumentalist response is too drastic to consider. Finally, an instrumentalist may seek to define "*e* is potential evidence that *h* saves the phenomena" in a way that is very different from what I have suggested, and that avoids both the realist presuppositions of that definition and the problems besetting other definitions I have considered. I invite the instrumentalist to provide such a definition.

9

Two Paradoxes of Evidence: Ravens and Grue

1. Introduction

Having developed my theory of evidence in the preceding chapters, two questions will be of concern to me in the chapters that follow. First, can the theory of evidence I am proposing cast light on important paradoxes and other issues that have been raised about evidence by philosophers? Second, can the theory be usefully applied to scientific cases?

The present chapter is devoted to Carl G. Hempel's paradox of the ravens[1] and Nelson Goodman's "new riddle of induction."[2] These paradoxes have been the bane of existence of philosophers of science for the past 50 years. In what follows I will use the theory of evidence I propose to offer solutions to each of these paradoxes.[3] Both solutions are based on the idea that whether the fact that all observed As are Bs counts as evidence that all As are Bs depends crucially on what selection procedure is employed in choosing As for observation.

Hempel and Goodman introduce these paradoxes using an objective concept of evidence. In what follows, except where otherwise indicated, my solution focuses on potential and veridical evidence.

2. Hempel's Paradox of the Ravens

This paradox is generated from three assumptions, each of which seems acceptable.

---

1. Carl G. Hempel, "Studies in the Logic of Confirmation," reprinted in Hempel, *Aspects of Scientific Explanation* (New York: Free Press, 1965), 3–51.
2. Nelson Goodman, *Fact, Fiction, and Forecast*, 4th ed. (Cambridge, MA: Harvard University Press, 1983).
3. For the ravens paradox I use some material from my *The Nature of Explanation* (New York: Oxford University Press, 1983), ch. 11.

First Assumption: If all the *A*s observed so far are *B*s (where these are numerous and varied), this fact is evidence that all *A*s are *B*s. (For example, if all the numerous and varied ravens so far observed are black, this is evidence that all ravens are black.)[4]

Second Assumption: If *e* is evidence that *h*, and if *h* is logically equivalent to *h'*, then *e* is evidence that *h'*.

Third Assumption: A hypothesis of the form "All *non-B*s are *non-A*s" is logically equivalent to "All *A*s are *B*s."

The paradox arises as follows. Each of the nonblack things we have observed (such as my brown shoes, the red rug on the floor, and the white walls in the room) is a nonraven. Therefore, by the First Assumption, the fact that all of these (and other such) nonblack things are nonravens is evidence that all nonblack things are nonravens. By the Third Assumption, the latter hypothesis is logically equivalent to the hypothesis that all ravens are black. Therefore, by the Second Assumption, the fact that all the observed nonblack things are nonravens is evidence that all ravens are black. This seems absurd, since it would mean that one could obtain evidence about the color of ravens without ever observing any ravens (or any birds, or any living things). Simply observing nonblack things such as my brown shoes, the red rug, and the white walls, and noting that they are nonravens, yields evidence that all ravens are black.

The solution I will offer is that normally when we obtain information of the sort just indicated concerning nonblack nonravens, our procedure for selecting nonblack things to observe is biased in a certain way. When it is biased in this way, then even if all the observed *A*s are *B*s (or all the observed *non-B*s are *non-A*s), this fact does not render the probability that all *A*s are *B*s (all *non-B*s are *non-A*s) greater than $\frac{1}{2}$. Hence, on my conception of (potential or veridical) evidence, this fact is not evidence that all *A*s are *B*s. In short, the First Assumption above is not universally valid.

I will speak of a *selection procedure* as a rule for determining how to test, or obtain evidence for or against, a hypothesis.[5] Here is a simple selection procedure for testing the hypothesis (*h*) that all ravens are black:

SP_1: Select ravens to observe from different locales and at different times of the year.

Some selection procedures are, I shall say, strongly biased. Here is a strongly biased selection procedure for testing *h*:

SP_2: Select ravens to observe from cages that contain only black birds.

4. This is a generalization of Hempel's original assumption, which he called Nicod's Criterion for Confirmation. That assumption is that an object confirms a universal conditional hypothesis of the form "for any *x*, if *x* has *P* then *x* has *Q*" if the object has both *P* and *Q*. This assumption allows a single instance of a generalization to confirm that generalization. If "confirmation" is construed as "evidence," and if the latter is understood as something that provides a good reason for believing the hypothesis, then Nicod's Criterion is much too strong. I wish to consider Hempel's paradox for a concept such as my potential evidence.
5. The idea of a selection procedure was introduced in the appendix to chapter 2.

This selection procedure is strongly biased in favor of black. Any ravens one observes using SP_2 are bound to be black. We can say

p(all observed ravens are black/ravens are selected for observation in accordance with SP_2) = 1,

where this probability, as well as others that follow, is construed as an objective epistemic one in the sense outlined in chapter 5. Disregarding any information about the color of ravens or about any selection procedure to be used, we might also say that the probability of the hypothesis h is much less than 1, that is,

p(all observed ravens are black) $<<$ 1.[6]

More generally, let us say that a selection procedure SP for choosing As to observe is strongly biased in favor of B if

(a) p(all observed As are Bs/As are selected for observation in accordance with SP) is close to, or equal to, 1;
(b) p(all observed As are Bs) is not close to, or equal to, 1.[7]

I will argue that if a selection procedure SP used for choosing As to observe is strongly biased in favor of B, then, in the case of many hypotheses of the form "All As are Bs,"

p(all As are Bs/all observed As are Bs) $< \frac{1}{2}$.

If this is so, then by my account of evidence (which requires a probability greater than $\frac{1}{2}$), when such a selection procedure is used, the fact that all observed As are Bs is not (potential or veridical) evidence that all As are Bs.

6. More precisely, using one of the special relativizations introduced in chapter 5, this can be written as

$$p_{d(c,s)}(\text{all observed ravens are black}) << 1,$$

where $d(c,s)$ means that facts about the color of ravens and about the selection procedure used are disregarded. As noted in chapter 5, this does not mean that we are supposing that ravens, or the ones observed, have no color or that no selection procedure was used in observing them. Nor does it mean that we are aware (or unaware) of such facts, or that all facts are being disregarded (which, I think, would result in an undefined probability). It is simply that whatever facts there are about the color of ravens or about the selection procedure used to observe them are being ignored.

It can still be the case that a conditional probability such as "$p_{d(c,s)}$(all observed ravens are black/3 observed ravens are black)" is defined. Here we disregard any actual facts about colors of ravens and determine how reasonable it is to believe that all ravens are black, *on the assumption that* 3 observed ravens are black.

7. As indicated in note 6, both of these probabilities involve a disregarding condition, which in (a) entails that facts about the color of ravens and about any actual selection procedures used are disregarded. (Although the probability in (a) is conditionalized to the assumption that selection procedure SP is being used, (a) does not claim that it is being used.)

There is a weaker notion of bias according to which SP is a selection procedure for choosing As to observe that is biased in favor of B if and only if p(all observed As are Bs/As are selected for observation in accordance with SP) > p(all observed As are Bs). That is, using SP increases the probability that all observed As are Bs. In what follows I will use the stronger notion, since in the ravens paradox the selection procedures we have in mind for choosing nonblack things to observe are strongly biased in favor of nonravens.

3. The General Argument

The argument proceeds by using Bayes' theorem in probability and making certain intuitive assumptions. First, we apply Bayes' theorem to hypotheses of the form "All As are Bs" and to As, selected in accordance with some selection procedure SP, that are found to be Bs. To simplify the notation, let

h = all As are Bs
e = all observed As are Bs
i = As are selected for observation in accordance with SP

According to Bayes' theorem,

(1) $p(h/e\&i) = \dfrac{p(h/i) \times p(e/h\&i)}{p(e/i)}$

One assumption we need to make is that $p(e/i)$ is not 0. This means that the selection procedure used to select As for observation cannot make it impossible that all the observed As are Bs.

Now h entails e, so that $p(e/h\&i) = 1$. Also, $p(h/i) = p(h)$, since the probability that all As are Bs should not be affected by what procedures are used to select As for observation, even though this can affect the probability that all observed As are Bs. Therefore, from Bayes' theorem (1),

(2) $p(h/e\&i) = p(h)/p(e/i)$.

Now suppose that SP is a selection procedure for choosing As that is strongly biased in favor of B. Then, by the definition of "strongly biased," p(all observed As are Bs/SP is used), that is, $p(e/i)$, is close to, or equal to, 1. So from (2),

(3) $p(h/e\&i)$ is approximately equal to $p(h)$.

That is, the probability that all As are Bs, given that all observed As are Bs and that the selection procedure used is strongly biased in favor of B, is approximately equal to the ("prior") probability that all observed As are Bs.

Suppose that h is a universal statement of the form "all As are Bs" whose prior probability is low, say significantly less than $\frac{1}{2}$, that is,

(4) $p(h) << \frac{1}{2}$.

This will frequently be the case, particularly when h is a nontautological generalization not restricted to some very small set of objects. For example, disregarding any information about the color of ravens, we might agree that the probability that all ravens are black is significantly less than $\frac{1}{2}$.

From (3) and (4) we obtain

(5) $p(h/e\&i) < \frac{1}{2}$.

That is, the probability that all As are Bs, given that all observed As are Bs and that As are selected for observation in accordance with a selection procedure strongly biased in favor of B, is less than $\frac{1}{2}$. If (5) is true, then, by my account of evidence, given that such a selection procedure is used (an assumption made as part of the background information), the fact that (e) all observed As are Bs is not evidence that (h) all As are Bs.

4. Application to the Ravens

We are now in a position to apply this to the ravens paradox. Let us return to the hypothesis that generated the paradox:

All nonblack things are nonravens.

Here is a selection procedure for nonblack things that is strongly biased in favor of nonravens:

SP_3: Select nonblack things to observe that are known in advance to be nonravens.

p(all observed nonblack things are nonravens/nonblack things are selected for observation in accordance SP_3) = 1, while the prior probability p(all observed nonblack things are nonravens) is not close to, or equal to, 1. Here are other selection procedures for nonblack things that are strongly biased in favor of nonravens:

SP_4: Select nonblack things to observe from inside your house.
SP_5: Select nonblack things to observe that are inanimate.

Given background information we possess indicating that it is extremely likely that nothing in your house, or inanimate, is a raven,

(6) p(all observed nonblack things are nonravens/nonblack things are selected for observation in accordance with SP_3 or SP_4 or SP_5) is close to, or equal to, 1.

Now,

p(all nonblack things are nonravens) = p(all ravens are black),

since the two hypotheses are logically equivalent. But assuming the prior probability p(all ravens are black) is significantly less than $\frac{1}{2}$, from (6) and the previous argument, where the selection procedure used is SP_3 or SP_4 or SP_5,

p(all nonblack things are nonravens/all observed nonblack things are nonravens) < $\frac{1}{2}$.

Accordingly, with selection procedures such as SP_3, SP_4, or SP_5 the fact that all the observed nonblack things are nonravens is not evidence that all nonblack things are nonravens. But in generating the paradox, I spoke of examining nonblack things such as my brown shoes, the red rug, and the white walls. These are items likely to be chosen by following strongly biased selection procedures such as SP_3 or SP_4 or SP_5, that is, by selecting nonblack things known in advance to be nonravens, or nonblack things inside your house, or nonblack things that are inanimate. With such selection procedures, the probability that all observed nonblack things are nonravens is close to 1, that is, (6) obtains.

This does not mean that all selection procedures for testing "all nonblack things are nonravens" are strongly biased in favor of nonravens. Consider

SP_6: Select nonblack things to observe from a large group consisting of nonblack birds of every species that has nonblack birds.

SP_7: Select nonblack things to observe from a group of nonblack birds that has been collected as follows: if nonblack ravens exist these comprise half the group; if they don't exist, then the group consists of nonblack birds other than ravens.

These selection procedures do not seem to be strongly biased in favor of nonravens, that is,

p(all observed nonblack things are nonravens/nonblack things are selected for observation in accordance with SP_6 or SP_7) is not close to, or equal to, 1.

For the sake of argument, assume that this is correct. However, even if SP_7 is free of bias, we are not in a position to follow it and know that we have. Suppose we obtain a group of nonblack birds with no ravens as part of the group. In the absence of information about the color of ravens, we do not know whether the group contains no ravens because no nonblack ones exist, or because the ones that do exist have not yet been found.

When we select nonblack things to observe in testing "all nonblack things are nonravens" it is much easier to follow, and know that we have followed, selection procedures such as SP_3, SP_4, and SP_5. It is easy to follow, and know that we have followed, a procedure that requires that we select a nonblack object known in advance to be a nonraven, or that we select nonblack things from inside one's house, or that are inanimate. But such selection procedures are strongly biased in favor of nonravens. So, using them, and obtaining the result that all the nonblack objects observed are nonravens, will not yield evidence that all nonblack things are nonravens.

By contrast, selection procedures SP_6 and SP_7 do not seem strongly biased. But using them requires much more knowledge than we are ever likely to have. Perhaps there are unbiased selection procedures that can be followed, and be known to be followed, for choosing nonblack things to observe in testing "all nonblack things are nonravens." What I am suggesting is that the ravens paradox, in which items such as brown shoes, the red rug, and the white walls yield evidence that all nonblack things are nonravens, is generated by using strongly biased selection procedures.

To be sure, in testing the hypothesis "all ravens are black" one can also employ a strongly biased selection procedure, such as

SP_2: Select ravens for observation from cages that contain only black birds.

If only SP_2 is used, then the fact that all the ravens observed are black would not count as evidence that all ravens are black. But there are also unbiased selection procedures, such as

SP_1: Select ravens to observe from different locales, and at different times of the year.

Not only is SP_1 unbiased, but it is a selection procedure that (unlike SP_6 and SP_7) we can follow and know that we have followed. That is why we do and should (normally) test the hypothesis "all ravens are black" by observing ravens to determine whether they are black rather than by observing nonblack things to determine whether they are nonravens.

Where e reports that all of the many and varied ravens observed are black, h is that all ravens are black, and SP_1 was used to obtain e, we can conclude that

$$p_{e \& SP_1 \text{ used to obtain } e}(h) > .5.$$

Even more strongly

$$p_{e \& SP_1 \text{ used to obtain } e}(\text{there is an explanatory connection between } h \text{ and } e) > .5,$$

since, given that (the unbiased) SP_1 was used to obtain e, it is very probable that the reason that all observed ravens are black is that all ravens are black; it is also very probable that some hypothesis (such as a genetic one) correctly explains both why all ravens are black and why the ones that happen to be observed are. Accordingly, e is potential evidence that h, given that SP_1 is used.

5. A Comparison with Hempel's Solution

How does this differ from Hempel's own solution to the paradox? According to Hempel, the reason that we do not take the fact that the brown shoes, the red rug, and the white walls are all nonravens as evidence that all nonblack things are nonravens is this: In selecting the brown shoes, the red rug, and the white walls for observation, we know in advance that these nonblack things are not ravens. So the information that they are nonravens adds nothing to what we already know. As Hempel puts it, "the outcome [of the observation] becomes entirely irrelevant for the confirmation of the hypothesis and thus can yield no new evidence for us."[8] Observing these items could provide evidence that all nonblack things are nonravens if it was not already known that they are nonravens. Suppose, for instance, that there is an opaque box with a sign on it saying "this box contains a nonblack item." And suppose, given the background information, there is no more reason to suppose it contains a raven than a nonraven. Then, opening the box and discovering that the nonblack item it contains is a nonraven (it might be a brown shoe) would, according to Hempel, provide some evidence that all nonblack items are nonravens.

Hempel's idea here is that something is evidence for a hypothesis only if it is "new evidence," only if it adds something to what we already know. Although Hempel does not employ probabilities to express this, his idea accords with the positive relevance conception of evidence discussed in chapter 3: e is evidence that h if and only if e increases h's probability. If background information b already includes the fact that this nonblack item is a brown shoe (and therefore a nonraven), the information (e) that it is a nonraven does not increase the probability that all nonblack things are nonravens over what it is on b alone.

In chapter 4 I gave reasons for rejecting the positive relevance definition as either necessary or sufficient for evidence. But my aim here is not merely to say that Hempel is espousing or suggesting an incorrect account of evidence. My aim is only to show that his resolution of the ravens paradox is different from mine.

---

8. Carl G. Hempel, *Aspects of Scientific Explanation*, p. 19.

I agree with Hempel that when selecting a nonblack item for observation in testing the hypothesis that all nonblack things are nonravens, we usually know more about it than that it is nonblack. (For example, we may know that it is a rug, and hence not a raven.) My point is not that this additional knowledge about the item makes the fact that it is a nonraven irrelevant since it adds nothing to the evidence (or because it fails to raise its probability). It is rather that in selecting things like shoes, a rug, or walls to observe we are probably following a selection procedure such as SP_3, SP_4, or SP_5 that is strongly biased in favor of nonravens. With such a selection procedure the hypothesis "all nonblack things are nonravens" fails to be more probable than not on the information that all the nonblack items selected for observation were nonravens. Hence the latter will not constitute (potential or veridical) evidence that this hypothesis is true.

6. Grue: Goodman's New Riddle of Induction

Goodman's great paradox begins with this piece of information

    e: All the emeralds so far examined are green.

The fact that this is so is surely evidence, indeed strong evidence, that

    h: All emeralds are always green.

Now define "grue" as follows:

    Definition: x is grue at time t if and only if t is prior to 2500 A.D. and x is green at t, or t is 2500 A.D. or later and x is blue at t.[9]

Since e is true, and since the emeralds so far examined have all been examined prior to 2500, the following is also true:

    e':   All the emeralds so far examined are grue.

So by parity of reasoning, e' should be evidence, indeed strong evidence, that

    h':   All emeralds are always grue.

But h' says that while emeralds before 2500 are green, emeralds beginning in 2500 are blue. And even though it is true that all emeralds observed so far are grue (because they are green and it is prior to 2500), this fact is not evidence that all emeralds are always grue, that is, green before 2500 and blue thereafter. That would be absurd! Can my theory of (potential) evidence help us to avoid this conclusion?

---

9. Goodman's original definition is that "grue" applies to all things examined before some specific time T just in case they are green, and to other things just in case they are blue. In our response to Goodman, Stephen Barker and I ("On the New Riddle of Induction," *Philosophical Review*, 69 (1960), 511–522) used the definition in the text (not Goodman's), except that (in 1960) we chose T to be the year 2000, which seems to have created a precedent for other writers on the subject. Since January 1, 2000 is now a memory, I have taken the liberty of pushing the date far into the future.

It can, if we can show that the probability of the grue hypothesis h', given the information e', is low, that is, $p(h'/e') < \frac{1}{2}$, while the probability of the green hypothesis, given the information e, is high, that is, $p(h/e) > \frac{1}{2}$. Note that in view of the definition of "grue," e and e' are equivalent: they are mutually entailing. So, in accordance with the rules of probability,

$p(h'/e') = p(h'/e)$.

Now, since h and h' are incompatible (assuming there are emeralds after 2500), if $p(h/e) > \frac{1}{2}$, then it follows that $p(h'/e) < \frac{1}{2}$. So if e is evidence that h, then e (and hence e') cannot be evidence that h'. Assuming that $p(h/e) > \frac{1}{2}$, is the explanation condition p(there is an explanatory connection between h and $e/h\&e) > \frac{1}{2}$ satisfied? Yes, it is. Given that all emeralds are always green and that all emeralds so far examined are green, the probability is very high that the reason that the latter is true is that the former is. So if $p(h/e) > \frac{1}{2}$, and the above explanation condition is satisfied by the green hypothesis, and the probabilities involved are sufficiently high so that their product is greater than $\frac{1}{2}$, then, since e does not entail h, e will be (potential) evidence that h. But if $p(h'/e) < \frac{1}{2}$, then even though the explanation condition is also satisfied by the grue hypothesis h', e will not be (potential) evidence that h'.

How can we show that $p(h'/e) < \frac{1}{2}$ while $p(h/e) > \frac{1}{2}$? One approach, suggested first by Carnap, and defended later by Barker and me, is that grue is a temporal property in the sense that a specific time, viz. 2500, is invoked in characterizing the property, whereas this is not so in the case of green. Moreover, induction works only for nontemporal properties, and not for temporal ones. In the present case this means that since h' attributes a temporal property (grue) to all emeralds, we cannot say that the probability of h', given that all examined emeralds satisfy h', is high. We cannot say that $p(h'/e') > \frac{1}{2}$. By contrast, since h attributes a nontemporal property (green) to emeralds, we can say that the probability of h, given that all examined emeralds satisfy h, is high, that is, $p(h/e) > \frac{1}{2}$. But if $p(h/e) > \frac{1}{2}$, then, since h' conflicts with h, $p(h'/e) < \frac{1}{2}$, and therefore $p(h'/e') < \frac{1}{2}$, since e is equivalent to e', given that all the emeralds so far examined have been examined before 2500. But on my account, if $p(h'/e') < \frac{1}{2}$, then e' is not (potential or veridical) evidence that h'.

This solution (which I no longer believe to be adequate) raises two important questions: What is a temporal property? And why should such properties not be "projected" (to use Goodman's term)? That is, why should we refuse to say that $p(h'/e') > \frac{1}{2}$? My answer to these questions, and my solution to the paradox, will be developed step by step in what follows.

7. My Solution to Grue: The First Step

To begin with, it is untrue that every property that mentions a specific time cannot be projected. Suppose, for instance, that there is a certain necktie produced by Harvard University emblazoned with the letters MCMLVI, and that all the owners of such ties that we have interviewed were graduated from Harvard in

1956. We might then legitimately infer that all owners of this type of tie were graduated from Harvard in 1956, despite the fact that "were graduated from Harvard in 1956" expresses a temporal property.

We need to be more selective with the temporal properties we say cannot be projected. Grue is a very special type of temporal property. It is a disjunctive one having this form:

(1) x has property P at time t if and only if x has property Q_1 at t and t is prior to a specific time T or x has property Q_2 at t and t is T or later.

This is not yet a sufficient characterization. Two provisos must be added. First, the properties Q_1 and Q_2 must be incompatible, for example, green and blue (something cannot have both at once). Second (this will become important later) the properties Q_1 and Q_2 (for example, green and blue) must not be thought of as disjunctive properties satisfying (1).[10]

Grue, and indeed any property of type (1), is a property of an even more general type that is not necessarily temporal. Consider disjunctive properties with this form:

(2) x has P if and only if x has Q_1 and condition C obtains or x has Q_2 and condition C does not obtain.

Again, the properties Q_1 and Q_2 must be incompatible. And they must not be disjunctive properties satisfying (2). In the grue case, Q_1 is green and Q_2 is blue. Condition C is that the time at which x has whatever color it has is before 2500. Goodman's paradox can be generated with respect to any property of type (2), whether temporal or not, if all the P's examined have been Q_1 and condition C obtains. For example, again let Q_1 be green, and Q_2 be blue, but let condition C be that x's temperature is less than some fixed value M. Suppose that all emeralds so far examined have been green and have been at temperatures below M degrees. Then if we project P with respect to emeralds, we generate a conclusion that entails that at temperatures reaching or exceeding M emeralds are blue. We generate the paradox even when the property P in question is nontemporal.[11]

How can Goodman's puzzle be resolved? Suppose that information e says that all P_1s so far examined are P_2, and hypothesis h states that all P_1s are P_2. Under what conditions can we infer that $p(h/e) > \frac{1}{2}$? That is, under what conditions can we project P_2 relative to P_1? Let us look at the grue case first.

We *can* project the property grue relative to the property of being an emerald, but only if evidence e reports on times both before and after T (2500), that is, only if e reports that the emeralds examined before T are grue (and hence green) and that emeralds examined at T or later are grue (and hence blue).

More generally, if P_2 (for example, grue) is a disjunctive property of types (1) or (2), with Q_1 and Q_2 (for example, green and blue) as disjuncts, and if P_1 (for example, being an emerald) is not a disjunctive property of types (1) or (2), and

---

10. A perceptive discussion of the disjunctive character of grue is found in David H. Sanford, "A Grue Thought in a Bleen Shade: 'Grue' as a Disjunctive Predicate," in Douglas Stalker, ed., *GRUE! The New Riddle of Induction* (Chicago: Open Court, 1994), 173–192.
11. See James Hullett and Robert Schwartz, "Grue: Some Remarks," *Journal of Philosophy*, 64 (1967), pp. 259–271.

if e reports that all the P_1s examined are P_2 in virtue of being Q_1 and none in virtue of being Q_2, then we cannot conclude that $p(h/e) > \frac{1}{2}$, where h is that all P_1s are P_2.

Consider the following two types of selection procedures for a hypothesis h of the form "All P_1s are P_2":

SP_1: Select P_1s to observe at times that are both before and after T. (Or more generally, for properties of form (2), select P_1s to observe that satisfy condition C and also ones that fail to satisfy C.)

SP_2: Select P_1s to observe at times that are only before T. (Select only P_1s to observe that satisfy C.)

Where P_1 and P_2 are properties of the sort described in the previous paragraph, $p(h/e) > \frac{1}{2}$ only if the selection procedure SP_1 is employed, not SP_2.

The basic idea derives from an injunction to "vary the instances." A disjunctive property P of the type depicted in (1) and (2) applies to two different sorts of cases: ones in which an item that is P (for example, grue) has property Q_1 (green) before time T (condition C is satisfied) and ones in which an item that is P has an incompatible property Q_2 (blue) at or after T (condition C is not satisfied). Since property P, when projected, is supposed to apply to items of both types, where these types are incompatible, items of both types need to be obtained as instances of the generalization. That is, SP_1 is to be followed, not SP_2.

For example, projecting the property grue, in the case of emeralds, requires that some emeralds be examined before 2500 to determine whether they are then green, and that some emeralds be examined after 2500 to determine whether they are then blue. Only if both of these determinations are made, and the emeralds examined before 2500 are green and those examined later are blue, can the resulting information e confer high probability on the hypothesis that all emeralds are grue.

Contrast this case with one in which the property green is projected with respect to emeralds. This property is not being construed as one that applies to two different sorts of cases: ones in which a green item has some nondisjunctive property Q_1 before time T and ones in which a green item has some incompatible nondisjunctive property Q_2 after T. So projecting the property green, in the case of emeralds, does not require that some emeralds be examined before T to determine whether they have such a property Q_1 and that some emeralds be examined after T to determine whether they have Q_2. Accordingly, it is not the case that only if both determinations are made and the emeralds examined before T have Q_1 while those examined after T have Q_2 can the resulting information (that all the examined emeralds are green) confer high probability on the hypothesis that all emeralds are green. In this case selection procedure SP_2 (as well as SP_1) can be used. One can select emeralds to observe at times that are only before some T, or at times that are before T and after.

The important point is not that grue, unlike green, is a temporal property. That is not enough to prevent grue from being projected. The important point is that grue is a certain type of disjunctive property, while green is not. To project this type of disjunctive property P with respect to some type of item, one needs to vary the instances observed by examining items that satisfy the condition C

and items that do not. (The Appendix to this chapter contains a probabilistic treatment of these claims.)

8. "I Object. Don't Forget Bleen!" (Nelson Goodman)

All of this is subject to what seems like a devastating objection. In characterizing a disjunctive property such as grue as one satisfying condition (1), or, more generally, (2), I indicated that the properties Q_1 and Q_2 must not be disjunctive properties satisfying (1) and (2). But there lies the rub, as Goodman gleefully points out. To illustrate the problem we define "bleen" as follows:

> Definition: x is bleen at time t if and only if t is prior to 2500 A.D. and x is blue at t or t is 2500 A.D. or later and x is green at t.

Now, thinking of grue and bleen as our basic properties, we can characterize the properties green and blue in a way that satisfies conditions (1) and (2):

(3) x is green at t if and only if t is prior to 2500 A.D. and x is grue at t or t is 2500 A.D. or later and x is bleen at t.

(4) x is blue at t if and only if t is prior to 2500 A.D. and x is bleen at t or t is 2500 A.D. or later and x is grue at t.

Looking at the properties green and blue this way and treating grue and bleen as our basic nondisjunctive properties, green and blue become disjunctive properties satisfying conditions (1) and (2). Accordingly, to project green with respect to emeralds, we need to examine emeralds both before and after 2500. We must use selection procedure SP_1 and not SP_2. This directly contradicts what was said earlier.

So where do we stand? Can the property grue be projected with respect to emeralds by examining only emeralds before 2500? In gathering information that will be evidence for or against the hypothesis that all emeralds are grue, can we use SP_2 and select emeralds to observe at times that are only before 2500? Similarly can the property green be projected with respect to emeralds only by examining emeralds both before and after 2500, that is, by following only SP_1?

My answer is that *for us*, that is, for normal human beings, green and blue are not disjunctive properties of types (1) and (2) subject to a temporal condition, while grue and bleen are. What I mean by this is explained as follows:

(a) For us, the properties green and blue are not defined in the disjunctive way given above. Our dictionaries do not define the terms "green" and "blue" in terms of "grue" and "bleen" and a specific time. Nor do dictionaries in other languages with words for the properties blue and green. Indeed, the dictionaries I own do not even contain the words "grue" and "bleen."

(b) When we attempt to ascertain whether something we are examining is green (or blue) at a certain time t we do not, and do not need to, ascertain whether it is grue at t and t is before 2500 or whether it is bleen at t and t is 2500 or later. For example, if it is within five minutes of midnight, one way or the other, December 31, 2499, but we do not know which, and we are presented with a colored object, we could examine it and determine whether it is then green (or blue) without knowing whether midnight has passed.

By contrast,

(c) For us, the properties grue and bleen are defined disjunctively in the manner of (1) and (2) and are subject to a temporal condition. We understand these properties only by reference to such definitions.

(d) When we attempt to ascertain whether something is grue (or bleen) at a certain time t, we need to ascertain whether it is green at t and t is before 2500 or whether it is blue at t and t is 2500 or later. For example, if it is within five minutes of midnight, one way or the other, December 31, 2499, but we do not know which, and we are presented with a colored object, by examining it we could not determine whether it is then grue (or bleen) without knowing whether midnight has passed.

We might, however, imagine some extraordinary group of individuals very different from us in the following respects:

(a)' For members of this group the properties grue and bleen are not defined disjunctively in the manner of (1) and (2). Their dictionaries do not define "grue" and "bleen" in terms of "green" and "blue" and a specific time. Indeed, their dictionaries do not even contain the words "green" and "blue."

(b)' When members of this extraordinary group attempt to ascertain whether something they are examining is grue (or bleen) at a certain time t, they do not, and do not need to, ascertain whether it is green at t and t is before 2500 or whether it is blue at t and t is 2500 or later. For example, if it is five minutes before or after midnight, December 31, 2499, but they do not know which, if they are presented with a colored object they could determine whether it is then grue (or bleen) without knowing whether midnight has passed.

(c)' For them, the properties green and blue are defined in the manner of (3) and (4). They understand these properties only by reference to such definitions.

(d)' When they attempt to ascertain whether something is green (or blue) at a certain time t, they need to ascertain whether it is grue at t and t is before 2500, or whether it is bleen at t and t is 2500 or later. If it is five minutes before or after midnight, December 31, 2499, but they do not know which, and they are presented with a colored object, by examining it they could not determine whether it is then green (or blue) without knowing whether midnight has passed.

It may be useful to draw an analogy with a different sort of case involving a disjunction that is nontemporal but is different from ones that can spawn Goodman's paradox. Suppose there is an extraordinary group of persons who have a word in their language for male robins and a different word for female robins, but no word for robins. (Perhaps they regard male robins and female ones as belonging to different species.) Using their words for "male robin" and "female robin" we can then define our word "robin" for them, as follows:

x is a robin if and only if x is a male robin or a female robin.

This is how they will understand the word "robin," which is new for them. Moreover, when members of this group attempt to ascertain whether something is a robin they will determine whether or not it is a male robin, and if it is not, whether or not it is a female robin. If it is one or the other it is a robin; if it is neither it is not a robin. For them, but not us, "robin" is a sex-linked term.

In the case of grue, what we are imagining is that for members of the extraordinary group the properties green and blue are disjunctive ones subject to a temporal condition, while grue and bleen are not. We have no idea how they do what they do, in particular how they determine whether something is grue at a certain time t without knowing whether t is before 2500 or later. Nor do we have any idea why, in order to determine whether something is green at a certain time t they need to know whether t is before or after 2500. We are imagining simply that these things are so.

My claim is that if there were (or could be) such extraordinary people, they would be justified in projecting the property grue with respect to emeralds after examining emeralds before 2500; they would not need to wait until 2500 to examine emeralds then as well. They would be justified in using selection procedure SP_2. And if there were such extraordinary people they would be justified in projecting the property green with respect to emeralds only by examining emeralds both before and after 2500, that is, by using SP_1.

However, the claim is not that there *are* people such as the extraordinary ones being imagined. Nor is it that there *could be*, in some robust sense of "could be." It may not be physically possible. The claim is only that it is *logically* possible.[12] There is no contradiction (or at least I have not found one) in imagining the existence of extraordinary persons satisfying conditions (a)′–(d)′. Accordingly, there is no contradiction in supposing the existence of extraordinary persons who are justified in projecting the property grue with respect to emeralds after examining only emeralds before 2500. However, we are not such extraordinary people and there is no reason to believe that any such people exist, or physically speaking, could exist.

We need to consider what all of this means for evidence. (Is the fact that all emeralds observed so far are green evidence that all emeralds are always green?) Before turning to this, however, I want to contrast what I have said so far with Goodman's resolution of the paradox.

9. A Contrast with Goodman's Solution

In offering his solution, Goodman, like me, allows the possibility (whether logical or physical) that persons exist who are justified in projecting the property grue with respect to emeralds after examining emeralds only before 2500.

Briefly, Goodman's solution is based on the idea that the term "green" is much better *entrenched* than "grue." The term "green" (as well as other terms true of the same class of things) has been used much more frequently than "grue" (or other coextensive terms) in hypotheses of the form "All *A*s are *B*s" that have actually come to be adopted. Goodman's question is this: When is a hy-

---

12. Here I disagree with Judith Thomson, who claims that it is not even logically possible. See Judith Jarvis Thomson, "Grue," *Journal of Philosophy*, 63 (1966), 289–309.

pothesis of the form "All *As* are *Bs*" projectible, that is, when is it confirmed by instances consisting of reports that particular *As* are *Bs*?

Suppose that two conflicting hypotheses "All *As* are *Bs*" and "All *As* are *Cs*" are such that all their examined instances are true. But suppose that the term *B* is much better entrenched than the term *C*. Then, according to Goodman, the hypothesis "All *As* are *Cs*" is not projectible. It receives no confirmation from its instances. Thus, although all the examined instances of the hypothesis "All emeralds are grue" are true, that hypothesis does not receive confirming support from those instances. The reason is that this hypothesis is "overridden" by the conflicting hypothesis "All emeralds are green" which (up to now) has equal numbers of examined instances but uses the better entrenched term "green" and conflicts with no hypotheses with still-better entrenched terms. Under these circumstances examined instances of green emeralds confirm the hypothesis that all emeralds are green, whereas examined instances of grue emeralds fail to confirm the hypothesis that all emeralds are grue.

On this solution it is at least logically, if not physically, possible that persons exist for whom examined instances of grue emeralds confirm the hypothesis that all emeralds are grue, whereas examined instances of green emeralds fail to confirm the hypothesis that all emeralds are green. For such persons "grue" would be a better entrenched term than "green." It would be used more frequently than "green" by such persons in hypotheses of the form "All *As* are *Bs*" that have actually come to be adopted by such persons. So, for such persons, the hypothesis "All emeralds are green" would be overridden by the hypothesis "All emeralds are grue" which (until now) has equal numbers of examined instances but uses what is for them the better entrenched term "grue" and conflicts with no hypotheses with still better entrenched terms.

Goodman's solution appeals to entrenchment. Although all the emeralds examined so far are both green and grue, "green" is a much better entrenched term. It appears much more frequently than "grue" in hypotheses of the form "All *As* are *Bs*" that we have come to accept. This claim I do not want to deny. My question, however, is why this is so. Why have we accepted generalizations of the form "All *As* are green" or "All green things are *B*" much more frequently than "All *As* are grue" and "All grue things are *B*"? My solution offers an answer. (Goodman simply accepts that this is so.)

The answer is that for us grue is a disjunctive property of types (1) and (2) of section 7, whereas green is not. (For us, conditions (a)–(d) of section 8 hold.) Accordingly, for us, to generalize from examined instances of grue to hypotheses of the form "All *As* are grue" and "All grue things are *B*" (where *A* and *B* are not for us disjunctive properties of types (1) and (2)), we need to examine *As* (for "All *As* are grue") and grue things (for "All grue things are *B*") *both before and after 2500*. Since for us grue is a disjunctive property of type (2), in order to generalize we need to vary the instances and examine both things that satisfy the condition *C* in a property of type (2) and things that fail to satisfy *C*. Since green is not for us a disjunctive property of types (1) and (2), in order to generalize from examined instances of green to hypotheses of the form "All *As* are green" and "All green things are *B*" (where *A* and *B* are not for us disjunctive properties of types (1) and (2)) we do not need to examine *As* (for "all *As* are green") and green things (for "All green things are B") both before and after 2500, or ones that satisfy some corresponding condition C and others that fail to.

Accordingly, my solution is not based on the idea of entrenchment, which is really an idea about terms used in generalizations we have come to accept. It is based on the idea that for us, because grue is a disjunctive property of a certain sort, whereas green is not, in order to generalize from examined cases of emeralds that are grue, we need to examine emeralds that satisfy one side of the disjunction and emeralds that satisfy the other.

10. So Is It Evidence Or Isn't It?

Is the fact that

(e) All emeralds examined so far are green

evidence that

(h) All emeralds are always green

or isn't it?

Consider the simplest case first, subjective evidence. We green speakers (the normal folks who satisfy (a)–(d) of section 8 with respect to green and grue) believe that e is (veridical) evidence that h and that h is true; our reason for believing that h is true is that e is. In short, the fact that e is true is our subjective evidence that h is true. Similarly, if grue speakers existed (extraordinary but imaginary beings who satisfy (a)'–(d)') the fact that all the emeralds examined so far are green (and hence grue) would be their subjective evidence that all emeralds are always grue.

In section 8 I claimed that a grue speaker is justified in projecting the property grue, while a green speaker is justified in projecting the property green. This is a case of ES-evidence, where the epistemic situation is understood as including a speaker's knowledge of definitions and of how to ascertain whether something is grue or green. A grue speaker's ES-evidence that all emeralds are always grue would be that all emeralds so far examined are grue. A green speaker's ES-evidence that all emeralds are always green is that all emeralds so far examined are green.

The crucial question concerns potential evidence. Is e potential evidence that h? Now, in fact, e can be understood in two ways: all the emeralds examined so far have been determined to be green; all the emeralds examined so far are in fact green (whether or not this has been determined). I discuss the first in the present section and the second in the one that follows.

Whether e, understood in the first way, is potential evidence that h depends on the selection procedure used. I have characterized a selection procedure as a rule for determining how to test, or obtain evidence for or against, a hypothesis. In the case of our hypothesis h a selection procedure might include a rule for selecting emeralds to observe. If, for instance, such a rule called for selecting emeralds only from a box containing green objects, then e would not be potential evidence that h. But a selection procedure for our hypothesis h may also include a rule for how to determine whether an emerald is green at a given time t. Many such rules are possible, but let me concentrate on two.

$SP(green)_1$: Determine whether an emerald is green at a time t simply by looking at it at t, in good light, at a distance at which it can be seen clearly (etc.), and ascertaining whether it looks green.

$SP(green)_2$: Determine whether an emerald is green at t by looking at it at t and ascertaining whether it looks grue at t (the way our imagined grue speaker does) and t is prior to 2500 or whether it looks bleen at t and t is 2500 or later.

Suppose that all the emeralds selected for observation so far (before 2500) have been determined to be green. Is that fact potential evidence that all emeralds are always green? That depends not only on which selection procedure was used to select emeralds for observation but on which one was used to determine whether an emerald is green. Suppose that $SP(green)_2$ was used (for example, by genuine grue speakers, who could not use $SP(green)_1$). Someone following $SP(green)_2$ and examining only emeralds before the year 2500 to determine whether they look grue and hence are green would need to wait until 2500 to examine emeralds to determine whether they look bleen after 2500 and hence are green. Such a person would need to do this in order to "vary the instances" to obtain genuine potential evidence that all emeralds are always green. If $SP(green)_2$ is really the selection procedure that was used for determining whether an emerald is green, then the probability that all emeralds are always green, given that all the emeralds examined so far are determined to be green, is not high. That is, $p(h/e\ \&\ SP(green)_2$ is used$) < \frac{1}{2}$. Accordingly, given the use of $SP(green)_2$, e is not potential evidence that h.

By contrast, someone following $SP(green)_1$ and examining only emeralds before 2500 would not need to wait until 2500 to examine emeralds to determine whether they look bleen after 2500 and hence are green. The date 2500 plays no role in following $SP(green)_1$ the way it does in following $SP(green)_2$. If $SP(green)_1$ were followed, then the probability that (h) all emeralds are always green, given that (e) all emeralds so far examined are determined to be green, would be high. Since, as noted in section 6, the explanation condition for potential evidence is satisfied, as is the condition that e does not entail h, if the probabilities are sufficiently high, then, assuming $SP(green)_1$ were used, e would be potential evidence that h.

Now, as Goodman loves to do, let us compare the situation with respect to the grue hypothesis. The question is whether

e': All emeralds examined so far are determined to be grue

is potential evidence that

h': All emeralds are always grue.

We need to say what selection procedure is being used. By analogy with the previous ones for green emeralds, we have

$SP(grue)_1$: Determine whether an emerald is grue at t simply by looking at it at t in good light (etc.) and ascertaining whether it looks grue.

$SP(grue)_2$: Determine whether an emerald is grue at t by looking at it at t and ascertaining whether it looks green at t, where t is prior to 2500, or whether it looks blue at t and t is 2500 or later.

If $SP(grue)_2$ is used in obtaining the result e', then e' is not potential evidence that h'. Someone (such as us) following this selection procedure and examining only emeralds before 2500 to determine whether they look green and hence are grue would need to wait until 2500 to examine emeralds to determine whether they look blue and hence are grue. Such a person would need to do this in order to "vary the instances" to obtain genuine potential evidence that all emeralds are always grue. Using $SP(grue)_2$, examining emeralds only before 2500 would not yield a sufficiently high probability for the grue hypothesis.

By contrast, a genuine grue speaker following $SP(grue)_1$ and examining emeralds only before 2500 would not need to wait until 2500 to examine emeralds to determine whether they look blue and hence are grue. Using $SP(grue)_1$, examining emeralds only before 2500 and determining that all of them are grue would yield a sufficiently high probability for $p(h'/e')$ to allow e' to be potential evidence that h'.

In short, (e) the fact that all emeralds observed so far are determined to be green is potential evidence that (h) all emeralds are always green, if selection procedure $SP(green)_1$ is used, but not $SP(green)_2$. And (e') the fact that all emeralds observed so far are grue is potential evidence that (h') all emeralds are always grue, if selection procedure $SP(grue)_1$ is used, but not $SP(grue)_2$. It should be emphasized that this is not to relativize the concept of potential evidence to a particular person or to a type of epistemic situation. Given that $SP(green)_1$ is used e is potential evidence that h, and given that $SP(grue)_1$ is used e' is potential evidence that h', independently of who believes what. Nor are e and e' just potential evidence for persons in epistemic situations of certain types.

Now, as a matter of fact, there are no grue speakers, that is, extraordinary persons who satisfy conditions (a)′–(d)′ of section 8 for defining and identifying grue and green properties. There are just ordinary, everyday people like us, who satisfy conditions (a)–(d). So, even if it is logically possible that a selection procedure such as $SP(grue)_1$ is followed in determining whether an emerald is grue, and that a selection procedure such as $SP(green)_2$ is followed in determining whether an emerald is green, this will never happen (we confidently believe). In any real-life situation in which selection procedures involve actual observations of emeralds $SP(green)_1$ and $SP(grue)_2$ will be followed, in which case the fact that all observed emeralds are determined to be green will be potential evidence that all emeralds are always green, and the fact that all observed emeralds are determined to be grue will not be potential evidence that all emeralds are always grue. If (as we also confidently believe) the green hypothesis is true and the reason that all observed emeralds have been determined to be green is that all emeralds are green, then the fact that all observed emeralds are determined to be green is veridical evidence for this hypothesis (in the strong sense of chapter 8). If the hypothesis turns out to be false, or even if true it is not the reason that all observed emeralds are determined to be green, then the fact about the observed emeralds is not veridical evidence that all emeralds are green.

11. Grue Looks

Let us turn to the second interpretation of

(e) All the observed emeralds examined so far are green,

according to which this is true even if the emeralds examined have not been determined to be green (for example, they were not examined for color). Suppose, indeed, that none of the ones examined was determined to be green, although all of the ones examined are in fact green. Accordingly, no selection procedure such as $SP(green)_1$ or $SP(green)_2$ was employed. Under these conditions, is the fact that e is true potential evidence that h?

Here we need to look at what lies behind selection procedures for green, such as $SP(green)_1$ and $SP(green)_2$. If the first of these can be used, that is because green things have green looks whose presence at a time can be ascertained (by certain individuals) without determining whether the time is before or after 2500. (These will be called "green nontemporal looks" or "green looks," for short.) If the second of these selection procedures can be used, that is because green things have grue looks whose presence at a time can be ascertained (by certain extraordinary individuals) without determining whether the time is before or after 2500. (These will be called "grue nontemporal looks" or "grue looks," for short.) Let us assume that an emerald is green at t if and only if it has a green ("nontemporal") look at t, or a grue ("nontemporal") look at t, where t is before 2500, and a bleen ("nontemporal") look, where t is 2500 or later. Then all emeralds examined so far (that is, before 2500) are green if and only if any emerald examined so far has a green look, or a grue look, or both.

Now consider three possibilities:

(1) All the emeralds examined so far have green ("nontemporal") looks, but no grue ("nontemporal") looks (whether or not anyone has in fact observed those "looks"). Let this be denoted by $Gn\&-Gr$. Then it is more reasonable than not to believe that all emeralds have green looks while none has a grue look, that is,

$p_{Gn\&-Gr}$(all emeralds have green looks and no emeralds have grue looks) $> \frac{1}{2}$.

Since the emeralds examined so far all have green looks and none has a grue look, there is no need to vary the instances to include emeralds examined before 2500 with grue looks and emeralds examined in 2500 or later with bleen looks. From the probability statement above it follows that

$p_{Gn\&-Gr}$(all emeralds have green looks) $> \frac{1}{2}$,

and hence that

$p_{Gn\&-Gr}$(all emeralds are green) $> \frac{1}{2}$.

In this case, where h is that all emeralds are green, it can also be concluded that $p_{Gn\&-Gr}$(there is an explanatory connection between h and $Gn\&-Gr$) $> \frac{1}{2}$. Given that all examined emeralds have green looks and none has a grue look, the probability is high that the reason for this is that all emeralds are green and none are grue. Accordingly, the fact that all examined emeralds have green looks and none have grue looks is potential evidence that all emeralds are green.

(2) All the emeralds examined so far have grue ("nontemporal") looks, but no green ("nontemporal") looks: $-Gn\&Gr$. This fact would not be potential evidence that all emeralds are green. The reason is that the examined instances are not varied with respect to "looks." If all the emeralds examined so far have grue looks, then none has a bleen look. But, we said, x is green at t if and only if either x has a green look at t, or x has a grue look at t, where t is prior to 2500, or a bleen look at t, where t is 2500 or later. If no emerald examined so far is green in virtue of having a green look, then such emeralds must be green in virtue of having a grue look at t where t is before 2500. (Here I exclude grue-like properties with switching dates different from 2500.) But then, since no examined emeralds are green in virtue of having bleen looks in 2500 or later, the instances must be varied to include emeralds examined in 2500 or later that have bleen looks. This must be so for the degree of reasonableness of belief to be high that all emeralds (including the unexamined ones) have grue looks if the date is before 2500 or bleen looks if the date is 2500 or later, and hence that all such emeralds are green. But since in this case the instances are not varied, the degree of reasonableness of belief is not high.

(3) All the emeralds examined so far have both green ("nontemporal") looks and grue ("nontemporal") looks: $Gn\&Gr$. Now it is impossible for all emeralds at all times to have both green and grue looks, since then some emeralds (viz. ones examined after 2500) would be both green and blue. That is,

$p_{Gn\&Gr}$(all emeralds have green and grue looks) = 0.

Since, under the present assumption, all the examined emeralds have both looks, we may conclude that

$p_{Gn\&Gr}$(All emeralds have green looks) = $p_{Gn\&Gr}$(all emeralds have grue looks),

which entails that both $p_{Gn\&Gr}$(all emeralds are green) and $p_{Gn\&Gr}$(all emeralds are grue) are less than or equal to $\frac{1}{2}$. Accordingly, the fact that $Gn\&Gr$, that is, that all examined emeralds have both green and grue looks, could not be potential evidence that all emeralds are green and it could not be potential evidence that all emeralds are grue.

So the question about potential evidence comes down to deciding which, if any, of the following is true.

(1) $Gn\&-Gr$: all emeralds examined so far have green looks, none have grue looks.
(2) $-Gn\&Gr$: all emeralds examined so far have grue looks, none have green looks.
(3) $Gn\&Gr$: all emeralds examined so far have both green and grue looks.

We (green-speakers) know that (2) is false, since we have examined at least some emeralds for color and observed their green looks. That leaves (1) and (3). We know that all the emeralds we have examined for color have green looks; let us suppose that all the emeralds we have examined have green looks (whether or not we have examined them all for color). The question then is this: do any of the ones examined by anyone (including grue speakers, if they exist) also have grue looks? We green speakers haven't observed any such looks. But they might still be there, even on the emeralds we have examined.

So all that can be said is this:

Gn is potential evidence that all emeralds are green, *if* −Gr is true. (I assume green speakers know that Gn is true.)

Is −Gr true? We have observed no grue ("nontemporal") looks. Nor have we observed any grue speakers who claim to have observed such looks, or act as if they do. Given our epistemic situation, the fact that all examined emeralds have green looks, and hence are green, is at least ES-evidence that all emeralds are green. If, as we strongly believe and are justified in believing, there are no ("nontemporal") grue looks associated with examined emeralds, the fact that all examined emeralds have green looks, and hence are green, is potential evidence that all emeralds are green. The ES-evidence claim here is relativized to our epistemic situation. But the potential evidence claim above is not. It is conditionalized not to an epistemic situation but to the fact that no examined emeralds have grue ("nontemporal") looks. If that claim is true, a corresponding unconditional potential evidence claim is also true. If it is false, such a potential evidence claim is false.

12. Green Meets Grue

Suppose that grue ("nontemporal") looks exist, as do grue speakers, and a green speaker meets a grue speaker and hires him as an assistant to examine emeralds to see if they are green. The grue speaker uses $SP(green)_2$ and determines that all emeralds so far examined are green. Despite the use of $SP(green)_2$ by the assistant, it would seem that the green speaker can use this report to project "green" and claim that the assistant's report is potential evidence that all emeralds are green. But if he can, isn't the green speaker's selection procedure for the property green that of the grue speaker, viz. $SP(green)_2$? And if it is, doesn't this mean that, contrary to what I said earlier, the green speaker cannot project "green" based on the assistant's report?[13]

To begin with, in the imagined case the green speaker is not using $SP(green)_2$ but something like

$SP(green)_3$: Determine whether an emerald is green at t by asking a reliable grue speaker to examine the emerald and determine whether it is green using $SP(green)_2$.

However, there is a deeper point to be made in response. As emphasized in the last section, it is possible to follow $SP(green)_1$ because emeralds have green nontemporal looks. Similarly, it would be possible to follow $SP(green)_2$ if emeralds have grue nontemporal looks. Suppose that emeralds have grue looks as well as green looks whose presence at a time can be ascertained without determining whether that time is before or after 2500. And suppose that 100 emeralds have been examined, 50 by a green speaker using $SP(green)_1$ and 50 by a grue speaker using $SP(green)_2$. Let e_1 be the information that the 50 emeralds examined by the

---

13. I owe this objection to a persistent critic.

green-speaker were determined to be green (and hence are grue). Let e_2 be the analogous information about the grue-speaker. Then, using epistemic probability and disregarding other information, we may conclude that

p(all emeralds are green/e_1 & $SP(green)_1$ is used to obtain e_1 & e_2 & $SP(green)_2$ is used to obtain e_2) = p(all emeralds are grue/e_1 & $SP(green)_1$ is used to obtain e_1 & e_2 & $SP(green)_2$ is used to obtain e_2).

That is, assuming the facts in question—e_1 & $SP(green)_1$ is used to obtain e_1 & e_2 & $SP(green)_2$ is used to obtain e_2—the green hypothesis would be no more probable than the grue one. Accordingly, if these facts obtained (which entail that all examined emeralds are determined to be green), they would not be potential evidence for either the green or the grue hypothesis.

Let $A_1 = SP(green)_1$ is used, and $A_2 = SP(green)_2$ is used. Consider the following probability schema:

(1) $p_{A1 \& A2}$(all emeralds are green/$e_1 \& e_2$) = r.

Suppose there are, and can be, no grue speakers because there are no grue looks whose presence at a time t can be ascertained without determining whether t is before or after 2500. Then A_2 is always false, and (1) is false for every r. The same holds for

(2) $p_{A1 \& A2}$(all emeralds are grue/$e_1 \& e_2$) = r.

We green speakers believe that there are no grue speakers who can use $SP(green)_2$, because we believe that there are no grue nontemporal looks. Moreover, we believe that there are no nontemporal "grue-type" looks (looks like putative grue looks except for a different temporal switching date).

So, to return to the original example in this section, a green speaker believes he could not hire a grue-speaking assistant. Suppose this belief is correct. Then

$p_{A2 \& e2}$(all emeralds are green) > $\frac{1}{2}$

is false, since $A_2 \& e_2$ are false for all grue-type properties. What is true is that

(3) $p_{A1 \& e1}$(all emeralds are green) > $\frac{1}{2}$.

So if, as we confidently believe, there are no grue (or grue-type) looks whose presence can be ascertained "nontemporally," and hence there can be no grue speakers who employ $SP(green)_2$, then the fact that A_1 and e_1 are true allows e_1 to be potential evidence that all emeralds are green. On the other hand, if A_2 and e_2 are true—if there are grue looks and grue speakers who ascertain that all the emeralds they have examined using $SP(green)_2$ are green, then e_1 is not potential evidence that all emeralds are green.

13. Ravens and Grue: Birds of a Feather?

In generating the ravens paradox a strongly biased selection procedure is used. When nonblack items observed around the house, such as my brown shoes, the red rug, and the white walls, are found to be nonravens, the selection procedure being used for nonblack things is strongly biased in favor of nonravens. Can a similar point be made in the case of grue?

Let the background information b contain the fact that all the emeralds examined so far are determined (in the normal way) to be green. Consider this selection procedure:

SP: Select emeralds for observation only before 2500.

Let us suppose that the following probability claims are true:

(1) p(all emeralds that have been or will be selected for observation are grue/emeralds are and will be selected for observation in accordance with SP, and b is the case) is close to, or equal to, 1.
(2) p(all emeralds that have been or will be selected for observation are grue/b is the case) is not close to, or equal to, 1.

According to claim (1), given that emeralds are and will be selected for observation only before 2500 and that all the emeralds selected so far are determined (in the normal way) to be green, the probability is close to 1 that all the emeralds that have been or will be selected for observation will be green and hence grue (since they will all be selected before 2500). Whereas, according to (2), given that all the emeralds selected so far are determined (in the normal way) to be green, but there is no time restriction on when other emeralds will be selected (such as in SP), the probability that all emeralds selected will be grue is not close to 1. Therefore, on the information b, the selection procedure SP is strongly biased in favor of grue.

Let us compare this to the green hypothesis, and assume that the following probability claims are true:

(3) p(all emeralds that have been or will be selected for observation are green/emeralds are and will be selected for observation in accordance with SP, and b is the case) is close to, or equal to, 1.
(4) p(all emeralds that have been or will be selected for observation are green/b is the case) is close to, or equal to, 1.

According to claim (3), given that emeralds are selected for observation only before 2500 and that all the emeralds selected so far are determined (in the normal way) to be green, the probability is close to 1 that all emeralds that have been or will be selected in this way will be green. Indeed, if (1) is true, so is (3). According to claim (4), however, given that all the emeralds selected so far are determined (in the normal way) to be green but there is no time restriction such as the one in SP, the probability that all emeralds selected will be green is still close to 1. So, given b, although SP is strongly biased with respect to grue, it is not strongly biased with respect to green.

To be sure, a grue speaker can respond by changing the background information by substituting the fact that (b') all the emeralds examined so far are determined to be green by the extraordinary method used by grue speakers (that is, by following $SP(green)_2$ of the previous section). Then, by analogous reasoning, it can be shown that, given b', SP above is strongly biased with respect to green but not grue.[14] But so far as we know, no one does or can determine whether emeralds are green in such an extraordinary way. Given the normal way of determining whether emeralds are green (and grue), and given that all the emeralds examined so far are determined to be green, the selection procedure SP is strongly biased in favor of grue but not green.

14. Appendix

Suppose we are dealing with a disjunctive property P of type (2), section 7:

> x has P if and only if x has Q_1 and condition C obtains or x has Q_2 and condition C does not obtain,

where Q_1 and Q_2 are incompatible and not themselves disjunctive properties of this type. Suppose that all items of a certain type A (such as emeralds) so far examined have property P. Under what conditions do examined As need to be varied, and what sort of variation is required, before we can project the property P to all As?

Let $p(P/A)$ represent the probability that an arbitrarily selected A has property P. If P is a disjunctive property of type (2), then

$$p(P/A) = p((Q_1 \& C) \text{ or } (Q_2 \& -C)/A),$$

where the probability on the right side is the probability that an arbitrary A either has Q_1 and condition C obtains or it has Q_2 and condition C does not obtain. Since these disjuncts are mutually exclusive,

$$p(P/A) = p(Q_1 \& C/A) + p(Q_2 \& -C/A).$$

By the multiplication rule of probability,

(i) $\quad p(P/A) = p(C/A) \times p(Q_1/C \& A) + p(-C/A) \times p(Q_2/-C \& A).$

Suppose that $p(C/A) = 1$, and hence $p(-C/A) = 0$. In this case, for an arbitrarily selected A, the probability that condition C obtains is 1. Here there is no possibility of varying the instances by examining As when condition C obtains and when it does not. In such a case, $p(P/A) = p(Q_1/C \& A)$. So if all the As observed so far have Q_1 and condition C obtains, we might conclude that $p(Q_1/C \& A)$ is close to 1, and hence so is $p(P/A)$. Here observing As only under condition C and finding them all to have Q_1 suffices to render it highly probable that As have the disjunctive property P.[15]

Suppose that $p(C/A) < 1$. Then we can show that $p(P/A) = p(Q_1/C \& A)$ only if $p(Q_1/C \& A) = p(Q_2/-C \& A)$. Let $a = p(C/A)$, so that $p(-C/A) = 1-a$. Let $b = p(Q_1/C \& A)$; let $c = p(Q_2/-C \& A)$. Then, from the multiplication rule (i),

---

14. In (2) if b is replaced by b' then the probability will be close to 1. In (4) if b is replaced by b' then the probability will not be close to 1. (1) and (3) remain the same if b' replaces b. So, given b', SP is strongly biased with respect to green but not grue.
15. I am assuming that the Q properties, such as green and blue, are determined to be present in the ordinary ways we employ, not the extraordinary ways employed by grue speakers and their ilk.

$p(P/A) = ab + (1-a)c = ab + c - ac$.

So $p(P/A) = p(Q_1/C\&A)$ only if $ab+c-ac=b$, that is, only if $b(1-a)=c(1-a)$. Now by hypothesis, $a<1$. So $p(P/A) = p(Q_1/C\&A)$ only if $p(Q_1/C\&A) = p(Q_2/-C\&A)$. So if $p(C/A) < 1$, and if $p(Q_1/C\&A)$ is not equal to $p(Q_2/-C\&A)$, then $p(P/A)$ is not equal to $p(Q_1/C\&A)$. This means that if it is not a certainty that an arbitrary A will satisfy C, then we cannot conclude that the probability that an arbitrary A has the disjunctive property P is just the probability that an A under condition C has property Q_1. We need to determine the probability that an arbitrary A *under a condition that violates C has property Q_2*. That is, we need to determine $p(Q_2/-C\&A)$. To determine the latter we must vary the conditions so that As will be examined when C is satisfied *and* when it is not.

Applying this to the grue case, let

P = the property grue
Q_1 = the property green
Q_2 = the property blue
A = the property of being an emerald
C = the condition that the time at which the arbitrary item is selected is before 2500.

In this case $p(C/A)$ is not 1, since it is not certain that given that an emerald is arbitrarily selected at some time, that time is before 2500. If the probability is greater than zero that emeralds exist after 2500, then $p(C/A) < 1$. Therefore, unless we can assume that the probability that an emerald arbitrarily selected before 2500 is green is equal to the probability that an emerald not selected before 2500 is blue, the probability that an arbitrarily selected emerald is grue, that is, $p(P/A)$, cannot be assumed equal to the probability that an emerald arbitrarily selected before 2500 is green, that is, to $p(Q_1/A\&C)$. Rather it is equal to (i) above, that is, to the sum of two products. The first is the probability that an arbitrarily selected emerald is selected before 2500 multiplied by the probability that an emerald arbitrarily selected before 2500 is green; the second is the probability that an arbitrarily selected emerald is not selected before 2500 multiplied by the probability that an arbitrary emerald not selected before 2500 is blue. To determine these probabilities, in particular $p(Q_1/A\&C)$ and $p(Q_2/A\&-C)$, we need to vary the instances observed. We need to examine emeralds before 2500 so that condition C is satisfied, and we need to examine emeralds in 2500 or later, so that $-C$ is satisfied.

10

Explanation versus Prediction: Which Carries More Evidential Weight?

1. The Historical Thesis of Evidence

According to a standard view, predictions of new phenomena provide stronger evidence for a theory than explanations of old ones. More precisely, a theory that predicts phenomena that did not prompt the initial formulation of that theory is better supported by those phenomena than is a theory by known phenomena that generated the theory in the first place and that the theory was used to explain. So say various philosophers of science, including William Whewell in the nineteenth century and Karl Popper in the twentieth, to mention just two.[1]

Stephen Brush takes issue with this on historical grounds.[2] He argues that, generally speaking, scientists do not regard new phenomena predicted by a theory, even ones of a kind totally different from those that prompted the theory in the first place, as providing better evidential support for that theory than is provided by already known phenomena explained by the theory. By contrast, Brush claims, there are cases, including general relativity and the periodic law of elements, in which scientists tend to consider known phenomena explained by a theory as constituting much stronger evidence than novel predictions.[3]

---

1. William Whewell, *The Philosophy of the Inductive Sciences* (New York: Johnson reprint, 1967; from the 1847 ed.) Karl Popper, *The Logic of Scientific Discovery* (London: Hutchinson, 1959).
2. Stephen Brush, "Prediction and Theory Evaluation: The Case of Light Bending," *Science*, 246 (1989), 1124–1129.
3. To what extent Brush wants to generalize this explanationist position is a question I leave for him to answer. There are passages that strongly suggest a more general position. For example: "There is even some reason to suspect that a successful explanation of a fact that other theories have already failed to explain satisfactorily (for example, the Mercury perihelion) is more convincing than the prediction of a new fact, at least until the competing theories have had their chance (and failed) to explain it" (p. 1127). In what follows, I consider a generalized explanationist thesis.

Both the predictionist and the explanationist are committed to an interesting historical thesis about putative evidence e and a hypothesis h, viz.

Historical thesis: Whether e if true is evidence that h, or how strong that evidence is, depends on certain historical facts about e, h, or their relationship.[4]

For example, whether e if true is evidence that h, or how strong it is, depends on whether e was known to be true before or after h was formulated. Various historical positions are possible, as Alan Musgrave noted years ago in a very interesting article.[5] On a simple predictionist view (which Musgrave classifies as "purely temporal"), e is evidence that h only if e was not known to be true when h was first proposed. On another view (which Musgrave attributes to Elie Zahar and calls "heuristic"), e is evidence that h only if when h was first formulated it was not devised in order to explain e. On yet a third historical view (which Musgrave himself accepts), e is evidence that h only if e cannot be explained by a "predecessor" theory, that is, by a competing theory which was devised by scientists prior to the formulation of h.

These are three examples of historical views concerned with the temporal order in which e and h were formulated or known to be true. But other historical positions are possible. For example, it might be held that whether e if true is evidence that h, and how strong that evidence is, always depends on historical facts concerning how the results reported in e were obtained, for example, what sampling methods were in fact used by those who reported that e (more on this in section 2).

The historical thesis is not that e or h are themselves propositions about particular historical events. (For example, e might be that light can be diffracted, and h might be that light is a wave motion.) Rather the thesis is that even if neither e nor h describes a particular event or set of events that occurred, whether e, if true, is evidence that h depends upon the occurrence of some particular event or set of events pertaining to when or how e, h, or the relationship between them came to be formulated or believed, or how the results in e were obtained.

Is the historical thesis true or false? It is clearly true in the case of subjective evidence. Whether e is some person's evidence that h depends on certain historical facts about e, h, and their relationship, viz. whether that person in fact has or had certain beliefs about e, h, and their relationship. However, defenders of the historical thesis are not speaking of subjective evidence. They are not speaking of something that someone *takes to be* evidence but of something that is (objective) evidence. In cases of the latter sort, I propose to argue that the historical thesis is sometimes true, and sometimes false, depending on the type of evidence in question. I will show how this comports with my own theory of potential and veridical evidence. Then I will consider what implications, if any, this has for the

---

4. e and h here are propositions. As noted in chapter 2, this is the customary practice of philosophers who speak of e as being evidence that h, even though it is the fact that e is true that is evidence. The historical thesis, then, concerns the propositions e and h or their relationship.
5. Alan Musgrave, "Logical versus Historical Theories of Confirmation," *British Journal for the Philosophy of Science*, 25 (1974), 1–23.

debate between Brush and the predictionists and for the problem of "old evidence" raised by Clark Glymour.

Before beginning, however, let me mention a curious but interesting fact about certain well-known philosophical theories or definitions of objective evidence (other than mine), including Carnap's theory of confirmation,[6] Hempel's satisfaction theory,[7] Glymour's bootstrap account,[8] and the usual hypothetico-deductive account. These theories are incompatible with the historical thesis.[9] They hold that whether, or the extent to which, e is evidence that h is an a priori fact about the relationship between e and h. It is in no way affected by empirical issues such as the time at which h was first proposed, or e was first known, or by the intentions with which h was formulated, or how information reported in e was obtained. Defenders of these views must reject both the predictionist and the explanationist claims about evidence. They must say that whether, or the extent to which, e supports h has nothing to do with whether e was first formulated as a novel prediction from h or whether e was known before h and h was constructed to explain it.

Accordingly, we have two extreme or absolutist positions. There is the position, reflected in the historical thesis, that evidence is always historical (in the sense indicated). And there is a contrasting position, reflected in a priori views, that evidence is never historical. Does the truth lie at either extreme? Or is it somewhere in the middle? In what follows I will pursue these questions with respect to my concepts of potential and veridical evidence, the ones most crucial to scientists (although what I will say is also applicable to ES- evidence).

2. Selection Procedures

Suppose that an investigator decides to test the efficacy of a drug D in relieving symptoms S. The hypothesis under consideration is

$h:$ Drug D relieves symptoms S in approximately 95% of the cases.

As I have emphasized in chapter 9, whether (and the extent to which) some test result is evidence that a certain hypothesis is true depends on the selection procedure used to obtain that result. Here are two of the many possible selection procedures for testing h:

$SP_1:$ Choose a sample of 2000 persons of different ages, both sexes, who have symptoms S in varying degrees; divide them arbitrarily into two groups; give one group drug D and the other a placebo; determine how many in each group have their symptoms relieved.

$SP_2:$ Choose a sample of 2000 females aged 5 all of whom have symptoms S in a very mild form; proceed as in SP_1.

---

6. Rudolf Carnap, *Logical Foundations of Probability*.
7. Carl G. Hempel, *Aspects of Scientific Explanation*.
8. Clark Glymour, *Theory and Evidence* (Princeton: Princeton University Press, 1980).
9. See Laura J. Snyder, "Is Evidence Historical?," *Scientific Methods: Conceptual and Historical Problems*, Peter Achinstein and Laura J. Snyder, eds. (Malabar, Florida: Krieger, 1994).

Now suppose that our investigator obtains the following result:

e: In a group of 1000 persons with symptoms *S* taking drug *D*, 950 persons had relief of *S*; in a control group of 1000 *S*-sufferers not taking *D* but a placebo none had symptoms *S* relieved.

If result *e* was obtained by following SP_2, then *e*, although true, would not be particularly good evidence that *h*, certainly not as strong as that obtained by following SP_1. The reason, of course, is that SP_1, by contrast with SP_2, gives a sample that is varied with respect to factors that may well be relevant: age, sex, and severity of symptoms. (Hypothesis *h* does not restrict itself to 5-year-old girls with mild symptoms, but asserts a cure rate for the general population of sufferers with varying degrees of the symptoms in question.)

This means that if the result described in *e* is obtained, then whether that result so described constitutes evidence that *h*, and how strong that evidence is, depends on a historical fact about *e*, viz. how in fact *e* was obtained. If *e* resulted from following SP_1, then *e* is pretty strong evidence that *h*; if *e* was obtained by following SP_2, then *e* is pretty weak evidence that *h*, if it is evidence at all. Just by looking at *e* and *h*, and even by ascertaining that *e* is true, we are unable to determine to what extent, if any, *e*, if true, supports *h*. We need to invoke "history."

Here is a third selection procedure:

SP_3: Choose a sample of 2000 persons all of whom have *S* in varying degrees; divide them arbitrarily into two groups; give one group drugs *D* and *D'* (where *D'* relieves symptoms *S* in 95% of the cases and blocks possible curative effects of *D* when taken together); give the other group a placebo.

If this was used to generate *e*, then result *e* fails to provide any support for *h*. And, again, that this is so can be ascertained only by learning a historical fact about *e*, viz. that what *e* (truly) reports was obtained by following SP_3.[10]

So far then we seem to have support for the historical thesis about evidence. Can we generalize from examples like this to all cases? Can we say that for any *e*, and any hypothesis *h*, whether, or to what extent, *e*, if true, is evidence that *h* depends upon historical facts about how *e* was obtained?

Consider another type of case, in which the aim is to test the hypothesis

h: John will win the lottery.

At the present time suppose that this can be done only "indirectly" by obtaining information about who bought tickets and how many. Two investigators proceed to obtain information of this sort, each one following a different selection procedure:

10. It might be objected that if this is so, then, contrary to what I have been saying, the concept of evidence involved—potential or veridical evidence—cannot be objective. Whether *e* is potential (or veridical) evidence that *h* in such cases will depend on what selection procedure was employed, which, in turn, depends on what some person(s) believed about *e* and *h*, viz. upon what was believed about how *e* was generated. My reply is that in the cases in question whether *e* is potential evidence that *h* depends on what selection procedure was in fact employed, not on what beliefs its employer(s) may have had about it.

SP_4: Determine who bought tickets, and how many, by asking lottery officials.

SP_5: Determine this by consulting the local newspaper, which publishes this information as a service to its readers.

These results are obtained:

e_1: Investigator 1 reports that in the lottery 1000 tickets were sold, of which John owns 999, and no further tickets will be sold.

e_2: Investigator 2 reports the same thing.

Whether e_1 or e_2 or both are evidence that h depends on what selection procedure was employed by the investigators in obtaining the information in their reports. Suppose that investigator 1 followed SP_4, while investigator 2 followed SP_5. And suppose that newspapers, by contrast to lottery officials, are usually unreliable in such reports. Then, although e_1, if true, is strong evidence that h, e_2 is not. In any case, to determine whether, or the extent to which, e_1 or e_2 supports h we again need to determine a historical fact concerning how the ticket information reported in e_1 and e_2 was obtained.

Now, however, let us distill the information reported in e_1 and e_2 into the following:

e: In the lottery 1000 tickets were sold, of which John owns 999, and no further tickets will be sold.

In this case, whether e, if true, is evidence that h, and how strong it is, does not depend on historical facts concerning how information reported in e was obtained, or concerning when or how e, h, or their relationship came to be formulated, believed, or known. To be sure, whether, or the extent to which, e, if true, supports h does depend upon other historical facts, such as whether the lottery is fair, whether certain conditions will "interfere," and so forth. But these are not historical facts of the kind relevant for the historical thesis of evidence. In this case, unlike the drug case, in order for e, if true, to be evidence that h, it is irrelevant how or when the information in e was obtained, or even whether it was obtained. Accordingly, we have a case that violates the historical thesis of evidence.

Since examples similar to each of the above can be readily constructed, we may conclude that there are cases that satisfy the historical thesis of evidence, and others that fail to satisfy it. With respect to a hypothesis h, if e speaks of observations or tests made that yield certain results, then whether e if true is evidence that h, and if so, how strong that evidence is, depends on what selection procedure was employed in making these observations or tests. That is a historical fact of the kind that conforms to the historical thesis of evidence. However, there are many cases, such as the last one noted, in which e does not speak of observations or tests, but of certain facts that are described independently of observations or tests. Such cases do not conform to the historical thesis.

What implications, if any, does this hold for whether predictions or explanations provide stronger evidence?

3. Predictions versus Explanations

Let us return to the original question proposed by Brush. Do novel facts predicted by a theory provide stronger evidence for that theory than known facts explained by the theory, as Whewell and Popper claim? Or is the reverse true?

A preliminary point is worth making. As should be obvious from discussions in earlier chapters, the mere fact that some theory or hypothesis h was (or indeed can be) used successfully to predict some novel e does not suffice to make e evidence that h. Nor does the fact that some theory or hypothesis h was (or can be) used to explain some known fact e suffice to make e evidence that h. Suppose there is a lottery whose 1 million tickets go on sale on Tuesday; only one ticket can be purchased by any given person; and the drawing will be made on Friday. My hypothesis is that you will win the lottery. Suppose I use this hypothesis to predict that you will buy 1 ticket. The fact that my prediction turned out to be correct is not evidence that you will win the lottery. Suppose that on Sunday it comes to be known that you are in a good mood. And suppose I use the hypothesis that you won the lottery to explain this known fact. This would not suffice to make the fact that you are in a good mood evidence that you won the lottery. It is readily shown that on the account of (potential and veridical) evidence I propose, neither the fact that e was (or can be) correctly predicted from a hypothesis h, nor the fact that h was (or can be) used to explain a known fact e, suffices to make e evidence that h.

Accordingly, what the "predictionist" and "explanationist" may want to say is this. Whether e is a novel fact correctly predicted from a hypothesis h, or whether e is a known fact that h was invoked to explain, is *relevant* for the question of whether e is evidence that h, and particularly for the question of how strong that evidence is. On the predictionist view, a novel fact correctly predicted by a hypothesis provides stronger evidence than does an already known fact explained by the hypothesis. For an explanationist it is the reverse. Which view is correct?

My answer is this: Neither one. *Sometimes a prediction provides better evidence for a hypothesis, sometimes an explanation does, and sometimes they are equally good. Which obtains has nothing to do with the fact that it is a prediction of novel facts or that it is an explanation of known ones.*

To show this, I will begin with a case that violates the historical thesis of evidence. Here it should be easy to show that whether the putative evidence is known before or after the hypothesis is formulated is irrelevant for whether it is evidence that h or how strong that evidence is. Let

h: This coin is fair, that is, if tossed in random ways under normal conditions it will land heads approximately half the time in the long run
e: This coin is physically symmetrical, and in a series of 1000 random tosses under normal conditions it landed heads approximately half the time.

Let us suppose that e is empirically complete with respect to h (that is, whether e is evidence that h, and how strong that evidence is, does not depend on empirical facts other than e).[11] In particular, whether e is evidence that h does not depend on when, how, or even whether, e comes to be known, or on whether e was known first and h then formulated, or on whether h was conceived first and e then stated as a prediction from it. Putative evidence e supports hypothesis h and does so (equally well) whether or not e is known before or after h was initially formulated, indeed whether or not h was ever formulated or e is ever known to be true or any selection procedure was used to obtain e.

The same holds in a case that is analogous except that e is not empirically complete with respect to h. Let

   e: In the lottery 1000 tickets were sold, of which John owned 999 at the time of the drawing.
   h: John won.

Whether e is evidence that h, and if so how strong it is, depends on the truth of certain facts other than e, such as whether the lottery was fair. But it does not depend on whether e was known before or after h was formulated, or on whether e was ever known, or on what selection procedure, if any, was employed to obtain e.

So let us focus instead on cases that satisfy the historical thesis of evidence. We might suppose that at least in such cases explanations (or predictions) are always better for evidence. Return once again to our drug hypothesis

   h: Drug D relieves symptoms S in approximately 95% of the cases.

Consider two evidence claims, the first a prediction about an unknown future event, the second a report about something already known:

   e_1: In the next clinical trial of 1000 patients who suffer from symptoms S and who take D approximately 950 will get some relief.
   e_2: In a trial that has already taken place involving 1000 patients with S who took D (we know that) approximately 950 got some relief.

On the prediction view, e_1 (if true) is stronger evidence for h than is e_2. On the explanation view, it is the reverse. And to sharpen the case, let us suppose that e_2, by contrast to e_1, was not only known to be true prior to the formulation of h, but that h was formulated with the intention of explaining e_2. Which view is correct? Neither one.

Let us take the prediction case e_1 first. Whether e_1 if true is evidence that h, and how strong it is, depends on the selection procedure to be used in the next clinical trial. Suppose this selection procedure calls for choosing just 5-year-old girls with very mild symptoms who in addition to D are also taking drug D' which ameliorates symptoms S in 95% of the cases and potentially blocks D from doing so. Then e_1 would be very weak evidence that h is true, if it is evidence at all. This is so, despite the fact that e_1 is a correct prediction from h, one not used in generating h in the first place. By contrast, suppose that the selection procedure used in the past trial mentioned in e_2 is much better with respect to h. For example, it calls for choosing humans of both sexes, of different ages, with

11. Microconditions are being disregarded. See chapter 5.

symptoms of varying degrees, who are not also taking drug D' or any other that can prevent D from working. Then e_2 would be evidence that h, indeed much stronger evidence than e_1. In such a case, a known fact explained by h would provide more support for h than a newly predicted fact would.

Obviously the situations here can be reversed. We might suppose that the selection procedure used to generate the prediction of e_1 is the one cited in the previous paragraph as being used to generate e_2 (and vice versa). In this situation a newly predicted fact would provide more support for h than an already explained one.

In these cases whether some fact is evidence, or how strong that evidence is, has nothing to do with whether it is being explained or predicted. It has to do with the selection procedure used to generate that evidence. In one situation—whether it involves something that is explained or predicted—we have a putative evidence statement generated by a selection procedure that is a good one relative to h; in the other case we have a flawed selection procedure. This is what matters for evidence, not whether the putative evidence is being explained or predicted.

4. A Response: Maher's Account

A response of the "predictionist" and "explanationist" will now be considered in this section and the two that follow. It involves formulating the information concerning whether e is a prediction or an explanation as part of the evidence statement itself or at least as part of the background information.[12]

In the example above, let

e: In a clinical trial of 1000 patients who suffer from symptoms S and who take drug D, approximately 95% got some relief
b_1: e is a prediction from h
b_2: h was devised to explain e
h: Drug D relieves symptoms S in approximately 95% of the cases.

According to the "predictionist,"

$e\&b_1$ is stronger evidence that h than is $e\&b_2$ (or e is stronger evidence that h given b_1 than it is given b_2).

According to the "explanationist,"

$e\&b_2$ is stronger evidence that h than is $e\&b_1$ (or e is stronger evidence that h given b_2 than it is given b_1).

Patrick Maher offers a predictionist response of this sort, which he formulates as follows.[13] Suppose that a hypothesis h is generated on some occasion by

---

12. See Eric Barnes, "Social Predictivism," *Erkenntnis*, 45 (1996), 69–89.
13. Patrick Maher, "Prediction, Accommodation, and the Logic of Discovery," *PSA*, 1988, vol. 1, 273–285; "How Prediction Enhances Confirmation," in J.M. Dunn and A. Gupta, eds., *Truth or Consequences* (Netherlands: Kluwer Academic Publishers, 1990), 327–343.

a method M. Let M_h be that M generated a hypothesis that entails h. Let O be that (as Maher puts it) e was available to M when it generated h. Then, according to Maher, the predictionist thesis is this:

(PT) $p(h/M_h\&e\&-O) > p(h/M_h\&e\&O)$.[14]

The probability on the left is conditional on the assumption that e was not used to generate h. (This is the "prediction" case.) The probability on the right is conditional on the assumption that e was used to generate h (which Maher calls "accommodation"). Maher does not claim that (PT) holds universally. But he does prove a theorem showing that it holds under certain conditions, which, he claims, usually obtain in science.[15]

Before discussing any of these conditions, I will give an example Maher himself offers, and then show how he would deal with the drug case above. Maher's example involves coin tossing and it purports to show that prediction provides stronger evidence than "accommodation."[16] He imagines two cases, as follows:

(a) *Accommodation.* A fair coin is randomly tossed 99 times by an experimenter, and the outcome of each toss is recorded in a sentence e. The hypothesis h is a conjunction of e with the proposition (h') that the coin will land heads on the 100th toss. Let b contain the information that the results of the 99 tosses were first recorded by the experimenter, following which the experimenter formulated the conjunctive hypothesis h to "accommodate" e. (This is the experimenter's method M of generating hypothesis h.) According to Maher,

(1) $p(h/e\&b) = p(h'/e\&b) = \frac{1}{2}$.

That is, given that on each of the first 99 tosses the coin landed in the manner described by e and given the "accommodation" described in b, the probability that the coin will land heads on the 100th toss and that in the first 99 tosses it landed in the manner described by e is just the probability that it will land heads on the 100th toss, given that it landed the way it did on the first 99 tosses and given "accommodation." Assuming the coin is fair and tossed randomly, the latter probability should be $\frac{1}{2}$.

(b) *Prediction.* Propositions e, h, and h' are as above, but b is changed. Instead of first tossing the coin 99 times and recording the results, the experimenter predicts the results of the first 99 tosses, viz. e, and also predicts that the 100th toss will yield heads. (Call this conjunctive fact $b'(e)$.)[17] Then the coin is tossed 99 times and the results e are exactly as predicted. Now, according to Maher,

(2) $p(h/e\&b'(e)) = p(h'/e\&b'(e)) =$ approximately 1.

14. Maher construes p here as the subjective probability for some rational person who has not yet learned the truth-values of $M_{h'}$, e, and O. See "How Prediction Enhances Confirmation," p. 327. With suitable relativization, probability in (PT) could also be understood in an objective epistemic sense of the sort introduced in chapter 5.
15. "How Prediction Enhances Confirmation," p. 328.
16. "Prediction, Accommodation, and the Logic of Discovery," p. 275. See also his "Howson and Franklin on Prediction," *Philosophy of Science*, 60 (1993), 329–340.
17. The experimenter is using some method, however random, for generating predictions. Although Maher does not say so explicitly, I assume he would claim that the method being used can but need not be given in $b'(e)$.

Let CP_i be that the experimenter correctly predicts the result of the i-th toss. Where h' is that the 100th toss will be heads, let $P(h')$ be that the experimenter predicts that the 100th toss will be heads. Then $e\&b'(e)$ is equivalent to $CP_1 \ldots CP_{99}\&P(h')$. So from (2) we have

(3) $p(h'/CP_1 \ldots CP_{99}\&P(h'))$ = approximately 1.

In this case, given that the experimenter successfully predicts the results of the first 99 tosses, viz. e, and predicts that the 100th toss will be heads, the probability that his prediction (h') of heads on the 100th toss will be correct is close to 1. The only difference between (1) and (3) is that in (1) the data recorded in e are "accommodated" by the experimenter using the conjunctive hypothesis h ($=h'\&e$), while in (3) the data recorded in e are correctly predicted by the experimenter in advance of the experiment.

Maher attributes the difference in probabilities to the fact that in the successful prediction case, but not the "accommodation" case, there is a strong reason to suppose that the experimenter has a reliable method of predicting coin tosses. This is the basis for his claim that predictions provide stronger evidence than accomodations.

5. Maher and the Drug Case

Returning now to my drug case, here is how Maher's account would proceed. (In what follows I alter the example to conform to this account.) Suppose there are 11 clinical trials, each to determine the effectiveness of drug D in relieving symptoms S. Let

- e_1: In the first 10 clinical trials involving 1000 S- sufferers who took drug D, 95% got relief in 9 out of the 10 trials. In one of the ten trials, 75% got relief.
- e_2: In the 11th trial involving 1000 S-sufferers who took drug D, 95% got relief.
- M: If the relative frequency of 95% success in ten clinical trials involving 1000 S-sufferers is .9, infer the same 95% success rate in another trial involving 1000 S-sufferers who took D.
- M_{e2}: M was used to generate a hypothesis that entails e_2.
- M_h: M was used to generate a hypothesis that entails h.
- $O_{e(i)}$: e_i was available to M when M generated e_i.
- O_h: e_i was available to M when M generated h.
- h: $e_1\&e_2$.

Now, according to the predictionist thesis,

(PT) $p(h/M_h\&e_1\&-O_h) > p(h/M_h\&e_1\&O_h)$

The probability on the left represents the "prediction" case, the one on the right represents "accommodation." That is, if e_1 is a prediction, then h has a higher probability than if e_1 is accommodated (that is, if e_1 was available to M when M generated h).

In the prediction case, Maher claims, the fact that e_1, which is entailed by h, was correctly predicted (indicated by $e_1 \& - O_h$) boosts the reliability of the method M used to generate h. This in turn boosts the probability of h. By contrast, in the accommodation case, the fact that e_1 was accommodated (indicated by $e_1 \& O_h$) does not boost the reliability of the method M used to generate h, or does not boost it as much as the prediction case does. Hence in this case h's probability is not as high as it is in the prediction case.

One of the central assumptions of Maher's general theorem concerns the lack of such a boost in the case of accommodation. Maher speaks of the "reliability" R of a method for generating a hypothesis. This is the (objective) probability that the hypothesis generated by this method will be correct. Let the expected value of the reliability of a method M for generating a hypothesis h be denoted by $E(RM(h))$.[18] Consider M above, a method that will generate e_2. Maher makes an assumption for his theorem which has the following consequence [p. 336, see (22)]:

(1) $E(RM(e_2)/M_{e2} \& e_2 \& O_{e2} \& e_1) = E(RM(e_2)/e_1)$.

He assumes that if e is simply accommodated (but not predicted), then the expected value of the reliability of method M with respect to e_2 is not increased.

Now in the drug case above, this assumption does not hold. According to e_2, the result in the 11th trial satisfies M. This should boost the expected value of the probability that a hypothesis generated by M is correct, whether the results in the 11th trial were accommodated or predicted.

Maher's response is to agree that in this sort of case the predictionist thesis (PT) does not hold.[19] In such a case, he says, it is certain what the method M would predict, that is,

(2) $p(M_{e2}/e_2 \& O_{e2} \& e_1) = p(M_{e2}/e_2 \& - O_{e2} \& e_1) = 1$.

This violates an assumption for his theorem from which (1) above follows. Although he agrees that (PT) does not hold when (2) is true, he claims that in the usual scientific cases (2) is false. Maher writes:

> Only in very special cases can we predict with certainty what hypotheses scientists will generate.[20]

In response, note first that (2) is concerned not with the probability that a scientist will generate the hypothesis e_2 but with the probability that method M will generate e_2. I agree that the former probability is not 1, but it is the latter probability that is of concern in (2). So is Maher claiming that in typical cases of prediction, where a scientist uses a method for generating a predictive hypothesis h, the probability that the method generates h is less than 1? The answer seems to be yes. The only justification Maher offers for this answer is that typically in pre-

---

18. If a function has a finite number of values, the expected value is the sum of all products consisting of a given value and the probability of that value. If a function (such as probability) has infinitely many values, the expected value is defined using an integral.
19. Personal correspondence. See his "Howson and Franklin on Prediction," *Philosophy of Science*, 60 (June 1993), 329–340; see pp. 339–340. I am indebted to Patrick Maher for help in trying to get me to express his views accurately in these sections; I hope I have done so.
20. "Howson and Franklin on Prediction," p. 340.

dictive cases the method being used is not understood well enough to allow it to be certain that the method generates the prediction in question.

As mentioned in note 14, Maher employs a subjective interpretation of probability in his predictionist thesis (PT). So his present claim would be that a typical scientist who makes a prediction using a method M does not understand M well enough to be completely sure that M yields that prediction.[21] The scope and interest of Maher's predictionist thesis (PT), then, depends on how typical it is for scientists using a method to predict the truth of a hypothesis h not to understand the method well enough to know for sure whether h is actually predicted by that method. To say the least, this is very different from the usual predictionist idea. More importantly, it is one whose wide applicability seems questionable. Frequently, a scientist will be in a situation in which the method of prediction consists in deriving the prediction deductively from a theory or hypothesis, where the scientist knows whether or not the derivation is correct (without necessarily knowing whether what is derived is true).

6. Balmer's Formula

It may be useful to cite a simple historical example, viz. Balmer's formula, that is similar in certain respects to one employed by Maher.[22] When light from hydrogen is analyzed using a spectroscope, it is seen to consist in series of sharp lines of definite wavelengths. In 1885, Johann Jakob Balmer introduced a general formula that entailed the wavelengths of the four lines known by him at that time. The formula can be represented as follows:

$$1/\lambda_n = R(\tfrac{1}{4} - 1/n^2)$$

where λ is the wavelength of a given line, R is a constant, and $n = 3, 4, 5, 6$ for the four lines. Balmer does not claim to be *explaining* why the lines occur or have the wavelengths they do, but simply to be "represent[ing] the wavelengths of the different lines in a satisfactory manner."[23] This seems to be a case satisfying Maher's notion of "accommodation."

21. With an objective concept of probability, of the sort I defend in chapter 5, this conclusion would not be permitted, unless a special relativization is introduced that entails disregarding information indicating whether M yields the prediction in question.
22. Maher cites predictions from Mendeleyev's periodic table of the elements. Mendeleyev placed the elements in groups based on atomic weights (rather than on other bases) and showed that periodic repetitions of properties occur in the groups and can be used to predict new elements. He did so explicitly claiming that atomic weights physically determine and can explain properties of the elements and compounds. (See his "The Relations between the Properties of Elements and their Atomic Weights," reprinted in Henry A. Boorse and Lloyd Motz, eds., *The World of the Atom* (New York: Basic Books, 1966, vol. 1), p. 306.) By contrast, Balmer makes no physical or explanatory claims regarding his formula. While Balmer's formula "accommodates" the data, it is more clearly nonexplanatory than Maher's example, while also being predictive. Stephen Brush, by contrast to Maher, claims that the periodic table was regarded by scientists as providing stronger explanatory than predictive evidence.
23. Johann Jakob Balmer, "The Hydrogen Spectral Series," reprinted in William Francis Magie, *A Source Book in Physics* (Cambridge: Harvard University Press, 1965), 360–365; quote on p. 360.

Now Balmer indicates that he used his formula to obtain the wavelength of a fifth line (letting $n = 7$). He says he knew nothing of such a line when he performed the calculation, and was later informed that it exists and satisfies the formula. So the fifth line was, from his standpoint, a prediction that turned out to be correct. Moreover, he reports being informed that "many more hydrogen lines are known, which have been measured by Vogel and Huggins in the violet and ultra-violet parts of the hydrogen spectrum and the spectrum of the white stars" (p. 362). What impresses Balmer, however, is not the fact that he has made a successful prediction, but simply the fact that all of the lines, whether accommodated or predicted, satisfy his formula. He writes:

> From these comparisons it appears that the formula holds also for the fifth hydrogen line.... It further appears that Vogel's hydrogen lines and the corresponding Huggins lines of the white stars can be presented by the formula very satisfactorily. We may almost certainly assume that the other lines of the white stars which Huggins found further on in the ultra-violet part of the spectrum will be expressed by the formula. (p. 362)

As far as Balmer is concerned, it is the fact that the various lines, whether first known and later accommodated or first predicted and later known, all satisfy his formula that provides strong evidence for the last claim in the above passage.

Let $B(i)$ mean that line i satisfies Balmer's formula. Balmer's claim is that

(3) $p(B(5)/B(1) \ldots B(4))$ is very high.

This has nothing to do with accommodation or prediction. However, let the method for generating hypotheses of the form $B(i)$ be as follows:

M: Use Balmer's formula to obtain $B(i)$.

The probability in which Maher is interested for "accommodation" is

(4) $p(h/M\&e\&O)$,

that for "prediction" is

(5) $p(h/M\&e\&-O)$,

where h is $B(5)\&e$, e is $B(1) \ldots B(4)$, and O says that e was available to M when it generated h.

If Maher's predictionist thesis (PT) holds for this case, then the probability in (5) should be greater than that in (4). But, I submit, they are equal. The relevant consideration is that in both cases lines 1 through 4 satisfy Balmer's formula, and that the hypothesis of interest concerning line 5 was generated using the Balmer formula. The truth-value of claims $B(1) \ldots B(4)$ is relevant, but it is irrelevant whether or not the truth-values of these claims were available to method M when it generated $B(5)$.

To be sure, in such a case Maher will say that Balmer's method M is such that whether $B(i)$ is a prediction using M is known for certain.[24] But I submit, it is just

---

24. Maher claims that his Mendeleyev example is one in which his present uncertainty thesis holds. He writes: "Nobody—without the benefit of hindsight—could be certain that Mendeleyev would propose the hypothesis he did" ("Howson and Franklin on Prediction,"

these kinds of predictions, viz. deductions from hypotheses and/or applications of quantitative formulas to new cases, that are frequently meant by those who defend a predictionist thesis. Moreover, I submit, they are not atypical in science.

7. Brush Redux

Brush is clearly denying a general predictionist thesis. By contrast, he cites cases in which scientists themselves regarded known evidence explained by a theory as stronger support for that theory than new evidence that was successfully predicted. And he seems to imply that this was reasonable. He offers an explanation for this claim, viz. that with explanations of the known phenomena, by contrast with successful predictions of the new ones, scientists had time to consider alternative theories that would generate these phenomena. Now, even if Brush does not do so, I want to extend this idea and consider a more general explanationist view that is committed to the following three theses that Brush invokes for some cases:

(1) A selection procedure for testing a hypothesis h is flawed, or at least inferior to another, other things being equal, if it fails to call for explicit consideration of competitors to h.
(2) The longer time scientists have to consider whether there are plausible competitors to h the more likely they are to find some if they exist.
(3) With putative evidence already known before the formulation of h scientists have (had) more time to consider whether there are plausible competitors to h than is the case with novel predictions.

I would challenge at least the first and third theses. In the drug example of section 2, selection procedure SP_1 for the drug hypothesis does not call for explicitly considering competitors to that hypothesis. Yet it does not seem flawed on that account, or inferior to one that does. However, even supposing it were inferior, whether or not a selection procedure calls for a consideration of competitors is completely irrelevant to whether the putative evidence claim is a prediction or a known fact being explained. In the case of a prediction, no less than that of an explanation, the selection procedure may call for a consideration of competitors.

For example, in our drug case, where h is "Drug D relieves symptoms S in approximately 95% of the cases," and e is the prediction "In the next clinical trial of 1000 patients suffering from symptoms S who take D, approximately 950 will

---

p. 340). But, again, this is not the probability of concern. The question is whether, with respect to a prediction Mendeleyev made using his periodic table, he was in any doubt about whether his table yields that prediction. After constructing his periodic table based on atomic weights, Mendeleyev writes: "... it appears to be certain when we look at the proposed table, that in some rows the corresponding members are missing; this appears especially clearly, e.g., for the row of calcium; in which there are missing the members analogous to sodium and lithium" (p. 310). Also, at the end of his paper he writes: "The discovery of numerous unknown elements is still to be expected, for instance, of elements similar to Al and Si having atomic weights from 65–75" (p. 312). Mendeleyev seems to have no doubt that these predictions are generated by his table.

get some relief," the selection procedure to be used for the next clinical trial might include the rule

> In conducting this next trial, determine whether the patients are also taking some other drug which relieves S in approximately 95% of the cases and which blocks any effectiveness D might have.

Such a selection procedure calls for the explicit consideration of a competitor to explain e, viz. that it will be some other drug, not D, that will relieve symptoms S in the next trial. This is so even though e is a prediction. Moreover, to respond to the third thesis about time for considering competitors, an investigator planning a future trial can have as much time as she likes to develop a selection procedure calling for a consideration of a competing hypothesis. More generally, in designing a novel experiment to test some hypothesis h, as much time may be spent in precluding competing hypotheses that will explain the test results as is spent in considering competing hypotheses for old data.

8. Thomson versus Hertz: The Wave Theory of Light

Let me invoke two scientific examples employed earlier. The first involves the dispute between Heinrich Hertz and J. J. Thomson over the nature of cathode rays, discussed in chapter 2 above. Recall that in 1883 Hertz observed that the cathode rays in his experiments were not deflected by an electric field. He took this to be strong evidence that cathode rays are not charged particles but some type of ether waves. In 1897 Thomson repeated Hertz's experiments but with a much higher evacuation of gas in the cathode tube than Hertz had been able to obtain. Thomson believed that when cathode rays pass through a gas they make it a conductor, which screens off the electric force from the charged particles comprising the cathode rays. This screening off effect will be reduced if the gas in the tube is more thoroughly evacuated. In Thomson's 1897 experiments electrical deflection of the cathode rays was detected, which Thomson took to be strong evidence that cathode rays are charged particles.

I want to consider the evidential report of Hertz in 1883, not of Thomson in 1897. Let

e = In Hertz's cathode ray experiments of 1883, no electrical deflection of cathode rays was detected.
h = Cathode rays are not electrically charged.

Hertz considered e to be strong (veridical) evidence that h. In 1897 Thomson claimed that Hertz's results as reported in e did not provide strong evidence that h, since Hertz's experimental set-up was flawed: He was employing insufficiently evacuated tubes. To use my previous terminology, Thomson was claiming that Hertz's selection procedure for testing h was inadequate, and hence that e is not (potential or veridical) evidence that h.

Here we can pick up on a point emphasized by Brush. Hertz, we might say, failed to use a selection procedure calling for considering a competitor to h to explain his results (viz. that cathode rays are charged particles, but that the tubes

Hertz was using were not sufficiently evacuated to allow an electrical force to act on these particles). But—and this is the point I want to emphasize—in determining whether, or to what extent, e is (potential or veridical) evidence that h, it is irrelevant whether Hertz's e was a novel prediction from an already formulated hypothesis h or an already known fact to be explained by h. Hertz writes that in performing the relevant experiments he was trying to answer two questions:

> Firstly: Do the cathode rays give rise to electrostatic forces in their neighbourhood? Secondly: In their course are they affected by external electrostatic forces?[25]

In his paper he did not predict what his experiments would show. Nor were the results of his experiments treated by him as facts known before he had formulated his hypothesis h. Once he obtained his experimental results he then claimed that they supported his theory:

> As far as the accuracy of the experiment allows, we can conclude with certainty that no electrostatic effect due to the cathode rays can be perceived. (p. 251)

To be sure, we might say that Hertz's *theory* itself predicted some such results, even if Hertz himself did not (that is, even if Hertz did not himself draw his conclusion before getting his experimental results). But even if we speak this way, Hertz did not claim or imply that his experimental results provide better (or weaker) evidence for his theory because the theory predicted them before they were obtained. Nor did Thomson in his criticism of Hertz allude to one or the other possibility. Whichever it was—whether a prediction or an explanation or neither—Hertz (Thomson was claiming) should have used a better selection procedure. This is what is criticizable in Hertz, not whether he was predicting a novel fact or explaining a known one.

The second example involves the nineteenth century argument for the wave theory of light given in chapter 7, section 7. The argument began with two claims: (i) that light travels from one point to another with a finite velocity; (ii) that in other known cases, such as sound waves, water waves, and projectiles, modes of transfer involving finite velocities are via classical waves in a medium or classical particles. From (i) and (ii), it is concluded that (h) light is either a wave motion in a medium or a stream of particles. In probabilistic terms,

(1) $p(h/(i)\&(ii))$ is close to 1.

Indeed, it might be claimed that

(2) $p(\text{there is an explanatory connection between } h \text{ and (i)}/(i)\&(ii)) > \frac{1}{2}$,

so that (since (i) and (ii) were deemed true in the nineteenth century), (i) constitutes potential evidence that h, given (ii). The claim that it does constitute evidence that light is either a classical wave or a classical stream of particles was made by wave theorists as part of their eliminative argument for the wave theory.

Now, what I believe can legitimately be said about this case is that although

---

25. Heinrich Hertz, *Miscellaneous Papers* (London: Macmillan, 1896).

(i) constitutes ES-evidence that h (where the epistemic situation is the one holding for wave theorists during the first four decades of the nineteenth century), it is not potential evidence that h. The reason it is not has nothing to do with whether wave theorists employed h to explain the already known fact that light travels from one point to another in a finite time or to predict new cases in which this is so. The reason is that the selection procedure is biased, although this was not known by wave theorists and could not be known until the twentieth century. The items selected to defend the claim that modes of transfer involve classical waves and particles—such as water waves, sound waves, and projectiles—are all items from the macroworld of classical waves and particles; there are none from the subatomic world subject to the laws of quantum mechanics which preclude classical waves and particles. Accordingly, (1) and (2) are criticizable, not the fact that wave theorists were explaining known facts (or predicting new ones).

I end this section with a quote from John Maynard Keynes, whose book on probability contains lots of insights. Here is one:

> The peculiar virtue of prediction or predesignation is altogether imaginary. The number of instances examined and the analogy between them are the essential points, and the question as to whether a particular hypothesis happens to be propounded before or after their examination is quite irrelevant.[26]

9. The Problem of "Old Evidence"

Years ago Clark Glymour raised a fundamental objection to the popular positive relevance definition of evidence.[27] Suppose that the probability of e is 1. If it is, then, assuming that the probability of h is greater than 0, $p(e/h) = 1$. Now, according to Bayes' theorem,

(1) $p(h/e) = p(h) \times p(e/h)/p(e)$.

Therefore, if $p(e) = 1$, then from (1) we obtain

(2) $p(h/e) = p(h)$.

That is, the probability of h, given e, is the same as its prior probability. Now on the positive relevance definition,

(3) e is evidence that h if and only if $p(h/e) > p(h)$.

So if $p(e) = 1$ and $p(h) > 0$, then from (2) and (3), e cannot be evidence that h.

Why is this a problem? Suppose, says Glymour, that prior to the introduction of hypothesis h, e was known with certainty to obtain. Glymour concludes that e's probability is then 1. (I will return to this claim in a moment.) If e's truth was known prior to the introduction of h, then e is "old evidence" with respect to h. It follows from the positive relevance definition of evidence that "old evidence" with respect to a hypothesis cannot be evidence that the hypothesis is true, since

---

26. J.M. Keynes, *A Treatise on Probability* (London: Macmillan, 1921).
27. Clark Glymour, *Theory and Evidence*, p. 86.

it cannot increase the probability of the hypothesis. This strikes many as absurd, since frequently phenomena considered evidence in favor of a theory were known with certainty to obtain prior to the formulation of the theory.

In section 3 of this chapter we considered whether predictions of novel facts provide stronger evidence for a theory than explanations of old ones, or whether the reverse is true. Predictionists (such as Whewell and Popper) might welcome Glymour's problem by responding that if e is already known to be true prior to the formulation of h, then it cannot be (very much) evidence for h, even if it is derivable from h. For them, in order that e be evidence that h, or at least substantial evidence, it must be a new prediction and not a fact known prior to the formulation of h.

Now to make this claim is to espouse the historical thesis of evidence formulated in section 1. To know whether e is a new prediction and not a fact known prior to the formulation of h one must know "historical" facts about e and h, viz. whether and when e is known and whether it is known prior to when h was proposed. In my earlier discussion I rejected the historical thesis as a *universal* principle. To be sure, there are cases in which determining whether, or to what extent, e supports h requires determining the truth of some historical fact about e, h, or their relationship. But there are also cases where this is not so at all. To claim that only predictions provide (substantial) evidence for a hypothesis is to espouse a mistaken historical thesis.

Glymour considered the problem of "old evidence" sufficient to show that one should not accept (3), the standard positive relevance definition of evidence. Since in chapter 4 I have given many other reasons to reject this definition, what I want to focus on here is not the question of whether Glymour's objection is devastating to the positive relevance account but on what one is to say about cases in which the probability of the putative evidence is 1.

However, before doing so I return to a central claim of Glymour's argument, viz. that if e is known to be true (whether or not this was prior to the formulation of h), then $p(e) = 1$. Is this true? It does, indeed, hold for subjective probability. If at a certain time t I know that e is true, then at t my subjective probability for e is maximal.[28] Does it hold for the concept of epistemic probability that I espouse?

My answer is that it depends on what relativization, if any, is being assumed in the probability statement. If e is known to be true, then it follows that it is true. And if the probability in question is relativized to e, or to the fact that e is known to be true, then, to be sure, the probability of e is 1 ($p_e(e) = 1$). But that is trivial and uninteresting. With other less trivial, more interesting, relativizations the probability is not 1, even though it is known that e is true. For example, suppose e is as follows:

e: This coin, which was tossed 100 times on January 1, 2000, landed heads 95% of the time.

---

28. Although subjectivists accept this, some wish to solve Glymour's problem by refusing to use one's actual degree of belief in e in determining $p(e)$ but some counterfactual degree of belief, such as one's degree of belief in e on everything one knows minus e. See Colin Howson and Peter Urbach, *Scientific Reasoning*, p. 404.

Suppose some person knows that e is true. We may want to consider the probability of e disregarding this knowledge, that is, $p_{d(Ke)}(e)$, in which "$d(Ke)$" means "disregard the fact that e is known to be true." In this case the probability may be less than 1, even though it is known that e is true.[29] Accordingly, with epistemic probability, from the fact that e is known to be true, it does not necessarily follow that the probability of e is 1. It depends on what is, or is not, being assumed.

I turn now to the question of whether Glymour's problem can be put in a form which does not introduce the idea that e is a fact (not) known prior to the formulation of h. Using epistemic probabilities, suppose that $p_b(e) = 1$. This is not a "historical" claim. It does not say that anyone in fact knows or came to know that e is true, or that anyone did so before or after the formulation of h. All it says is that, in view of b, the degree of reasonableness of believing e is maximal. Could it be the case that, given b, e is potential or veridical evidence that h? Or does the fact that $p_b(e) = 1$ preclude this possibility?

My definitions of potential and veridical evidence do *not* preclude this possibility. It is possible for the following conditions for potential evidence to be satisfied:

1. e and b are true
2. e does not entail h
3. $p_b(h/e) \times p_b$(there is an explanatory connection between h and $e/h\&e$) $> \frac{1}{2}$,

even if $p_b(e) = 1$.

To demonstrate this suppose the background information b contains the fact that there is a lottery consisting of 1000 tickets one of which will be drawn at random, and that John's wife Mary purchased 950 of these tickets as a present for John, who deposited them in his safe deposit box with a legal document saying that the tickets belong to him. Let e be that John owns 950 tickets in this lottery. Let h be that John will win. If e and b are true, then the above conditions for e's being potential evidence that h are satisfied. Yet $p_b(e) = 1$, since in this case the background b (which we may suppose contains standard legal principles of ownership) entails e. Given b, the degree of reasonableness of believing e (that John owns 950 tickets in this lottery) is maximal. Yet given b, e is potential evidence that h. And, if h is true, it is veridical evidence as well. The fact that, in the light of the background information b, it is maximally reasonable to believe e does not at all count against e's being evidence that h. Indeed, in this case, in the light of b, e is very strong evidence that h.

Now, let us consider what, if anything, happens if it is unreasonable to believe e. If $p(e) = 0$, then $p(h/e)$ is undefined. So let us consider a case in which e has low, but nonzero, probability. "Predictionists" champion such cases, because the lower the probability of the "prediction" e, the higher the posterior probability of h on e.

Let the background information b be that a fair coin will be randomly tossed 100 times. Let e be that the coin will land heads 100 times in a row. Let h be the hypothesis that the Devil will intervene after each random toss, causing the coin

---

29. This epistemic probability is analogous to a subjective one that would be used by subjectivists offering a counterfactual solution to Glymour's problem mentioned in the previous footnote.

to land heads each time during the first 100 tosses. In this case the hypothesis h entails (the prediction) e. Assuming b, the probability of e is very low: $p_b(e) = (\frac{1}{2})^{100}$. Yet, given b, it seems far-fetched to say that e, if true, is potential evidence that h. Indeed, the definition of potential evidence precludes this, even if e and b are both true, since (given normal background information) the probability of the devil hypothesis h, even on the assumption of e, is extremely low. The fact that the coin will land heads 100 times in a row, even if true, is not potential evidence that the Devil will intervene.

Can we follow the "predictionist" and say at least this: Where h entails some "prediction" e, the lower the probability of e the stronger the evidence that e confers upon h? No, we cannot. All we can say is that the lower the probability of e, in such a case, the higher the probability of h on e. But it does not follow from this that the lower the probability of e the stronger the evidence that e confers upon h, since e may confer no evidence upon h. Thus, in the previous example, let us change e to

e': The coin will land heads 1000 times in a row.

Let h and b be the same as before. Now we have $p_b(e') = (\frac{1}{2})^{1000}$, which is a much lower probability than $p_b(e) = (\frac{1}{2})^{100}$. Yet that does not make e' stronger evidence that h than e is, since, on my conception, e' is not evidence that h. The threshold for high probability required for evidence has not been reached.

What can be said is this. *If e is evidence that h*, then the lower the probability of e the stronger is the evidence that e confers upon h. For example, let b contain the information that this coin is perfectly symmetrical. Let h be the hypothesis that it will land heads approximately half the time. Let e_1 be the information that when tossed randomly the first 100 times it landed heads between 45 and 55 times. Let e_2 be the information that when tossed randomly the next 1000 times it landed heads exactly 500 times. We might say that, given b, both e_1 and e_2 count as evidence that h. Now in this case $p_b(e_2) < p_b(e_1)$, so that, indeed, $p_b(h/e_2) > p_b(h/e_1)$. But in this case also, in the light of b, e_2 is stronger evidence that h than is e_1.

Accordingly, whether e has very high or very low probability does not necessarily affect whether, or the extent to which, e is evidence that h. Nor is it in general true that if h "predicts" e, the lower is e's probability the stronger is e's evidence that h.

10. Conclusions

1. According to the historical thesis of evidence, whether e if true is evidence that h, or how strong that evidence is, depends on certain historical facts about e, h, or their relationship (for example, on whether e was known before or after h was formulated). Although this thesis holds for subjective evidence, it does not hold universally for the concepts of objective evidence I have introduced. Focusing on potential and veridical evidence, depending on the particular evidence claim and the selection procedure employed, in some cases the historical thesis holds, in others it does not.

2. Sometimes a novel fact that is predicted provides better evidence for a hypothesis than a known fact that is explained. Sometimes the reverse is true. Which obtains has nothing to do with whether it is a prediction or an explanation, but rather, in the cases in question, with the selection procedure used to generate the evidence. This is illustrated in the case of Hertz's claim that the results of his 1883 cathode ray experiments provide evidence that cathode rays are not electrically charged. It is also illustrated in the wave theorists' argument that light consists of classical waves or classical particles.

3. A response of Patrick Maher is examined which involves formulating the information concerning whether e is a prediction or a fact explained as part of the evidence statement itself or as part of the background information. This new formulation will not suffice to establish the superior power of prediction over explanation (or "accommodation"), or vice versa, in typical scientific cases of interest to predictionists and explanationists.

4. Glymour raises "the problem of old evidence" in order to reject the positive relevance definition. According to that definition, e cannot be evidence that h when e is "old evidence" known to be true. Predictionists may welcome this result, since for them, in order for e to be substantial evidence that h, e must be a new prediction, not a fact known prior to the formulation of h. To say this is to espouse the historical thesis of evidence as universal, which I do not. I discuss Glymour's problem in a form that does not commit one to the historical thesis. If the degree of reasonableness of believing the putative evidence e is maximal (corresponding to "old evidence"), can e be potential (or veridical) evidence that h? I demonstrate that it can be. So there is no comfort here for the predictionist. (Nor does the fact that the probability of e is maximal, where e is explainable by derivation from h, guarantee that e is potential evidence that h; so there is no comfort here for the explanationist either.) A second question is this: If h entails some "prediction" e, does the lower the probability of e mean the stronger the evidence that e confers upon h? The answer again is no. So again there is no comfort for the predictionist.

11

OLD-AGE AND NEW-AGE HOLISM

1. Old-Age Holism

There is a fundamental objection, urged by Duhem and Quine, against any project, such as mine, that seeks to define evidence in a way that allows what Duhem calls an "isolated" hypothesis to be confirmed or disconfirmed by empirical evidence. The objection derives from a very general thesis about confirming or disconfirming evidence. Here is Quine's well-known formulation of this thesis:

> ... our statements about the external world face the tribunal of sense experience not individually but only as a corporate body.[1]

Duhem's ideas, although formulated about physics, are usually extended to the sciences generally by his admirers. Here are two passages from Duhem:

> An experiment in physics can never condemn an isolated hypothesis but only a whole theoretical group.[2]
>
> No isolated hypothesis and no group of hypotheses separated from the rest of physics is capable of an absolutely autonomous experimental verification. (p. 258)

In terms of evidence, we might put the Duhem-Quine claim like this:

Evidential Holism: Nothing can be evidence for or against an individual, isolated hypothesis, but only for or against some theoretical group of hypotheses.

The question of what constitutes a "corporate body" for Quine, or a "theoretical group" for Duhem, is not really answered. A very robust holism, sometimes suggested by these authors, would assert that a corporate body or theoretical group includes all of the propositions generally accepted in a science such as physics.

---

1. W. V. Quine, *From a Logical Point of View* (Cambridge, MA: Harvard University Press, 1953), p. 41.
2. Pierre Duhem, *The Aim and Structure of Physical Theory* (Princeton: Princeton University Press, 1991; translated from the second edition published in 1914), p. 183.

A less demanding holism would say simply that there is always some set of hypotheses, and not an individual one, that evidence supports or disconfirms. The discussion that follows will be applicable to both versions. If evidential holism is correct, then, it might be claimed, no definition of evidence of the sort I propose is possible, since that definition allows something to be evidence for or against individual, isolated hypotheses. Whether evidential holism is incompatible with my concept of potential evidence is a question I will discuss in section 5, after considering various arguments for this doctrine.

What are the arguments for evidential holism? Here is Duhem's argument for the claim that it is impossible to "condemn an isolated hypothesis but only a whole theoretical group":

> A physicist decides to demonstrate the inaccuracy of a proposition; in order to deduce from this proposition the prediction of a phenomenon and institute the experiment which is to show whether the phenomenon is or is not produced, in order to interpret the results of this experiment and establish that the predicted phenomenon is not produced, he does not confine himself to making use of the proposition in question; he makes use also of a whole group of theories accepted by him as beyond dispute. The prediction of the phenomenon, whose nonproduction is to cut off debate, does not derive from the proposition challenged if taken by itself, but from the proposition at issue joined to that whole group of theories; if the predicted phenomenon is not produced, not only is the proposition questioned at fault, but so is the whole theoretical scaffolding used by the physicist. The only thing the experiment teaches us is that among the propositions used to predict the phenomenon and to establish whether it would be produced, there is at least one error; but where this error lies is just what it does not tell us. The physicist may declare that this error is contained in exactly the proposition he wishes to refute, but is he sure it is not in another proposition? If he is, he accepts implicitly the accuracy of all the other propositions he has used, and the validity of his conclusion is as great as the validity of his confidence. (p. 185)

Duhem's point is quite simple. A physicist, or more generally, a scientist, in order to "demonstrate the inaccuracy" of some hypothesis, deduces from it, *together with other assumptions*, some "prediction." If the latter turns out to be false, that does not disconfirm the hypothesis, but only the conjunction consisting of that hypothesis together with all the other assumptions used to arrive at the false prediction. As a point of logic, Duhem's reasoning is impeccable. If hypothesis h together with assumptions A entails proposition p, but h by itself does not, and if p is false, then so is the conjunction $h \& A$. We cannot conclude that h itself is false. Construed as a claim about *evidence*, however, whether Duhem's reasoning is impeccable is another matter, which will be considered in section 2.

Quine's argument for evidential holism is different. It derives from the claim that "individualism" (the idea that evidence can support individual, isolated hypotheses) is committed to a form of reductionism—a dogma that Quine rejects:

> But the dogma of reductionism has, in a subtler and more tenuous form, continued to influence the thought of empiricists. The notion lingers that to each statement, or each synthetic statement, there is associated a unique range of possible sensory events such that the occurrence of any of them

would add to the likelihood of truth of the statement, and that there is associated also another unique range of possible sensory events whose occurrence would detract from that likelihood. This notion is of course implicit in the verification theory of meaning.

The dogma of reductionism survives in the supposition that each statement, taken in isolation from its fellows, can admit of confirmation or infirmation at all. (pp. 40–41)

In this passage Quine makes several important claims about evidence. First, he seems to espouse a positive relevance account, according to which evidence for something increases its probability ("likelihood"), and evidence against it decreases it. Second, he seems to be saying that if you accept individualism, you are committed to the view that associated with any sentence (perhaps as part of its meaning) is some unique set of "sensory events" any of which would, in the positive relevance sense, confirm or disconfirm the sentence. But, he argues, there is no reason to suppose that each sentence has such an associated set. To suppose it does is a dogma. This associationist claim, Quine thinks, is a less radical version of what he calls "radical reductionism," according to which each meaningful statement is translatable into a statement about immediate experience (p. 38).

2. Are These Arguments for Evidential Holism Reasonable?

Duhem's holism, construed as a claim about evidence, is defended by appeal to the idea that evidence against (and perhaps also for) a hypothesis h is produced by, and only by, deducing some consequence e from h together with other assumptions b. Not-e is evidence against $h\&b$ if and only if e is false, while e is evidence in favor of $h\&b$ if and only if e is true. Such a view of evidence is hypothetico-deductive:

> e is evidence that the conjunction $h\&b$ is true if and only if $h\&b$ entails e (but b itself does not) and e is true.
>
> not-e is evidence that the conjunction $h\&b$ is false if and only if $h\&b$ entails e, and e is false.

Duhem's holism, then, understood as a claim about evidence, is the idea that evidence e (or not-e) is, in general, evidence for the truth (or falsity) not of some single, isolated hypothesis h, but for the truth of some set or conjunction of hypotheses $h\&b$. And his argument for this claim, which is based on a hypothetico-deductive account of evidence, is that in science an evidence report e is not generally derivable from an isolated hypothesis but only from a system of hypotheses.

Now, as I have argued in chapter 7, section 2, the hypothetico-deductive view of evidence is fundamentally flawed. It provides neither a necessary nor a sufficient condition for evidence. The fact that (e) John owns 950 tickets in the lottery is evidence that (h) he will win, given (b) there are 1000 tickets in the lottery all of which have been sold, even though $h\&b$ does not entail e. And although the hypothesis (h') that officials invalidated 998 lottery tickets before the drawing but not the two that Bill owned, together with (b), entails (e') that Bill won, the fact that Bill won (assuming that he did) is not evidence that h' is true; nor is it evidence that $h'\&b$ is true. On the theory of evidence I have espoused, a deductive connection between h and e is neither necessary nor sufficient for e to be evidence that h. So Duhem cannot reject my theory simply on the ground that in science an isolated hypothesis rarely entails a prediction. Later I will consider how Duhem might respond "holistically" to the particulars of my theory.

Turning to Quine's argument, my first response is to reject the positive relevance criterion of evidence he suggests, as I have done in chapter 4. Increase in probability is neither necessary nor sufficient for evidence. My purchasing 1 ticket in a 1 million ticket lottery increases the probability I will win, but it is not evidence that I will. Examples were given in chapter 4 in which e fails to increase h's probability, and even decreases it, yet e is evidence that h.

My second response to Quine is to reject the idea that to be an individualist about evidence you must be committed to the claim that associated with any sentence is some set of "sensory events" the occurrence of any of which would confirm (or disconfirm) that sentence. If this were true, and if the association in question is supposed to be an association in virtue of the meaning of the sentence, evidence statements would all be a priori, an idea I have rejected in chapter 2. But even if the association is not to be understood in terms of meaning, to reject evidential holism in favor of individualism, one need not espouse the view that it is only "sensory events" (whatever they are) that provide evidence for hypotheses. Such a view is indeed a holdover from the kind of positivism and reductionism that Quine deplores. But it is not a requirement of any definition of evidence I have considered earlier, including my own.

So let us broaden the view Quine is attacking to be that

(1) Associated with any ("isolated") hypothesis h there is some set of sentences $e_1, e_2, \ldots$, such that if any one of the latter were true, the fact that it is would be evidence that h.

In putting the claim this way we need not subscribe to the positive relevance (or indeed any other) definition of evidence. Nor need we be committed to the view that only "sensory events" confirm or disconfirm, or to the thesis that there is an analytic-synthetic distinction (the "first dogma" of empiricism that Quine rejects). Quine seems to be saying that (1) is mistaken, even if (1) is divorced from any positivistic or reductionist claims about the meaning of sentences.

How does Quine argue against this claim? He doesn't. He simply asserts that it is false. Recall what he says:

> The dogma of reductionism survives in the supposition that each statement, taken in isolation from its fellows, can admit of confirmation or infirmation at all. My countersuggestion . . . is that our statements about the external world face the tribunal of sense experience not individually but only as a corporate body. (p. 41)

To the claim that individual statements can be confirmed, Quine offers a "countersuggestion": they cannot. We need to find some arguments for rejecting individualism in favor of holism.

Now in fact the individualism to which I subscribe, and that Quine and Duhem will reject, is weaker than (1). We need not say that associated with *any* hypothesis *h* there is some set of sentences of the sort in question, but simply that

(2) There are individual, "isolated" hypotheses which are such that if certain statements are true, the fact that they are is evidence that these hypotheses are true.

What argument can be mustered against this thesis?

3. The Background Assumption Argument

I believe that the best and most interesting argument against individualism is that, in general, and particularly in the sciences, evidence claims require the use of background assumptions. This use is such that the putative evidence is evidence for (or against) the hypothesis *together with these background assumptions*.

For example, in Book III of the *Principia* Newton derives his law of gravity from the six "Phenomena" he cites concerning the observed motions of the planets and their satellites. (Phenomenon 1, for example, is that the satellites of Jupiter describe areas proportional to the times of description and their periods are proportional to the 3/2th power of their distance from Jupiter.) Newton regards the law of gravity as being "deduced from phenomena and made general by induction, which is the highest evidence a proposition can have in this [experimental] philosophy."[3] He clearly regards the six Phenomena that he cites as being evidence, or a large part of the evidence, that the law of gravity is true. However, in order to "deduce" the law of gravity from the Phenomena, Newton makes use of his three laws of motion in Book I and various theorems that follow from these. So, we might say, following Duhem and Quine, Newton's six Phenomena are really evidence not for his law of gravity but for his entire theoretical system, which includes his three laws of motion.

An obvious response to this is to say that Newton is indeed providing evidence for his law of gravity, but that he is doing so *relative to his laws of motion and theorems*. He is making a claim of the form "*e* is evidence that *h*, given *b*." He is not making the claim "*e* is evidence that *h&b*." Is there any difference between these claims? Are they equivalent? No, they are not. The former claim does not entail the latter, at least on my definition of (potential and veridical) evidence. To see this we can concentrate just on the condition that if *e* is evidence that *h*, given *b*, then $p(h/e\&b) > \frac{1}{2}$. The latter can be true, even if $p(h\&b/e) < \frac{1}{2}$. For example, let

b = There are 100 tickets in a fair lottery, one of which will be drawn at random.
e = John owns 95 tickets in this lottery.
h = John will win the lottery.

---

3. H. S. Thayer, *Newton's Philosophy of Nature* (New York: Hafner, 1953), p. 6. For a discussion of Newton's ideas about deduction, induction, and evidence, see my *Particles and Waves* (New York: Oxford University Press, 1991), chapter 2.

$p(h/e\&b) = .95$. Yet it does not follow that $p(h\&b/e) > \frac{1}{2}$. Indeed it can be much less.[4]

More generally, the following is provable: For any h, b, and e, if $p(b/e) < 1$, then $p(h\&b/e) < p(h/e\&b)$.[5] So the fact that $p(h/e\&b) > \frac{1}{2}$ does not require that $p(h\&b/e) > \frac{1}{2}$. Therefore, using my definition of evidence, the fact that e is evidence that h, given b, does not require that e is evidence that $h\&b$.

Accordingly, even if it were true that all evidence statements claiming that e is evidence that h need to be understood as relativized to some background assumptions b, on the definitions of evidence I propose it would not follow that such statements are true only if e is evidence that $h\&b$. It would not follow that, in effect, no e is evidence for an "isolated" hypothesis h, but only for a conjunction of h with some set of background assumptions.

Finally, it should be noted that the background assumption argument is incompatible with the second conjunction condition advocated in chapter 8, section 6, according to which if e is evidence that $h_1\&h_2\&\ldots\&h_n$, then e is evidence that h_1, and e is evidence that $h_2, \ldots,$ and e is evidence that h_n. This condition implies that if e is evidence for a "holistic" conjunction then it is evidence for each "isolated" conjunct. In other words, if the present conjunction condition is valid, then a "holistic" evidence claim presupposes a "nonholistic" one.

A holist must reject this conjunction condition and argue that (i) evidence that h always requires the use of background assumptions, and (ii) this requires that evidence is always evidence for h in conjunction with these background assumptions without its being evidence for each isolated background assumption. In the present section I have argued that (i) does not imply (ii). In the next section I turn to claim (i).

4. New-Age Holism

Suppose we agree that "isolated" hypotheses can receive evidence. How might we reformulate "evidential holism" so as to do some justice to the Duhem-Quine

---

4. Let b' be that the lottery mentioned in b was itself chosen at random by John from a set of 100 lotteries only one of which contains 100 tickets; the rest contain more tickets. In this case $p(h/e\&b\&b') = .95$, but $p(h\&b/e\&b') = p(b/e\&b') \times p(h/e\&b\&b') = .01 \times .95 = .0095$.

5. Proof:

1. $p(h\&b/e) = \dfrac{p(h\&b\&e)}{p(e)}$

2. $p(h/e\&b) = \dfrac{p(h\&b\&e)}{p(e) \times p(b/e)}$

3. From 1 and 2,
$$\dfrac{p(h\&b/e)}{p(h/e\&b)} = \dfrac{p(h\&b\&e)}{p(e)} \times \dfrac{p(e) \times p(b/e)}{p(h\&b\&e)}$$
$= p(b/e)$, which, by hypothesis, is less than 1

4. From 3,
$$\dfrac{p(h\&b/e)}{p(h/e\&b)} < 1$$

5. From 4,
$p(h\&b/e) < p(h/e\&b)$. QED

thesis? One way is to say simply that all evidential claims are to be understood as relativized to a set of background assumptions:

> *Relativized Evidential Holism:* Where h is some "isolated" hypothesis and e is some "isolated" fact, it is never true that e is evidence that h. Rather e is evidence that h, relative to some set of assumptions b.

The thesis of holism is retained in the idea that evidential claims are always to be understood as relativized to a system of assumptions.

In response to this thesis, it is a simple matter to construct true nonrelativized evidential claims by taking any background assumptions and putting them into the e-statement. For example, the previous lottery case can be reformulated as follows:

e': John owns 95 of the 100 tickets in a fair lottery, one ticket of which will be drawn at random
h: John will win the lottery.

Intuitively, e' is (strong) evidence that h. (On my definition of evidence, this is so since $p(e'$ correctly explains why h is true/$e') > \frac{1}{2}$.) Moreover, this is true without the need to relativize the evidential claim "e' is evidence that h" to any set of assumptions.[6] To be sure, in this example e' combines the information e (that John owns 95 tickets) with b (that there are 100 tickets in the lottery, one of which is to be drawn at random). In the first example, we relativized the evidential claim by saying that e is evidence that h, given b. In the second example we avoid relativization by combining e and b to form e', which is also evidence that h, a claim that now becomes a priori true. So, a holist might claim, e' is not an "isolated" fact, but a conjunction of "isolated" facts.

The problem with this response is that the holist fails to say what counts as "isolated," leaving that idea to be intuitively clear. Just about any sentence a holist might regard as nonconjunctive can be reformulated in equivalent terms as a conjunction, no conjunct of which is equivalent to the original. For example, we can formulate

e: John owns 95 tickets in the lottery

as the conjunction

John owns some tickets, and there are 95 of them, and they are in the lottery.

A more promising response of the holist is to note that the unrelativized claim above that e' is evidence that h is a priori, not empirical. He may now agree that if an evidential claim is a priori true, then no relativization is required. What he really means to assert is this: If an unrelativized evidential claim is empirical, then it needs to be understood as relativized to a set of empirical assumptions whose truth is logically necessary and sufficient for the truth of the evidential claim. So if a sentence of the form

(1) e is evidence that h

---

6. If one of the background assumptions is that there are no "interference" conditions of some type, this can be added to e'. See chapter 5.

is empirical, then it is incomplete, or at least it is not formulated in a maximally perspicuous way that will allow it to be defended properly. To complete it, or to express it in its most perspicuous way, it needs to be formulated as follows:

(2) e is evidence that h, given b.

In (2), b is some conjunction of empirical sentences whose truth is logically necessary and sufficient for the truth of (1). This is not to say that any such conjunction will suffice. (To avoid trivialization, a defender of this view will need to exclude from b evidential sentences such as "e is evidence that h" and perhaps others.) The claim is only that some set b of the sort described is required. We may call a sentence of form (2) that meets these conditions a *completely relativized* evidential claim. Such a sentence will be a priori; it will be complete, perspicuous, and defensible.

Holism, then, becomes the following view:

Completely Relativized Evidential Holism: In its most complete and most perspicuous form (which will allow it to be defended properly) an otherwise empirical evidential claim needs to be "completely relativized."

This will entail relativization to a (fairly large) set of empirical assumptions. (Either that, or else, if it is not relativized in this way, the complete set of empirical assumptions needs to be expressed in the e-position.) Without such "holistic" relativization we have one or both of two charges: lack of completeness and lack of perspicuity.

My objection is that if these charges were fair, then they could be made against any empirical claim whatever, whether or not that claim is an evidential one of form (1). Suppose I make some nonevidential empirical claim, such as

(3) Neutrinos have mass.

The charge could be made that this claim is not complete or sufficiently perspicuous unless additional information b is provided whose truth is logically necessary and sufficient for the truth of (3). In effect, what is needed to replace (3) is a claim of the form

(4) Neutrinos have mass, given b

that is a priori. This type of claim is complete and perspicuous; it can be defended properly.

My response is that to say that completeness and perspicuity always require providing such additional information is to make an unreasonably high demand, for two reasons. First, we ought to be able to separate an empirical claim such as (3) from any defense of that claim that might be offered. The claim may be complete and perspicuous in reasonable senses of these terms even if an argument for (or against) it is not given or available. Otherwise the idea of investigation, so crucial in science, becomes virtually impossible. Frequently, one seeks to investigate, question, or find evidence for or against, a claim already made which lacks any (or any sufficient) argument and indeed for which no method of verification is yet known. Unless one adopts the old positivist line that the meaning of a statement is its method of verification (a line that Quine, of course, rejects), it seems possible to understand a claim for which no argument is given.

Second, even if a defense of an empirical claim is offered it need not be a priori. I may defend my claim that (3) is true by appeal to the fact that this was recently announced in *The New York Times* as a result of experiments by Japanese physicists, even though the latter fact does not provide an a priori justification for (3). To be sure, my argument in this case can itself be defended by appeal to other empirical facts, such as ones having to do with the credentials of the physicists or with their experiments. But the fact that my argument can itself be so defended does not necessarily show that it is incomplete or not perspicuous.

Elsewhere I have argued that, in a case parallel to that of evidence, viz. explanation, what counts as complete and perspicuous is contextual: it depends, at least in part, on the knowledge and interests of the intended audience.[7] My explanation that John will win the lottery because he owns 95 tickets is less complete and perspicuous than it should be for an audience that has no idea of the number of tickets sold. For an audience that knows that the lottery contains 100 tickets, one of which will be drawn at random, this explanation has no such defects. The same is true for defending, or providing information that constitutes evidence for, my claim that John will win the lottery by citing the fact that he owns 95 tickets.

5. Holism and Potential Evidence

We can now return to a question raised at the beginning of this chapter. Is evidential holism incompatible with the concept of potential (or veridical) evidence? Let me focus on the stronger ("old-age") holism. The central condition for potential evidence is

> e is potential evidence that h, given b, only if p(there is an explanatory connection between h and $e/e\&b$) $> \frac{1}{2}$.

In chapter 7 it was proved that this is equivalent to

> e is potential evidence that h, given b, only if p(there is an explanatory connection between h and $e/h\&e\&b$) $\times p(h/e\&b) > \frac{1}{2}$.

This requires that both probabilities exceed $\frac{1}{2}$. Now an evidential holist could accept this condition but claim that it is satisfied only when h is a "theoretical group" (Duhem) or a "corporate body" (Quine), and not an "isolated" hypothesis. What the holist might say is that if h is an isolated hypothesis, then, for any e and b, either

(a) $p(h/e\&b) \leq \frac{1}{2}$

or

(b) p(there is an explanatory connection between h and $e/h\&e\&b$) $\leq \frac{1}{2}$.

7. Peter Achinstein, *The Nature of Explanation*.

Only when h is a "theoretical group" or "corporate body," $h_1 \& h_2 \& \ldots \& h_n$, will the probabilities in (a) and (b) both be greater than $\frac{1}{2}$.

Now, in fact, if $p(h_1 \& h_2 \& \ldots \& h_n/e\&b) > \frac{1}{2}$, then, from the probability rules, it is provable that for any isolated h_i in this group of hypotheses $p(h_i/e\&b) > \frac{1}{2}$. So condition (a) above cannot be satisfied. That is, if a theoretical group has a high probability, then so must any isolated hypothesis in this group. Accordingly, the evidential holist who employs my concept of potential evidence will need to focus on condition (b), according to which if h is isolated, then, for any e and b, the probability of an explanatory connection between h and e, given $h\&e\&b$, is less than $\frac{1}{2}$. Only when h is a theoretical group can this probability be greater than $\frac{1}{2}$.

So indeed it is possible to express evidential holism using the concept of potential evidence I have developed. The two are not incompatible. The holist would simply say that e is potential evidence that h only when h is a theoretical group. It must be noted, however, that this commits the holist not just to a holism about evidence but to one about explanation as well. If, for an isolated h and any e and b, the probability in condition (b) is always less than or equal to $\frac{1}{2}$, this must be due to a fact about explanation rather than probability. It must be due to the fact that only theoretical groups, not isolated hypotheses, can (correctly) explain.

Such a holism about explanation is possible but demanding. According to the account of correct explanation I outlined in chapter 7,

> If p is a complete content giving proposition with respect to the question Q, then p is a correct explanation of q if and only if p is true.

So, for example, I would count

> p: the reason that Al won the lottery is that he owned all the tickets

as a correct explanation of

> q: why Al won the lottery

in virtue of the fact that p is a complete content giving proposition with respect to the question Q (why did Al win the lottery?) and p is true. An explanation holist who accepts my account of explanation will need to say either that p is false, or that it has no truth-value, or that it is true but elliptical for a much more complex sentence involving a theoretical group of hypotheses about lotteries and how they work.

In defense of his position an explanation holist may say that although p is a correct explanation it is not a good one, or at least not a good scientific one. The latter requires a "theoretical group." In chapter 7 I distinguished a correct explanation, which (depending on the context) may or may not be a good one, from a good one, which may or may not be correct. The concept of a correct explanation is involved in my account of potential evidence. Admitting that isolated hypotheses can correctly explain (even if such explanations are not always good) is all that is required.

The holist may reply that even if *p* above is true—and hence that Al's owning all the tickets correctly explains why he won—the explanation needs to be defended by appeal to a "theoretical group." On my view, whether an explanation needs to be defended, and how, is a contextual matter that depends on the knowledge and interests of the intended audience. But even if *p* above needs to be defended by appeal to some "theoretical group," that would not show that *p* is false or elliptical for something more complex. A holist needs to present a different argument for such claims. What this is I leave to him.

Finally, as is the case with evidential holism, what counts as a "theoretical group" or "corporate body" must be indicated. (Surely not just any conjunction of hypotheses will do.) What this is again I leave for the holist to say. My aim here is simply to show how an evidential holist who, despite my criticisms in earlier sections, still wants to defend his position, can employ my concept of evidence by invoking a demanding explanation-holism according to which only "theoretical groups" or "corporate bodies," not isolated hypotheses, can correctly explain.

6. Conclusions

1. Evidential holism is the view that nothing can be evidence for or against an "isolated" hypothesis but only for or against some group of hypotheses. Evidential individualism is the view that there can be evidence for or against "isolated" hypotheses.

2. Duhem's argument for evidential holism is rejected, since it is based on a faulty hypothetico-deductive account that provides neither a necessary nor a sufficient condition for evidence. Quine's argument for evidential holism and against individualism invokes a positive relevance criterion of evidence, which I reject, as well as the dubious idea that only "sensory events" provide evidence for hypotheses. But Quine seems to be rejecting even a broader individualism that avoids positive relevance and "sensory events" views of evidence, as well as positivistic and reductionist claims about the meaning of sentences. However, he offers no argument against this broader individualism, but simply asserts that holism is true.

3. An argument that might be used to defend epistemic holism is that evidence claims require the use of background assumptions; accordingly, an evidential claim must be understood as saying that the putative evidence is evidence for or against the hypothesis together with these background assumptions. My reply is to distinguish claims of the form "*e* is evidence that *h*, given *b*" from ones of the form "*e* is evidence that *h&b*." Using my theory of evidence, it is provable that claims of the former sort do not imply ones of the latter sort. So even if it were true that all evidence claims need to be relativized to background assumptions ("new-age holism"), this would not establish that ("old-age") holism is true.

4. Various forms of relativized evidential holism ("new-age") are considered and rejected.

5. Finally, I argue that, contrary to what might be thought, evidential holism does not preclude a definition of potential evidence of the sort I propose. A holist can use this definition and simply claim that e is potential evidence that h only when h is a theoretical group. This requires a holism with respect to correct explanations. What is needed is a thorough and convincing defense of such a view. It is not established by arguments of Duhem or Quine or by other arguments I consider.

12

EVIDENCE FOR MOLECULES:
JEAN PERRIN AND MOLECULAR REALITY

The final two chapters of this book invoke case histories in physics: the discovery of the electron, credited to J. J. Thomson in 1897, and Jean Perrin's argument for the existence of molecules in 1908 and 1909 on the basis of experiments involving Brownian motion. Both cases illustrate use of the concepts of evidence I have defined. But each one also involves related philosophical and historical issues I propose to discuss.

In the case of Perrin my philosophical concern is with the question of how observable results Perrin obtained from his experiments could provide evidence that unobservable molecules exist, since the argument he uses seems to presuppose that molecules exist. I will also discuss the historical question of why as late as 1908 Perrin should have thought it necessary to argue that molecules exist. In the case of Thomson my philosophical questions are these: What constitutes a scientific discovery? What sort of epistemic situation does such a discovery require? What sort of evidence? The historical question is whether, in the case of the electron, Thomson was in such an epistemic situation. Did he have such evidence?

1. Introduction

In 1908 Jean Perrin conducted a series of experiments on Brownian motion from which he drew two conclusions of particular importance: (1) that molecules exist, and (2) that Avogadro's number N, the number of molecules in a substance whose weight in grams equals its molecular weight, is approximately 6×10^{23}. Perrin's experimental work and conclusions were set forth in a series of papers published in 1908 and 1909, the most famous of which is his "Brownian Movement and Molecular Reality."[1] An expanded version of his results appeared in his book *Atoms*, published in 1913.[2]

In 1926 Perrin received the Nobel Prize in physics primarily for his work on Brownian motion. Despite his considerable success, philosophers and historians of science who read his articles and book should find some of his key arguments puzzling. For one thing, why in 1908, after nineteenth century successes in the kinetic-molecular theory of gases, should Perrin have thought it necessary to argue that molecules exist? Yet argue for this he did.

A second puzzling fact is that Perrin's argument for the reality of molecules seems circular. In brief, from assumptions in kinetic theory involving the existence of molecules, Perrin derives a formula, the "law of atmospheres," that governs a volume of gas. The law relates the number of molecules per unit volume of a gas at a height above some reference plane to Avogadro's number. He then assumes that a slightly modified version of this same law can be applied to the distribution of much larger, microscopic particles (Brownian particles) suspended not in a gas but in a fluid. With this assumption he proposes a formula that relates the number of suspended Brownian particles per unit volume at a height above a reference plane to Avogadro's number and to various experimentally measurable quantities of the visible particles, including their mass, density, and numbers at different heights. He performs experiments measuring these quantities, for different Brownian particles, different liquids, and different temperatures. Each of these measurements, when combined with the law of atmospheres, yields approximately the same value for Avogadro's number. From this fact he concludes that molecules exist. The apparent circularity is that to reach this conclusion Perrin begins by *assuming* that molecules exist. That is an assumption presupposed by the law of atmospheres.

In this chapter I will examine Perrin's reasoning to see whether it is in fact circular. I believe that it is not, and indeed that it conforms with a valid pattern of reasoning frequently used by scientists to infer the existence of "unobservables." I will show how Perrin's experimental results provide veridical (and hence potential) evidence that molecules exist. I will discuss why, even in 1908, it was reasonable for Perrin to employ the pattern of reasoning he does in arguing for the existence of molecules. Finally, I will discuss the relationship between Perrin's reasoning and the debate between realists and antirealists regarding unobservable entities.

2. Perrin's Determination of Avogadro's Number and His Argument for Molecular Reality

Perrin's strategy is first to derive the law of atmospheres for gases.[3] He considers a volume of gas contained in a thin cylinder of unit cross-sectional area and small elevation h. The density of molecules making up the gas will be greatest at the bottom of the cylinder and decreases exponentially with increasing height.

1. Reprinted in *The Question of the Atom*, ed. Mary Jo Nye (Los Angeles: Tomash Publishers, 1986), 507–601.
2. Jean Perrin, *Atoms* (Woodbridge, CT: Ox Bow Press, 1990; translation of the original *Les Atomes*, 1913, by D. Ll. Hammick).
3. Here I follow the argument in *Atoms*, pp. 90ff; a briefer version is found in "Brownian Movement and Molecular Reality," pp. 529–530.

The pressure p at the bottom of the cylinder is more than the pressure p' at the top (just as air pressure at the bottom of a mountain is greater than at the top). The very small difference in pressure $p - p'$ balances the downward force of gravity gm_c on the mass m_c of gas in the cylinder. So

(1) $p - p' = gm_c.$

Now the mass m_c of the gas is to its volume $1 \times h$ as the gram molecular weight M (the mass in grams equal to the molecular weight of the gas) is to the volume v occupied by a gram molecular weight of the gas. That is,

(2) $\dfrac{m_c}{1 \times h} = \dfrac{M}{v}$

From (1) and (2) we get

(3) $p - p' = \dfrac{Mgh}{v}$

Now Perrin invokes the perfect gas law for one gram molecular weight of a gas

$pv = RT,$

where R is the gas constant and T is absolute temperature, and substitutes $v = RT/p$ for v in (3), obtaining

$p - p' = \dfrac{Mghp}{RT},$

or

(4) $\dfrac{p'}{p} = 1 - \dfrac{Mgh}{RT}$

The pressure of a gas is proportional to its density, and hence to the number of molecules per unit volume. So the ratio p'/p can be replaced by the ratio n'/n, where n' and n are the number of molecules per unit volume at the upper and lower levels, respectively. We obtain

(5) $\dfrac{n'}{n} = 1 - \dfrac{Mgh}{RT}$

If m is the mass of a molecule of gas and N is Avogadro's number, then

(6) $M = Nm;$

that is, the gram molecular weight M of a gas is equal to the number of molecules in a gram molecular weight multiplied by the mass of a molecule. So from (5) and (6) we obtain

(7) $\dfrac{n'}{n} = 1 - \dfrac{Nmgh}{RT}$

in which Avogadro's number N appears.[4]

Perrin proposes to use Eq. (7) to determine a value for Avogadro's number experimentally. The problem is that the molecular quantities n, n', and m are not directly measurable. So he makes a crucial assumption, viz. that visible particles comprising a dilute emulsion will behave like molecules in a gas with respect to their vertical distribution. In 1827 the English botanist Robert Brown discovered that small, microscopic particles suspended in a liquid do not sink but exhibit rapid, seemingly haphazard motions—so-called Brownian motion. Following Leon Gouy, Perrin assumed that the motions of the visible particles are caused by collisions with the molecules making up the liquid in which the particles are suspended. He also made the assumption that just as invisible molecules that comprise a gas obey the gas laws, so do the visible particles exhibiting Brownian motion in a liquid. Among other things, he assumed that the law (7) derived for molecules in a cylinder of gas could be extended to Brownian particles distributed in a dilute emulsion.

This means that just as the molecules comprising a gas are all identical in mass and volume, so will the Brownian particles have to be. However, in the latter case, the gravitational force acting on a particle will not be its weight mg, but its "effective weight," that is, the excess of its weight over the upward thrust caused by the liquid in which it is suspended. This is

$$(8) \quad mg - \frac{mdg}{D} = mg(1 - \frac{d}{D})$$

where D is the density of the material comprising the particles and d is the density of the liquid. Replacing the weight mg in (7) by the expression in (8), we obtain

$$(9) \quad \frac{n'}{n} = 1 - \frac{Nmg(1 - d/D)h}{RT}$$

In this equation, n' represents the number of Brownian particles per unit volume at the upper level and n the same at the lower level; m is the mass of a Brownian particle; N is Avogadro's number.[5] Equation (9) contains quantities for the suspended particles (not molecules) which Perrin attempted to determine experimentally.

This required the careful preparation of emulsions containing particles equal in size, and determining the density of the material comprising the particles, the

---

4. A mathematically more rigorous derivation using differential calculus yields

$$\frac{n'}{n} = e^{-Nmgh/RT}$$

where e is the natural log base. For tiny particles and small h the exponent becomes much smaller than 1, and the exponential factor can be expanded in a series whose first two terms are given on the right side of Eq. (7). In his 1909 article, by contrast to his book, Perrin employs the more rigorous derivation.

5. Strictly speaking, in Eq. (9) N represents a number for Brownian particles; that is, any quantity of these particles equal to their molecular weight will contain the same number N of particles. This number N will, according to Perrin's assumptions, be the same as Avogadro's number N for molecules.

mass of the particles, and (with microscopes) the number of suspended particles per unit volume at various heights—all difficult procedures. Experiments were performed with different emulsions, particles of different size and mass, different liquids, and different temperatures. With various values obtained experimentally for the quantities n, n', m, h, and T in Eq. (9), Perrin could use (9) to determine whether Avogadro's number is really a constant, and if so what its value is. He writes:

> In spite of all these variations, the value found for Avogadro's number N remains approximately constant, varying irregularly between 65×10^{22} and 72×10^{22} [that is 6.5×10^{23} and 7.2×10^{23}]. (*Atoms*, p. 105)

Immediately after this sentence Perrin draws a broader conclusion:

> Even if no other information were available as to the molecular magnitudes, *such constant results would justify the very suggestive hypotheses that have guided us, and we should certainly accept as extremely probable the values obtained with such concordance for the masses of the molecules and atoms.*[6]

Perrin's "suggestive hypotheses" include, of course, the assumption that molecules exist. He continues by noting that the values for Avogadro's number obtained through his experiments agree with the number (6.2×10^{23}) given by kinetic theory from considerations of viscosity of gases. And he concludes:

> *Such decisive agreement can leave no doubt as to the origin of the Brownian movement.* . . . The objective reality of the molecules therefore becomes hard to deny. (*Atoms*, p. 105, italics his)

Perrin's conclusions concerning the value of Avogadro's number and the reality of molecules are drawn from his experiments on Brownian particles suspended in a column of fluid. After drawing them Perrin goes on to consider the theory of Brownian motion developed by Einstein in 1905 which generates an equation relating Avogadro's number to the mean square of the displacement of the Brownian particles in a given direction during a given time. Perrin conducted experiments on such displacement, and using Einstein's equation he generated a value for N close to that achieved by his law-of-atmosphere experiments.[7]

At the end of his book Perrin notes that the value(s) he determined for Avogadro's number approximated ones obtained by a variety of different methods, including ones from experiments on radioactivity, blackbody radiation, and the motions of ions in liquids. And he writes:

> Our wonder is aroused at the very remarkable agreement found between values derived from the consideration of such widely different phenomena. See-

---

6. *Atoms*, p. 105, italics his. Using Avogadro's number N and the known molecular weights of substances, the mass of molecules is readily determined: mass of a molecule of substance S is equal to the molecular weight of S divided by Avogadro's number.
7. A useful discussion of these experiments and the statistical reasoning involved is found in Deborah Mayo, "Cartwright, Causality, and Coincidence," *PSA*, 1986, vol. 1, 42–58, and Mayo, "Brownian Motion and the Appraisal of Theories," in A. Donavan et al., eds., *Scrutinizing Science* (Dordrecht: Kluwer, 1988), 219–243.

ing that not only is the same magnitude obtained by each method when the conditions under which it is applied are varied as much as possible, but that the numbers thus established also agree among themselves, without discrepancy, for all the methods employed, the real existence of the molecule is given a probability bordering on certainty. (*Atoms*, 215–216; see also his 1909 paper, 598–599)

3. Is the Argument for Molecules Circular?

The basic structure of Perrin's reasoning seems to be this.

1. From various assumptions, including that molecules exist, and that gases containing them satisfy the ideal gas law, Perrin derives Eq. (7), which relates the number of molecules at a height h in a container of gas to Avogadro's number and to other quantities including the mass of a gas molecule and the temperature of the gas.
2. Perrin then claims that this formula, or a variation of it, can be applied to visible Brownian particles suspended in a fluid, yielding Eq. (9).
3. Next he devises ways to experimentally measure the quantities in Eq. (9) (other than Avogadro's number), and he conducts various experiments using different fluids and particles.
4. Each of these measurements, when combined with Eq. (9), yields approximately the value 6×10^{23} for N.
5. This approximate value for N is also obtained from experiments other than those involving particles suspended in a fluid.
6. From steps 4 and 5 Perrin concludes ("with a probability bordering on certainty") that molecules exist.

The apparent circularity consists in the fact that in step 1 Perrin is making the crucial assumption that molecules exist. Without this assumption he cannot derive Eq. (7), which gives a ratio of the number of molecules per unit volume at the height h to the number at the bottom of the cylinder. Is a charge of circularity warranted? In what follows I will consider some attempts to understand Perrin's reasoning so that circularity is avoided.

4. A Common-Cause Interpretation

Wesley Salmon urges that Perrin's reasoning to the reality of molecules is an example of a legitimate common-cause argument. The basic idea of such an argument is this. If very similar effects have been produced, and if it can reasonably be argued that none of these effects causes any of the others, then it can be concluded that these effects all result from a common cause. This, claims Salmon, is how Perrin argues for the reality of molecules: Perrin notes that experiments on various phenomena—including Brownian motion, alpha particle decay, X-ray diffraction, blackbody radiation, and electrochemical phenomena—all yield approximately the same value for Avogadro's number. Salmon asks us to imagine five different scientists engaged in experiments on the five phenomena mentioned. He writes:

These experiments seem on the surface to have nothing to do with one another [so that it is unlikely that one phenomenon studied causes the other]. Nevertheless, we ask each scientist to fill in the blank in this statement: On the basis of my experiments, assuming matter to be composed of molecules, I calculate the number of molecules in a mole [gram molecular weight] of any substance to be ———. When we find that all of them write numbers that, within the accuracy of their experiments, agree with 6×10^{23}, we are as impressed by the "remarkable agreement" as were Perrin and Poincaré. Certainly, these five hypothetical scientists have been counting entities that are objectively real.[8]

Later he says,

> Remember, for instance, the victims of mushroom poisoning; their common illness arose from the fact that each of them consumed food from a common pot. Similarly, I think, the agreement in values arising from different ascertainments of Avogadro's number results from the fact that in each of the physical procedures mentioned, the experimenter was dealing with substances composed of atoms and molecules—in accordance with the theory of the constitution of matter that we have all come to accept. The historical argument that convinced scientists of the reality of atoms and molecules is, I believe, philosophically impeccable.[9]

Since Salmon claims Perrin's argument is "philosophically impeccable," he would deny any circularity charge. According to him, the argument goes like this:

1. If molecules exist, then from experiments on Brownian motion we get a value for Avogadro's number of $N = 6 \times 10^{23}$.
2. If molecules exist, then from Rutherford's experiments on alpha particle decay, we get a similar value for Avogadro's number. The same is true for experiments involving X-ray diffraction, blackbody radiation, and electrochemical phenomena.
3. There is no reason to suppose that Brownian motion's resulting in a value of $N = 6 \times 10^{23}$ causes alpha particle decay to yield the same value, nor vice versa. The same applies to other cases.
4. So probably each phenomenon's yielding a similar value for N has a common cause, viz. the existence of molecules.
5. So probably molecules exist.

This argument is not circular, since no assumption is made in premises 1 and 2 that molecules do in fact exist. All that is assumed is a conditional: if molecules exist, then. . . . Salmon himself recognizes this when he writes:

---

8. Wesley Salmon, *Scientific Explanation and the Causal Structure of the World* (Princeton: Princeton University Press, 1984), p. 221.
9. *Ibid.*, p. 223. Nancy Cartwright, *How the Laws of Physics Lie* (Oxford: Oxford University Press, 1983), 82–85, offers a somewhat similar analysis of Perrin's reasoning. She takes it to be an argument to the "most probable cause." Experiments on seemingly unrelated phenomena all yield the same calculation for Avogadro's number. "Would it not be a coincidence if each of the observations was an artefact, and yet all agreed so closely about Avogadro's number? The convergence of results provides reason for thinking that the various models used in Perrin's diverse calculations were each good enough. It thus reassures us that those models can legitimately be used to infer the nature of the cause from the character of the effects" (pp. 84–85). For a criticism of Cartwright's account, see Deborah Mayo, "Cartwright, Causality, and Coincidence," *PSA*, vol. 1, pp. 42–58.

On the basis of my experiments, *assuming matter to be composed of molecules*, I calculate the number of molecules in a mole of any substance to be ———.
(p. 221, italics mine)

This corresponds to premises 1 and 2 above. The problem, however, is that this conditional assumption is too weak to yield the strong conclusion 4. The most that premises 1, 2, and 3 warrant is the conditional

4′. *If molecules exist*, then probably each phenomenon's yielding a similar value for N has a common cause.

But 4′ is much less than Perrin himself claims.

To generate the conclusion that Perrin wants, Salmon might alter the argument by adding an additional premise, viz. "molecules exist." But now the argument becomes clearly circular. A more promising approach is to delete the antecedent "if molecules exist" from premises 1 and 2 and assert simply that on each of the varied experiments in question physicists calculated the value of N to be 6×10^{23}, where the latter claim does not presuppose that this is the correct value or even that molecules exist. On this interpretation, we have similar effects (similar calculations of a number that is supposed to represent a number of molecules); and these effects do not cause one another. So by the common-cause principle, we may infer a common cause (without the antecedent assumption that molecules exist). The problem is that a common-cause argument by itself (even assuming its validity)[10] does not permit us to infer what that common cause is but only that there is one. Additional facts must be cited to show that it is the existence of molecules, and not something else, that is the common cause.[11]

One strategy for doing so would be to argue for two points: (a) that the existence of molecules can cause, or be a causal factor in producing, similar calculations of N from experiments on Brownian motion, alpha particle decay, X-ray diffraction, etc. (this could be done by showing how molecular processes can be involved in, or related to, the phenomena in question); and (b) that other possible causes do not produce these effects. Both before and after giving his common-cause argument involving the five hypothetical scientists, Salmon in fact goes some way toward arguing for points (a) and (b). He considers how N is related to

---

10. For a general criticism of common cause arguments, see Bas van Fraassen, *The Scientific Image* (Oxford: Oxford University Press, 1980), and Frank Arntzenius, "The Common Cause Principle," *PSA*, 1992, vol. 2, 227–237.
11. In explicating the idea of a common cause, Salmon employs Reichenbach's notion of a conjunctive fork defined probabilistically in terms of these 4 conditions: (i) $p(A\&B/C) = p(A/C) \times p(B/C)$; (ii) $p(A\&B/-C) = p(A/-C) \times p(B/-C)$; (iii) $p(A/C) > p(A/-C)$; (iv) $p(B/C) > P(B/-C)$. These conditions are satisfied, Salmon argues, if A and B represent experimental results from two different phenomena yielding the same value for Avogadro's number, and C includes the assumption that molecules exist. But as Salmon himself recognizes (pp. 167–168), these are not sufficient conditions for C to be a common cause of A and B. Nor, indeed, does the satisfaction of these conditions make it highly probable that C is a common cause of A and B. Incompatible C's could satisfy these conditions. (For an example, see my coin-tossing case, in chapter 6, section 10.) Yet Perrin (as well as Salmon) wants to conclude that, in all probability ("bordering on certainty"), molecules exist. The satisfaction of the conjunctive fork conditions will not yield such a conclusion.

the five phenomena cited. And he discusses one rival to molecular theory, viz. energeticism, which, he argues, is incapable of explaining the experimental results. But if indeed it is possible to defend points (a) and (b), then a common-cause argument is both unnecessary and unproductive. It is unnecessary because if (a) and (b) can be successfully defended, then the existence of molecules is shown to be probable without invoking a common-cause argument. It is unproductive because a common-cause argument does not by itself make probable the existence of molecules, contrary to what Salmon claims is shown by his "five hypothetical scientists" argument.

5. A Hypothetico-Deductive Interpretation

A different interpretation is to suppose that Perrin is engaging in a form of hypothetico-deductive reasoning: From the hypothesis that molecules exist and have properties he attributes to them he draws deductive conclusions regarding observable phenomena, including Brownian motion. He tests these conclusions experimentally and finds they are correct. From this he infers that his molecular hypotheses are probable, or at least that they are confirmed or supported by observations. This is no more circular than any use of hypothetico-deductive reasoning. In its simplest form it is just: O is derivable from T; O is true; hence, T is confirmed or probable. T is not being assumed to be true or probable at the outset.

From what hypothesis or set of hypotheses is Perrin supposed to have derived observational conclusions, and what observational conclusions does he derive? The following hypothesis is clearly among those from which Perrin derives consequences:

h = Chemical substances are composed of molecules, the number N of which in a gram molecular weight of a substance is the same for all substances.

A claim (indeed the most important one) that Perrin establishes experimentally that he takes to confirm h is this:

C. The calculation of N done by means of Perrin's experiments on Brownian particles using Eq. (9) is 6×10^{23}, and this number remains constant even when values for n', n, and so on, in Eq. (9) are varied.

Proposition C might well be called "observational." But it is not something that Perrin derives from his theoretical hypothesis h, nor from h together with other hypotheses he employs about molecules and Brownian particles. What Perrin does is to derive Eq. (9), not proposition C, deductively from such hypotheses. Then he uses Eq. (9) *together with results from various carefully designed experiments*, to establish C, which he regards as confirming molecular theory. But this is not the procedure advocated by hypothetico-deductivists. Contrary to the hypothetico-deductive view, the conclusion whose establishment is being claimed to confirm the theory is not derived from that theory.

Even if Perrin does not derive C from his theory, could he have done so? Is C derivable from the theoretical assumptions Perrin in fact makes? No, because even though one of the hypotheses Perrin was using is that N is a constant, he did not begin with any theoretical postulate concerning the numerical value of this constant. As noted, there were experiments on phenomena other than Brownian motion from which N was calculated to be approximately 6×10^{23}. But C is not derivable from this fact. Nor, in order to obtain his result C, did Perrin assume that these other experimental values for N were correct.

Finally, and perhaps most importantly, as I will argue in sections 7 and 8, Perrin's approach to confirming molecular theory is richer than that suggested by a hypothetico-deductive approach. He does not in fact defend this theory simply on the grounds that it entails true "observational" conclusions (whether or not these include C). Nor, therefore, is he subject to criticisms of the dubious hypothetico-deductive view of confirmation, according to which if h entails e, then e confirms h (see above, chapter 7).

6. Bootstrapping

Clark Glymour's idea of bootstrapping looks more promising than the hypothetico-deductive account because it uses experimental results together with hypotheses in a theory to confirm those hypotheses.[12] To invoke Glymour's own simple example, consider the ideal gas law expressed as

(10) $\quad PV = kT,$

where P represents the pressure of a gas, V its volume, T its absolute temperature, and k an undetermined constant. We suppose that we can experimentally determine values for P, V, and T, but not for k. The hypothesis (10) can be "bootstrap confirmed" by experimentally obtaining one set of measurements for P, V, and T, and then employing Eq. (10) itself to compute a value for k. Using this value for k, together with a second set of values for P, V, and T, we can instantiate this equation.

Glymour himself cites Perrin's reasoning in determining a value for Avogadro's number as an example of this type of confirmation.[13] Although he does not spell out the Perrin example, presumably what Glymour will say is this: Perrin's Eq. (9) relates Avogadro's number to measurable quantities n', n, m, etc. Using one set of measurements for these quantities, Perrin employed Eq. (9) itself to compute a value for N. Using this value for N, together with a second set of values for n', n, m, etc., Perrin instantiated, and thus "bootstrap confirmed," Eq. (9).

In other writings I have criticized Glymour's general account of bootstrap confirmation on the grounds that it allows the confirmation of equations containing completely undefined or obviously meaningless terms.[14] This objection is related to the point I now want to make.

---

12. Clark Glymour, *Theory and Evidence*.
13. Glymour, "Relevant Evidence," reprinted in Peter Achinstein, *The Concept of Evidence* (Oxford: Oxford University Press, 1983), p. 130, fn. 12.
14. Peter Achinstein, *The Nature of Explanation*, chapter 11; see note 16 below.

When the ideal gas equation $PV = kT$ gets confirmed in the manner indicated by Glymour—simply by experimentally determining two sets of values for P, V, and T—the term k (at least in Glymour's example) simply represents a constant, that is, a number. This constant can be given a molecular interpretation.[15] But it need not be; it can simply be construed as a constant of proportionality, that is, that number by which T needs to be multiplied to yield the same number as the product PV. This is the way that Glymour seems to be treating it. The value of that constant is to be determined experimentally.

Now in Perrin's Eq. (9) the constant N can be construed in a manner exactly analogous to the way Glymour seems to be treating k in (10), that is, as a numerical constant relating the other physical quantities in (9). Indeed, nothing in Glymour's theory of confirmation requires us to interpret N in Eq. (9) as a number *of anything*, let alone a number *of molecules*.[16] (The other quantities in (9) are physically interpreted.) Equation (9) would be "bootstrap confirmed" by two sets of measurements of the quantities n, n', etc., if N represents the number of angels on the head of a pin, or if N is just like a constant of proportionality. So "bootstrap confirming" Eq. (9), or any other equation (such as (7)) containing the constant N, does not confirm the existence *of molecules*. (No one, not even Glymour, takes "bootstrap confirming" the ideal gas Eq. (10) to be confirming the existence of molecules, even though k in (10) can be given a molecular interpretation.) But when Perrin determined Avogadro's number from his experiments using Eq. (9) he took his results to confirm the existence of molecules. Either he was mistaken in doing so, or else Glymour's "bootstrap confirmation" of Eq. (9) does not capture, at least not completely enough, the logic of Perrin's reasoning.

15. When Glymour writes Eq. (10) he does not make clear which constant he has in mind by k. Perhaps he means what physicists usually call R, the universal gas constant, which is experimentally determined in the manner he notes. Perhaps he means Boltzmann's constant k, which is equal to R divided by Avogadro's number N. In either case, Glymour's k in Eq. (10) can be related to molecular quantities.

16. It is instructive to look at an example which Glymour does work out involving 6 equations: (1) $A_1 = E_1$; (2) $B_1 = G_1 + G_2 + E_2$; (3) $A_2 = E_1 + E_2$; (4) $B_2 = G_1 + G_2$; (5) $A_3 = G_1 + E_1$; (6) $B_3 = G_2 + E_2$. The As and Bs are directly measurable quantities, the Es and Gs are "theoretical" quantities whose values can be determined indirectly through the theory by determining the values of the As and Bs. (Glymour identifies none of these quantities.) According to Glymour, here is how we can confirm hypothesis (1). We determine a value for A_1 directly by experiment, since this quantity is directly measurable. We obtain a value for the theoretical quantity E_1 by obtaining values for the observables B_1, B_3, and A_3, and then by using hypotheses (2), (5), and (6), mathematically computing a value for E_1. If this value is the same as the one determined for A_1 directly by experiment, then we have confirmed hypothesis (1). Note that we can confirm (1) without assigning any physical meaning to E_1. Only the observable quantities (the As and Bs) need be given any physical meaning.

In a simple counterexample to Glymour's system I develop in *The Nature of Explanation*, chapter 11, let A = the total force acting on a particle; B = the product of the particle's mass and acceleration; C = the quantity of God's attention focused on a particle. The theory consists of two equations: (i) $A = C$; (ii) $B = C$. On Glymour's account, these equations can be bootstrap-confirmed by measuring A and B and using equations (i) and (ii) in the manner Glymour proposes. Surely one cannot conclude that anything about God has been confirmed by such a procedure!

7. A Solution

I have noted three ways of construing Perrin's reasoning to the reality of molecules from his experimental determination of Avogadro's number on the basis of Brownian motion. Salmon's common-cause idea, as he formulates it, is not sufficient or necessary to yield the desired conclusion. The hypothetico-deductive account does not adequately represent Perrin's reasoning, since his calculation of N, from which he infers the existence of molecules, is not derived or derivable from the theoretical assumptions Perrin makes. Nor does Glymour's bootstrapping approach to Eq. (9) permit us to see how Perrin legitimately could have inferred the existence of molecules.

I will now suggest a way to understand Perrin's reasoning that avoids circularity and yields an argument free from problems of the previous interpretations. I will also show how his reasoning yields evidence that satisfies my conditions for potential and veridical evidence.

In section 3 I took the first step in Perrin's reasoning to be (in part) this: From various assumptions, including that molecules exist, and that gases containing them satisfy the ideal gas law, Perrin derives Eq. (7). This certainly is part of Perrin's reasoning, but it is not his "first premise." Both in his 1909 article and in his 1913 book, long before he begins to derive Eqs. (7) and (9), he offers a general discussion of the atomic theory and the existence of atoms and molecules comprising chemical substances. In his book 82 pages are spent developing atomic theory and giving chemical evidence in its favor before he turns to a discussion of Brownian motion in chapter 3 on p. 83. In his article he begins with a description of Brownian motion and then offers arguments that Brownian motion is caused by the agitation of molecules and hence that molecules exist. (This is also how he begins chapter 3 of his book.) Let me briefly mention two such arguments.

First, he writes:

> it was established by the work of M. Gouy (1888), not only that the hypothesis of molecular agitation gave an admissible explanation of the Brownian movement, but that no other cause of the movement could be imagined, which especially increased the significance of the hypothesis.[17]

Perrin notes that Gouy's experiments established that known "external" causes of motion in a fluid, including vibrations transmitted to the fluid by external causes, convection currents, and artificial illumination of the fluid, do not produce the Brownian motion. When each of these known causes was reduced or eliminated the Brownian motion continued unabated. So, Perrin concludes,

> it was difficult not to believe that these [Brownian] particles simply serve to reveal an internal agitation of the fluid, the better the smaller they [the Brownian particles] are, much as a cork follows better than a large ship the movements of the waves of the sea. (p. 511)

Perrin offers a second argument that he regards as stronger than the first. When a fluid is disturbed, the relative motions of its small but visible parts are ir-

17. "Brownian Movement and Molecular Reality," pp. 510–511.

regular. (This can be seen when colored powders are mixed into the fluid.) However, this irregularity of motion does not continue as the visible parts get smaller and smaller. At the level of Brownian motion an equilibrium is established between what Perrin calls "coordination" and "decoordination": if certain Brownian particles stop, then other Brownian particles in other regions assume the speed and direction of the ones that have stopped. From this Perrin draws the following conclusion:

> Since the distribution of motion in a fluid does not progress indefinitely, and is limited by a spontaneous recoordination, it follows that the fluids are themselves composed of granules or *molecules*, which can assume all possible motions relative to one another, but in the interior of which dissemination of motion is impossible. If such molecules had no existence it is not apparent how there would be any limit to the de-coordination of motion. On the contrary if they exist, there would be, unceasingly, partial re-coordination; by the passage of one near another, influencing it (it may be by *impact* or in any other manner), the speeds of these molecules will be continuously modified, in magnitude and direction, and from these same chances it will come about sometimes that neighboring molecules will have concordant motions. (514–515; italics his)

Perrin is arguing that irregular motions of the parts of a fluid become regular at the level of Brownian particles, in such a way that the total momentum of these particles is conserved. This strongly suggests that the Brownian particles (which are not responsible for their own motion) are being subjected to the influence of smaller particles still—molecules—which exhibit an equilibrium between coordination and decoordination of motion. He concludes:

> In brief the examination of Brownian movement alone suffices to suggest that every fluid is formed of elastic molecules animated by a perpetual motion. (p. 515)

In both arguments Perrin appears to be using eliminative-causal reasoning of the following sort:

A: (1) Given what is known, the possible causes of effect E (for example, Brownian motion) are $C, C_1, \ldots, C_n$ (for example, the motion of molecules, external vibrations, convection currents).
    (2) $C_1, \ldots, C_n$ do not produce effect E.
So probably,
C produces E.

A premise of type (1) may be defended by appeal to the fact that similar known observed effects are produced by and only by one of the types of causes on the list. Alternatively, it may be defended by appeal to more general established principles mandating one of these causes for an effect of that type. (Such principles may also provide a mechanism by means of which E can be produced by one or more of these causes.) A premise of type (2) may be defended by appeal to the fact that effect E is achieved in the absence of $C_1, \ldots, C_n$, or even when these causes are varied. In the first argument, for example, Perrin claims that Gouy's experiments take into account known external causes of motion in a fluid, and that molecular motion can in principle produce Brownian motion, since "the incessant movements of the [postulated] molecules of the fluid, which striking unceasingly the observed [Brownian] particles, drive about these particles irregularly through the fluid, except in the case where these impacts exactly counterbalance one another" (p. 513). This causal possibility, he obviously believes, can be defended by appeal to more general mechanical principles.[18] In addition, in the first argument Perrin defends premise (2) by appeal to the fact that the motion of the Brownian particles exists whether or not there is external agitation, convection currents, or the like.

Although causation is invoked in these arguments, this is different from Salmon's version of the common-cause argument. For one thing, a common cause of different phenomena is not inferred here. For another, in a common-cause argument no premises of types (1) and (2) need appear. Nor do they in Salmon's "five hypothetical scientists" argument cited earlier. (Nor are they required by the conjunctive fork conditions; see note 11.)

Even if arguments of type A do not establish the existence of molecules with certainty—since other possible causes cannot be precluded with certainty—Perrin believes that his arguments make it likely that Brownian motion is caused by the motion of molecules that make up the fluid. Accordingly, before any discussion of his own experimental results leading to his determination of Avogadro's number, and then to his claim that molecules exist, Perrin presents preliminary reasons to believe the latter claim. (In his article, as well as in his book, chemical arguments are also presented, for example, from combinations of elements and compounds). In addition, he presents reasons for believing that Avogadro's number exists, that is, that the number of molecules in a substance whose weight in grams is its molecular weight is the same for all substances.[19] Finally, as Perrin notes, values for Avogadro's number determined by experiments on phenomena other than Brownian motion yield approximately 6×10^{23}.[20] Accordingly, independently of his own experimental results with granules in an

---

18. At this point, however, Perrin does not present such principles or show quantitatively exactly how molecular motion can cause Brownian motion. This is important because some earlier investigators claimed that Brownian motion was incompatible with standard assumptions of the kinetic-molecular theory. One objection, first raised by Karl Nägeli in 1879, is that kinetic theory calculations show that the velocity that would be imparted to a Brownian particle by a collision with molecules would be much too small to observe, contrary to what is the case. A second objection is that Brownian motion, if produced by molecular collisions, would violate the second law of thermodynamics. (For a discussion of the validity and impact of these objections, see Mary Jo Nye, *Molecular Reality: A Perspective on the Scientific Work of Jean Perrin* (London: Macdonald, 1972), pp. 25–27, 101–102; Stephen G. Brush, "A History of Random Processes," *Archive for History of the Exact Sciences*, 5, 1968, pp. 10ff; Roberto Maiocchi, "The Case of Brownian Motion," *British Journal for the History of Science*, 23 (1990), pp. 261ff.

    Although Perrin in the qualitative discussion of Guoy does not present a quantitative mechanical explanation of Brownian motion, he does later in his article when he gives the theoretical explanation offered by Einstein in 1905. He responds to the second objection by defending a statistical, by contrast to a universal, interpretation of the second law of thermodynamics—one that he regards as established by Clausius, Maxwell, Helmholtz, Boltzmann, and Gibbs (see Perrin 1984, p. 512).

19. See "Brownian Movement and Molecular Reality," 515–516; *Atoms*, 18ff.
20. *Atoms*, 105, 215; "Brownian Movement . . . ," 521–524, 583–598.

emulsion, Perrin clearly believed that there was information (call it background information b) available to him and to other physicists and chemists that supported the following theoretical proposition:

> T = Chemical substances are composed of molecules, the number N of which in a gram molecular weight of any substance is (approximately) 6×10^{23}.[21]

Perrin believed that, on the basis of b alone, T was at least more probable than not, and hence that

(i) $p(T/b) > \frac{1}{2}$.[22]

He gives arguments (some noted above) to support (i).

Now, as indicated earlier, the experimental result achieved by Perrin on the basis of which he most firmly concludes that molecules exist is

> C = The calculation of N done by means of Perrin's experiments on Brownian particles using equation (9) is 6×10^{23}, and this number remains constant even when values for n', n, etc. in equation (9) are varied.

Proposition C is not a deductive consequence of T. Even if T is true, C could be false, since the particular experimental assumptions and conditions introduced by Perrin are not required by T to be appropriate to test T. To be sure, we might add experimental assumptions to T that would yield C as a deductive consequence. Such assumptions would include that the Brownian particles of gamboge employed by Perrin all have the same mass and volume, that such particles can be treated like large molecules obeying the standard gas laws, and that Stokes law is applicable to Brownian particles. Perrin gave empirical arguments for each of these and other assumptions he made.[23] But suppose that instead of adding these assumptions to T, we simply add to the background information b Perrin's experimental results which do not deductively entail these assumptions, but (Perrin believed) made them probable. If so, then, even if T together with all the information we are now counting as part of the background information b (including other determinations of N) does not entail the experimental result C,

---

21. Note that T, unlike h in section 5, gives a specific value for Avogadro's number. (h is simply "Chemical substances are composed of molecules, the number N of which in a gram molecular weight of a substance is the same for all substances.") The point of the difference is this. Perrin's derivation of Eq. (7), and then of Eq. (9), proceeds from h, not from the stronger T (which contains a specific value for N). Otherwise in Eqs. (7) and (9) he could have substituted a specific value for N. So from the perspective of the h-d account, which considers what theoretical hypotheses Perrin starts with, from which he derives conclusions (7) and (9), it is appropriate to choose h rather than T. However, my point in the present section is not to consider Perrin's initial theoretical hypotheses, but rather hypotheses he takes to be proved or made likely by his arguments. Clearly the stronger hypothesis T (which entails the weaker h) is what Perrin seeks to establish.
22. Probability is here construed in the objective epistemic sense of chapter 5.
23. For example, in "Stokes' Law and Brownian Motion," *Comptes Rendus* (1908), 147ff, he gives experimental arguments for the applicability of Stokes' law. For a discussion and criticism of these arguments, see Roberto Maiocchi, "The Case of Brownian Motion," 257–283; see 278–279.

it does make C probable. At least C becomes more probable given the truth of T than without it. That is, it is more likely that Perrin's experiments will yield $N = 6 \times 10^{23}$ given the assumption that there are molecules whose number $N = 6 \times 10^{23}$, than it is without such an assumption. So,

(ii) $p(C/T\&b) > p(C/b)$.

Now the following is a theorem of probability:

(iii) If $p(C/T\&b) > p(C/b)$, and if $p(T/b) > 0$ and $p(C/b) > 0$, then $p(T/C\&b) > p(T/b)$.[24]

Theorem (iii) states that if T increases C's probability on b, and if both T's and C's probability on b are greater than zero, then C increases T's probability on b.

Now, from (i), T's probability on b is not zero. And, since b contains the information that other experimental determinations of N yield approximately 6×10^{23}, C's probability on b is also not zero. So, it follows from (ii) and (iii) that Perrin's experimental result C increases the probability of the theoretical assumption T, that is

(iv) $p(T/C\&b) > p(T/b)$.

Finally, if T's probability on b alone is greater than $\frac{1}{2}$, that is, if (i) is true, then we can conclude that T's probability on $C\&b$ is at least as high, that is,

(v) $p(T/C\&b) > \frac{1}{2}$.

8. Evidence

Result (iv) will be of interest to those who consider increase in probability as sufficient as well as necessary for evidence. According to this positive relevance view,

(11) e is evidence that h, given b, if and only if $p(h/e\&b) > p(h/b)$.

On this conception, Perrin's experimental result C concerning the calculation of Avogadro's number from experiments on Brownian motion counts as evidence in favor of his theoretical assumption T, which postulates the existence of molecules. Moreover, the greater the boost in probability that T gives to C in (ii), the greater the boost in probability that C gives to T in (iv). Assuming these boosts in probability are high, on the positive relevance conception (11), we can conclude that Perrin's experimental result C provides strong evidence for his theoretical assumption T. (In section 9 I discuss Perrin's reasons for believing that his evidence was so strong.)

However, as I argue in chapter 4, positive relevance is neither necessary nor sufficient for evidence. In its place I advocate this definition in chapter 8:

24. Proof: According to Bayes' theorem, $p(T/C\&b) = p(T/b) \times p(C/T\&b)/p(C/b)$. By the assumptions of the theorem, $p(T/b) > 0$ and $p(C/b) > 0$. So $p(T/C\&b)/p(T/b) = p(C/T\&b)/p(C/b)$. Since, by the assumptions of the theorem, $p(C/T\&b) > p(C/b)$, it follows that $p(T/C\&b) > p(T/b)$.

(PE) e is potential evidence that h, given b, if and only if
  (a) e and b are true;
  (b) e does not entail h;
  (c) $p(h/e\&b) \times p(\text{there is an explanatory connection between } h \text{ and } e/h\&e\&b) > \frac{1}{2}$.

On this conception of evidence, what is important is result (v), not (iv). Increase in probability is neither necessary nor sufficient for evidence. But, it follows from condition (c) in (PE) that high probability is necessary. (c) also requires the high probability of an explanatory connection between h and e, given the truth of h&e&b. But this is also satisfied in the case of Perrin. Given T, that chemical substances are composed of molecules, the number N of which in a gram molecular weight of a substance is 6×10^{23}, and given C, that the calculation of N from Perrin's experiments using Eq. (9) is 6×10^{23}, and given the information noted in b, the probability is high that the reason C obtains is that T is true. The probability is high that Perrin's experimental calculation yielded 6×10^{23} because chemical substances are composed of molecules, the number N of which in a gram molecular weight is 6×10^{23}. So

  (vi) $p(\text{there is an explanatory connection between } T \text{ and } C/T\&C\&b) > \frac{1}{2}$.

If, as seems reasonable, we may also suppose that the probabilities in (v) and (vi) are sufficiently high to allow their product to be greater than $\frac{1}{2}$, then condition (c) in (PE) is satisfied. Assuming that C is true, as are the facts reported in b, and that C does not entail T, it follows that Perrin's experimental result C is potential evidence for his theoretical claim T. Since T is true, and since there is an explanatory connection between T and C, the experimental result C is also veridical evidence that T.

Perrin's reasoning, so represented, reaches the conclusion that his experimental result from Brownian motion constitutes evidence for the truth of a theoretical claim involving the existence of molecules. It does so on the conceptions of potential and veridical evidence I defend. Moreover, it does so without circularity. There is no undefended assumption at the outset that molecules exist. The claim in (i) that molecules probably exist is based on reasons for their existence cited in the background information b in (i). Perrin does not begin by assuming without argument that molecules probably exist. He begins by providing a basis for this assumption that includes experiments other than the one of concern to him in C.[25]

The pitfalls of the three interpretations in previous sections are avoided. Unlike the common-cause and bootstrap accounts, we end up by confirming a claim entailing that molecules exist. Unlike the hypothetico-deductive account, we need not suppose that Perrin's experimental result expressed in C is a deductive consequence of his theoretical assumptions about molecules. Nor need we accept the dubious hypothetico-deductive view of evidence.

25. As indicated in note 18, it also includes the theoretical assumption, later justified, that a quantitative mechanical explanation of Brownian motion can in principle be obtained from kinetic theory assumptions.

9. Why Argue for Molecules in 1908?

Suppose we agree that Perrin's reasoning has been adequately represented and that it is not circular. Why in 1908, eleven years after the discovery of the electron, let alone in 1913 when his book appeared, should Perrin have thought it necessary or even useful to present an argument for the existence of molecules?

To begin with, extending into the first decade of the twentieth century, serious opposition to any atomic-molecular theory had been expressed by some physicists and chemists. French and German positivists, including Mach, Duhem, and Poincaré, for whom unobservable entities underlying the observed phenomena were anathema, rejected any realist interpretation of atomic theory. At best, atoms and molecules, if invoked at all, were to be construed simply as instrumental, conceptual devices. The German physical chemist Friedrich Wilhelm Ostwald until at least 1908 rejected atomic theory in favor of the doctrine of "energetics." His grounds for doing so were partly philosophical (a repudiation of any form of unverifiable materialism) and partly based on scientific reasons (including the belief that atomic theories, being purely mechanical, should always entail reversible processes, something incompatible with observed thermodynamic phenomena). Indeed, in the preface to his book *Atoms*, Perrin explicitly mentions Ostwald's rejection of hypotheses about unobserved atomic structure. He describes Ostwald as advocating the "inductive method" which he (Perrin) takes to be concerned only with inferring what is observable from what is observed. By contrast, Perrin is employing what he calls the "intuitive method," which attempts "*to explain the complications of the visible in terms of invisible simplicity*" (*Atoms*, p. vii; his italics).

Although Ostwald rejected atomic theories well into the first decade of the twentieth century, he changed in views by 1909, as a result of the work of Thomson and Perrin.[26] At the end of his 1913 book Perrin refers to the recent controversy over atomic theory and boldly claims that as a result of his (and other) determinations of Avogadro's number,

> the atomic theory has triumphed. Its opponents, which until recently were numerous, have been convinced and have abandoned one after the other the sceptical position that was for a long time legitimate and no doubt useful. (*Atoms*, p. 216)

Perrin was very conscious of the controversies over atoms extending into the twentieth century and felt the need to settle the issue on the side of the atomists.[27]

---

26. "*I have convinced myself that we have recently come into possession of experimental proof of the discrete or grainy nature of matter, for which the atomic hypothesis had vainly been sought for centuries, even millenia*. The isolation and counting of gas ions on the one hand . . . and the agreement of Brownian movements with the predictions of the kinetic hypothesis on the other hand, which has been shown by a series of researchers, most completely by J. Perrin—this evidence now justifies even the most cautious scientist in speaking of the *experimental* proof of the atomistic nature of space-filling matter" (Friedrich Wilhelm Ostwald, *Grundriff der Allgemeinen Chemie* (Leipzig: Engelmann, 1909; quoted in Brush, "A History of Random Processes,"; italics Ostwald's).

27. For a very informative extended discussion, see Mary Jo Nye, *Molecular Reality: A Perspective on the Scientific Work of Jean Perrin* (London: Macdonald, 1972).

As noted earlier, independently of his 1908 experiments, Perrin believed that available information provided at least some support for the existence of atoms and molecules. Yet he regarded his own experimental results as particularly important in this connection. Why?

Although he considered previous arguments to be supportive, he believed that they did not supply sufficiently direct evidence for molecules. Interestingly, in 1901 he regarded the evidence for the existence of electrons—evidence he himself had helped to develop (see chapter 2)—to be more direct than that for molecules:

> It is remarkable that the existence of these corpuscles [following J. J. Thomson, Perrin used this term for electrons], thanks to the strong electric charges which they carry, is demonstrated in a more direct manner than that of atoms or molecules, which are much larger.[28]

In 1901, although neither molecules nor electrons were visible as discrete particles observable with a microscope, the effects of electrons were more directly observable than those of molecules. Cathode rays, that is, streams of negatively charged electrons produced in cathode tubes, were observed to produce fluorescence in the glass of the tube, as well as on zinc sulfide screens, and to be deflected by magnetic and electric fields. Neutral molecules and atoms were not known to have these or analogous observable effects. Prior to his experiments, or at least prior to the study of Brownian motion, Perrin regarded the evidence for molecules to be less direct, based as it was in chemistry on the regularities of chemical composition and proportion, and in physics, especially in the kinetic theory of gases developed by Maxwell, on phenomena of heat transfer, on the success of mechanical theories in general, and on the ability of chemical theory as well as the kinetic theory to explain a range of observable phenomena.[29] For Perrin, Brownian motion was for molecules what cathode rays were for electrons. Both phenomena provided a relatively direct link between the postulated entities and their observable effects.

Second, Perrin regarded his evidence for molecules as providing more precise and certain quantitative information about molecules than was previously available. He considered his determination of Avogadro's number and of the masses and diameters of molecules and atoms to be more accurate than previous estimates. He writes:

> this same equation [one corresponding to (9) above] affords a means for determining the constant N, and the constants depending on it, which is, it appears, *capable of an unlimited precision*. The preparation of a uniform emulsion and the determination of the magnitudes other than N which enter into the equation can in reality be pushed to whatever degree of perfection [is] desired. It is simply a question of patience and time; nothing limits *a priori* the accuracy of the results, and the mass of the atom can be obtained, if desired, with the same precision as the mass of the Earth.[30]

---

28. Quoted in Nye, *Molecular Reality*, 83–84; from "Les Hypotheses Moleculaires," *Revue Scientifique*, 15 (1901), 449–461; quotation, p. 460.
29. For Maxwell's arguments for molecules, see my *Particles and Waves*, essays 7 and 8.
30. "Brownian Movement...," 555–556.

Finally, Perrin regarded his experimental results on Brownian motion as important in the confirmation of atomic theory for another reason as well. These results included not only a determination of Avogadro's number from law-of-atmosphere experiments—which has been the focus of attention in this chapter. They also included such a determination from Einstein's theory of Brownian motion, which from kinetic theory assumptions generates a formula relating the displacement of Brownian particles to N. Perrin considered Einstein's theory crucial in providing what he called a "mechanism" by which an equilibrium is reached in molecular situations such as those governed by the law of atmospheres. Prior to his 1908 experiments Perrin considered Einstein's theory to be experimentally unverified.

Let us return now to the probabilistic reconstruction of Perrin's reasoning. We formulated his major experimental result as

> C = The calculation of N done by means of Perrin's experiments on Brownian motion using Eq. (9) is 6×10^{23}, and this number remains constant when values for n', n, etc. in Eq. (9) are varied.

A theoretical claim for which C is supposed to provide evidence is

> T = Chemical substances are composed of molecules, the number N of which in a gram molecular weight of any substance is (approximately) 6×10^{23}.

Perrin believed that T's probability is increased by establishing C, that is,

(iv) $p(T/C\&b) > p(T/b)$.

Indeed, because of the two facts noted above concerning the evidence reported in C—its directness and precision—by contrast to other evidence for molecules contained in b, Perrin believed that C gave a substantial boost to the probability of T. If this is right, then for those who adopt the positive relevance account of evidence, and the associated idea that the bigger the increase the stronger the evidence, Perrin's experimental result C provided substantial evidence for the theoretical claim T. More precisely, on this account of evidence, C and b together count as stronger evidence for T than b by itself if and only if (iv) obtains.[31] So on this view of evidence, we can understand at least one reason why Perrin regarded his experimental result C as important. Not only did it, together with b, provide evidence for T, but also it provided stronger evidence for T than b alone, that is, than information available before Perrin's experiments.

Matters are not so simple on the account of evidence I propose. Although the latter sanctions the conclusion that, given b, Perrin's experimental result C is potential and veridical evidence for theoretical claim T, the question of the strength of that evidence is not settled by (PE). Nevertheless, this much can be said. If, in accordance with (PE), both e_1 and e_2 are potential evidence that h, given b, and if e_1 reports a higher frequency of the property in question than does e_2, or if e_1 contains a larger, more varied, or more precisely described sample than e_2, or if

---

31. On a standard view, e_1 provides stronger evidence than e_2 for h if and only if $p(h/e_1) - p(h) > p(h/e_2) - p(h)$, that is, if and only if $p(h/e_1) > p(h/e_2)$. See chapter 3.

it describes items more directly associated with those in h than does e_2, then e_1 is stronger evidence for h than e_2 *in one or more of these respects*. In Perrin's case, I have argued, definition (PE) is satisfied. His experimental result C counts as evidence for T, given the background information b in question. It can also be argued that b itself contains evidence for T. But C is stronger evidence for T than b in several respects, including precision and directness. This is among the reasons why Perrin believed his evidential claim was worth making in 1908.

10. Perrin and Realism

Salmon urges that Perrin's argument is an argument for scientific realism. Moreover, unlike the usual philosophical arguments for realism, it is empirical rather than a priori. He writes:

> In an effort to alleviate this intellectual discomfort [produced by philosophical arguments between realists and anti-realists], I decided to try an empirical approach to the philosophical problem. Since it seemed unlikely that scientists would have been moved by the kinds of arguments supplied by philosophers, I felt that some insight might be gained if we were to consider the evidence and arguments that convinced scientists of the reality of unobservable entities.[32]

Salmon believes that Perrin's argument to the reality of molecules provides a "clear and compelling example" of a scientific argument for the existence of unobservable entities. As noted, he considers Perrin's argument to be of the common-cause variety. Even if we reject this interpretation, the question remains as to whether Perrin's argument is, or is best construed as, an argument for scientific realism, and whether Perrin himself understood it in this way.

An antirealist might provide a very different interpretation of Perrin's conclusion. Instead of claiming that the theoretical claim T is true, all that Perrin is doing, or at least all that he is entitled to do as a scientist, is infer that T is empirically adequate, that it "saves the phenomena." T can accomplish the latter without being true. This suggests two questions, one philosophical, one historical. First, is the antirealist correct in supposing that it is possible to have a valid argument to the probable conclusion that some theory saves the phenomena that is not also a valid argument to the probable conclusion that the theory is true? Second, even if it is possible, is it historically plausible to construe Perrin's reasoning in this way?

The answer to the first question is yes, in accordance with the account of "saving the phenomena" provided in chapter 8, section 7. There I showed that it is possible for it to be highly probable that a theory saves the phenomena while it is highly improbable that the theory is true.

Applying this to Perrin's reasoning, he conducted a series of experiments on Brownian motion with different values for n', n, and so on, in Eq. (9). Let C_i be the proposition that the calculation of N done by means of Perrin's ith experiment on Brownian particles using Eq. (9) is (approximately) 6×10^{23}. As before, let

---

32. Salmon, *Scientific Explanation* . . . , 213–214.

> T = Chemical substances are composed of molecules, the number N of which in a gram molecular weight of any substance is (approximately) 6×10^{23}.

In section 7 above, I noted the possibility of taking Perrin's background information b to contain assumptions that together with T entail C.[33] On an antirealist interpretation, Perrin could be arguing that it is highly probable that T saves the C-phenomena without arguing that T is probably true. That is, Perrin could be arguing that

(12) $\lim_{m,n\to\infty} p(T \text{ saves } C_{n+1} \ldots C_{n+m}/C_1 \ldots C_n \& b) = 1$

without supposing that $p(T/C_1 \ldots C_n \& b)$ is high, or that

(13) $\lim_{n\to\infty} p(T/C_1 \ldots C_n \& b) = 1.$

Accordingly, an antirealist has a way of understanding Perrin's reasoning that does not commit the antirealist, or Perrin, to drawing the conclusion that the theory itself is true or highly probable, and hence to drawing the conclusion that molecules are real. As argued in chapter 8, such an antirealist interpretation would preclude our saying that Perrin's experimental results C constitute (potential or veridical) evidence that theory T is true, because the explanatory connection condition would be violated. Nor, I argued, can the antirealist maintain that C constitutes potential or veridical evidence that T *saves the phenomena*—at least not without abandoning antirealism or supplying some yet to be defined concept of evidence that discards an "explanatory connection" condition and avoids the problems that I believe require such a condition.

Even if an antirealist reconstruction is possible—even if all the antirealist wants is to show that Perrin's argument establishes (12) and not that Perrin's experiments provide evidence that molecules exist or evidence that the molecular theory saves the phenomena—is this antirealist reconstruction historically plausible in the case of Perrin?

Admittedly, Perrin makes some remarks which may suggest antirealism. For example, near the end of his paper he writes:

> Lastly, although with the existence of molecules or atoms the various realities of number, mass, or charge, for which we have been able to fix the magnitude, obtrude themselves forcibly, it is manifest that we ought always to be in a position to express all the visible realities without making any appeal to the elements still invisible. But it is very easy to show how this may be done for all the phenomena referred to in the course of this Memoir. (p. 599)

Perrin argues that one can take various laws which relate Avogadro's number to measurable quantities and derive a new equation containing only measurable quantities. If the two laws governing different phenomena are expressible as $N = f(A,B,C)$ and $N = g(D,E,F)$, where A–F are measurable quantities, we can write $f(A,B,C) = g(D,E,F)$, in which (as Perrin puts it) "*only evident realities occur*" (p. 600, his italics). But Perrin does not conclude from this that the most we can

---

33. Although this possibility was noted, in that section I supposed only that T increases C's probability on b.

say is that the molecular hypothesis is empirically adequate or has only instrumental value. His main point seems to be that by expressing an equation of the last form above we obtain a result that "expresses a profound connection between two phenomena at first sight completely independent, such as the transmutation of radium and the Brownian movement" (p. 600).

Most of his comments strongly suggest an attitude of realism. For example, "The real existence of the molecule is given a probability bordering on certainty."[34] Elsewhere, he writes, "Thus the molecular theory of the Brownian motion can be regarded as experimentally established, and, at the same time, *it becomes very difficult to deny the objective reality of molecules.*"[35] These are more typical passages.

Finally, Perrin's argument as I have reconstructed it probabilistically in steps (i)–(v) in section 7 is not an antirealist argument. It is not an argument simply to the conclusion that T saves the phenomena, or to (12), a probabilistic version of this. Step (v) asserts the high probability of T itself.

Accordingly, an antirealist must show not simply that Perrin's reasoning *can be* reformulated in an antirealist way to the conclusion (12), but that such a reformulation is required or desirable for historical or logical reasons. The historical grounds for such a reformulation are dubious at best. On logical grounds, considering my probabilistic reconstruction, the antirealist would need to show that there are invalid steps in the argument that can be removed only by adopting an antirealist conclusion such as (12) rather than a stronger realist conclusion such as (13). He must show that Perrin's preliminary arguments leading to step (i)—for example, eliminative-causal arguments of type A in section 7 from Brownian motion (not appealing to his own experimental results), arguments from chemical combinations, from kinetic theory, and from other determinations of Avogadro's number—do not give T a high probability. The antirealist needs to show not simply that these arguments *can be* reformulated as arguments to the conclusion that T saves the phenomena, but that something is faulty with these arguments themselves, that they fail to confer high probability on T (they fail to establish step (i)). He must show that Perrin's scientific reasoning is erroneous. This is not demonstrated by showing simply that the antirealist conclusion "T saves the phenomena" is possible, or even that it is more probable than T itself, since it commits one to much less than T. In the absence of arguments against specific steps in Perrin's reasoning, one can conclude, with Salmon, that Perrin supplies a reasonable empirical argument for the reality of molecules.[36]

---

34. *Atoms*, 215–216.
35. "Brownian Movement...," 554, italics his.
36. For very helpful criticisms of material in this chapter, I am indebted to Laura J. Snyder, Robert Rynasiewicz, Nancy Cartwright, Michael Redhead, students in my 1994 graduate seminar at Johns Hopkins, and four excellent referees for *Perspectives on Science* in which a version of this chapter initially appeared.

13

WHO REALLY DISCOVERED THE ELECTRON?

In this final chapter I return once again to the electron, a subject that figured prominently in chapter 2. The year 1997 marked the 100th anniversary of the discovery of the electron, perhaps the most important particle in physics. J. J. Thomson is credited with that discovery. Did he really discover the electron? If so, or if not, what did he do? What constitutes a discovery? What sort of evidence, if any, does it require? Did Thomson have such evidence?

1. Two Problems with Identifying J. J. Thomson as the Discoverer

Heroes are falling in this age of revisionist history. Thomas Jefferson, according to one recent authority, was a fanatic who defended the excesses of the French Revolution. Albert Einstein was not the saintly physicist we were led to believe, but was mean as hell to his first wife. And, more to the present purpose, J. J. Thomson really didn't discover the electron. Or so claim two recent authors, Theodore Arabatzis, in a 1996 article on the discovery of the electron,[1] and Robert Rynasiewicz, at a February 1997 AAAS symposium in honor of the 100th anniversary of the discovery.

I would like my heroes to retain their heroic status. However, my aim in this chapter is not to defend Thomson's reputation, but to raise the more general question of what constitutes a discovery. My strategy will be this. First, I want to say why anyone would even begin to doubt that Thomson discovered the electron. Second, I want to suggest a general view about discovery. Third, I will contrast this with several opposing positions, some of which allow Thomson to retain his status, and others of which entail that Thomson did not discover the

---

1. Theodore Arabatzis, "Rethinking the 'Discovery' of the Electron," *Studies in History and Philosophy of Science*, 27B (December 1996), 405–435.

electron; I find all of these opposing views wanting. So who, if anyone, discovered the electron? In the next part of the chapter I will say how the view of discovery I develop applies to Thomson, and also ask why we should care about who discovered the electron, or anything else. Finally, I will consider whether the discoverer of something must have evidence that it exists, whether Thomson in fact had such evidence in the case of electrons, and if so, whether it was, and had to be, evidence of all four types distinguished in chapter 2.

Let me begin, then, with two problems with identifying Thomson as the discoverer of the electron. The first is that before Thomson's experiments in 1897 several other physicists reached conclusions from experiments with cathode rays that were quite similar to his. One was William Crookes. In 1879, in a lecture before the British Association at Sheffield, Crookes advanced the theory that cathode rays do not consist of atoms,

> but that they consist of something much smaller than the atom—fragments of matter, ultra-atomic corpuscles, minute things, very much smaller, very much lighter than atoms—things which appear to be the foundation stones of which atoms are composed.[2]

So 18 years before Thomson's experiments, Crookes proposed two revolutionary ideas essential to Thomson's work in 1897: That cathode rays consist of corpuscles smaller than atoms, and that atoms are composed of such corpuscles. Shouldn't Crookes be accorded the title "discoverer of the electron"?

Another physicist with earlier views about the electron was Arthur Schuster. In 1884, following his own cathode ray experiments, Schuster claimed that cathode rays are particulate in nature and that the particles all carry the same quantity of electricity.[3] He also performed experiments on the magnetic deflection of the rays, which by 1890 allowed him to compute upper and lower bounds for the ratio of charge to mass of the particles comprising the rays. Unlike Thomson (and Crookes in 1879), however, Schuster claimed that the particles were negatively charged gas molecules.

Philipp Lenard is still another physicist with a considerable claim to be the discoverer of the electron. In 1892 he constructed a cathode tube with a special window capable of directing cathode rays outside the tube. He showed that the cathode rays could penetrate thin layers of metal and travel about half a centimeter outside the tube before the phosphorescence produced is reduced to half its original value. The cathode rays, therefore, could not be charged molecules or atoms, since the metal foils used were much too thick to allow molecules or atoms to pass through.

Other physicists as well, such as Hertz, Perrin, and Wiechert, made important contributions to the discovery. So why elevate Thomson and say that *he* discovered the electron? Why not say that the discovery was an effort on the part of many?

---

2. William Crookes, "Modern Views on Matter: The Realization of a Dream" (an address delivered before the Congress of Applied Chemistry at Berlin, June 5, 1903), *Annual Report of the Board of Regents of the Smithsonian Institution* (Washington, D.C.: Government Printing Office, 1904), p. 231. In this paper Crookes quotes the present passage from his 1879 lecture.
3. See Arthur Schuster, *The Progress of Physics* (Cambridge: Cambridge University Press, 1911), p. 61.

The second problem is this. Even assuming that Thomson discovered something, was it really the *electron?* How could it be, since Thomson got so many things wrong about the electron? The most obvious is that he believed that electrons are particles or corpuscles (as he called them), *and not waves*. In a marvelous twist of history, Thomson's son, G. P. Thomson, received the Nobel Prize for experiments in the 1920s demonstrating the wave nature of electrons. Another mistaken belief was that electrons are the *only* constituents of atoms. Still others were that the charge carried by electrons is not the smallest charge carried by charged particles, and that the mass of the electron, classically viewed, is entirely electromagnetic, a view Thomson came to hold later. Why not deny that Thomson discovered anything at all, since nothing exists that satisfies his electron theory?

To deal with these issues something quite general needs to be said about discovery.

2. What Is Discovery?

The type of discovery with which I am concerned is discovering some physical thing or type of thing (such as the electron, the Pacific Ocean), rather than discovering some abstract object (such as a proof), or discovering *that* something is the case (such as that the electron is negatively charged). Later I will consider a sense of discovering some thing X that requires a knowledge that it is X, as well as a sense that does not.

My view has three components, the first of which is ontological. Discovering something requires the existence of what is discovered. You cannot discover what doesn't exist—the ether, the Loch Ness monster, or the fountain of youth—even if you think you have. You may discover the *idea* or the *concept* of these things. Everyone may think you have discovered the things corresponding to these ideas or concepts. They may honor you and give you a Nobel Prize. But if these things don't exist, you haven't discovered them.

The second component of discovery is *epistemic*. A certain state of knowledge is required. If you are to be counted as the discoverer of something, not only must that thing exist but also you must know that it does. Crookes in 1879 did not discover electrons because he lacked such knowledge; his theoretical claim that cathode rays consist of subatomic particles, although correct, was not sufficiently established to produce the knowledge that such particles exist. However, not just any way of generating knowledge will do for discovery. I may know that something exists because I have read that it does in an authoritative book. Discovery, in the sense we are after, requires that the knowledge be first-hand, as it were.

What counts as "first-hand" can vary with the type of object in question. Since my concern is with discovering physical objects one might offer this rough characterization: knowledge that the objects exist is generated, at least in part, by observing those objects or their direct effects. This knowledge may require rather strenuous inferences and calculations from the observations. (Scientific discovery is usually not like discovering a cockroach in the kitchen or a nail in your shoe.) As noted, discovery involves not just any observations that will produce knowledge of the object's existence, but observations of the object itself or its direct effects. I may come to know of the existence of a certain library book by observing a computer screen in my office which claims that the library owns it. I may discover *that* the book exists by doing this. But I may never discover the book itself if I can't find it on the shelf. In discovering the book at least among the things that make me know that it exists is my seeing it. Finally, for discovery, the knowledge in question involves having as one's reason, or at least part of one's reason, for believing that X exists the belief that it is X or its direct effects that have been observed. My knowledge that electrons exist may come about as a result of my reading the sentence "Authorities say that electrons exist" on my computer screen. What is on my screen is a direct effect of electrons. But in such a case, I am supposing, my reason for believing that electrons exist does not include the belief that I have observed electrons or their direct effects on the screen.[4]

Putting together these features of the second (epistemic) component of discovery, we can say that someone is in an *epistemic situation necessary for discovering* X if that person knows that X exists, observations of X or its direct effects caused, or are among the things that caused, that person to believe that X exists, and among that person's reasons for believing that X exists is that X or its direct effects have been observed. More briefly, I will say that such a person knows that X exists from observations of X or its direct effects.

The third component of discovery is priority. If I am the discoverer of something, then the epistemic situation I have just described must be a "first." I put it this way because it is possible to relativize discovery claims to a group or even to a single individual. I might say that I discovered that book in the library last Tuesday, meaning that last Tuesday is the first time *for me*. It is the first time I knew the book existed by observing it, even though others knew this before I did. I might also make a claim such as this: I was the first member of my department to discover the book, thereby claiming my priority over others in a certain group. Perhaps it is in this sense that we say that Columbus discovered America, meaning that he was the first European to do so. And, of course, the relevant group may be the entire human race. Those who claim that Thomson discovered the electron mean, I think, that Thomson was the first human to do so.

There is a rather simple way to combine these three components of discovery, if we recognize that knowing that something exists entails that it does, if, as already indicated, we confine our attention to discovering physical objects, and if we employ the previously introduced concept of an epistemic situation necessary for discovery. The simple way is this:

> P discovered X if and only if P was the first person (in some group) to be in an epistemic situation necessary for discovering X.

That is, P was the first person (in some group) to know that X exists, to be caused to believe that X exists from observations of X or its direct effects, and to have as a reason for believing that X exists that X or its direct effects have been observed.

---

4. I am indebted to Kent Staley for this example and this point.

3. Clarifying Points

Before contrasting this with opposing views, and applying this to Thomson and the electron, some points need clarification.

1. On this account, to discover X, you do not need to observe X directly. It suffices to observe certain causal effects of X that can yield knowledge of X's existence. If I see a cloud of dust moving down the dirt road that is obviously being produced by a car approaching, then I can discover a car that is approaching even though I cannot see the car itself, but only the cloud of dust it is producing as it moves. However, it is not sufficient to come to know of X's existence via observations of just any sort. If I read a letter from you saying that you will be driving up the dirt road to my house at noon today, and I know you to be someone who always keeps his word, that, by itself, does not suffice for me to say that I discovered a car that is approaching at noon, even if I know that the car is approaching. Discovering the car requires observations of the car or its direct effects.

2. This will prompt the question "What counts as observing direct effects?" Some physicists want to say that the tracks produced by electrons in cloud chambers are direct effects, because electrons, being charged, ionize gas molecules around which drops of water condense forming the tracks. By contrast, neutrons, being neutral, cannot ionize gas molecules, and hence do not leave tracks. They are detected by bombarding charged particles that do leave tracks. More recently detected particles, such as the top quark, involve many different effects that are less direct than these.[5] This is a complex issue that cannot be quickly settled.[6] What seems to be involved is not some absolute idea of directness, but a relative one. Given the nature of the item whose effects it is (for example, if it is a neutron it cannot produce a track but must interact with charged particles that do produce tracks), this degree of directness in detecting its effects not only yields knowledge that the item exists but also furnishes the best, or one of the best, means at the moment available for obtaining that knowledge.

3. On this account, the observations of X or its effects need not be made by the discoverer, but by others. What is required is only that the discoverer be the first to know that X exists from such observations. The planet Neptune was discovered independently by Adams and Leverrier from observations of the perturbations of Uranus caused by Neptune. These observations were made by others, but complex calculations enabled these astronomers to infer where the new planet could be observed in the sky. The first actual telescopic observation of Neptune was made not by either of these astronomers but by Johann Galle at the Berlin Observatory. Although Galle may have been the first to see Neptune, he is not its discoverer, because he was not the first to come to know of its existence from observations of Neptune or its effects.

4. To discover X it is not sufficient simply to postulate, or speculate, or theorize that X exists. In 1920 Rutherford theorized that neutrons exist. But Chad-

---

5. For an illuminating discussion, see Kent Staley, "Over the Top: Experiment and the Testing of Hypotheses in the Search for the Top Quark" (Ph.D. dissertation, Johns Hopkins University, 1997).
6. For more discussion, see my *Concepts of Science* (Baltimore: Johns Hopkins University Press, 1968), ch. 5.

wick in 1932, not Rutherford in 1920, is the discoverer of these particles. Not before 1932 were there experimental results that allowed the existence of this particle to be known.

In connection with the electron, there are two physicists whose names I have not mentioned so far: Larmor and Lorentz. Both had theories about what they called electrons. Setting aside questions about whether they were referring to what we call electrons, one reason these physicists are not the discoverers of electrons is, I think, epistemic. Although their theories explained experimental results, such results were not sufficiently strong to justify a knowledge-claim regarding the electron's existence. Their claims about electrons were primarily theory-driven.

5. This view allows there to be multiple, independent discoverers, as were Adams and Leverrier. They came to be in the appropriate epistemic situations at approximately the same time. It allows a cooperative group of scientists to be the discoverers—as in the recent case of the top quark. And it allows scientists to make contributions to the discovery of X without themselves being discoverers or part of a group that discovered X. Plücker did not discover the electron, though in 1859 he made a crucial contribution to that discovery, viz. the discovery of cathode rays.

6. We need to distinguish two ways of understanding the phrase "knowing that X exists" in my definition of discovery, and hence two senses of discovery. Suppose that while hiking in the Rockies, I pick up some shiny stones. You inform me that I have discovered gold. This could be true, even if I don't know that it is gold. In this case by observing the stones I have come to know of the existence of something that, unbeknownst to me, is gold. That is one sense in which I could have discovered gold. Of course, I might also have come to know of these objects that they are gold. That is another sense in which I could have discovered gold.[7]

The same applies to discovering the electron. To say that Thomson discovered the electron might mean only that by suitable observations he came to know of the existence of something that happens to be the electron, even if he didn't realize this. Or it might mean something stronger to the effect that he came to know that the thing in question has the electron properties (whatever those are). I shall speak of the latter as the "stronger" sense of discovery and the former as the "weaker."

4. Contrasting Views of Discovery

The present view of discovery will now be contrasted with several others, including ones suggested by two historians of science who have discussed the history of the discovery of the electron. Although the primary focus of these authors is historical and not philosophical, what they claim about Thomson

7. In philosophical jargon this distinction corresponds to that between referential transparency and opacity in the expression "discovering X" (and in "knowing that X exists"). In the referentially transparent sense, but not the referentially opaque sense, if I have discovered X, and if $X = Y$, then it follows that I have discovered Y.

suggests more general views about what counts as a discovery. These more general views provide sufficient conditions for discovery, or necessary ones, or both. I want to indicate how these views conflict with mine, and why I reject them both as generalizations about discovery and as particular views about what made, or failed to make, Thomson the discoverer of the electron.

Manipulation-and-measurement view. At the end of her important 1987 paper on Thomson, Isobel Falconer writes:

> In the light of this reinterpretation of Thomson's work is there any sense remaining in which he can be said to have "discovered the electron"? Arriving at the theoretical concept of the electron was not much of a problem in 1897. Numerous such ideas were "in the air." What Thomson achieved was to demonstrate their validity experimentally. Regardless of his own commitments and intentions, it was Thomson who began to make the electron "real" in Hacking's sense of the word. He pinpointed an experimental phenomenon in which electrons could be identified and methods by which they could be isolated, measured, and manipulated.[8]

Several things are suggested here, but one is that Thomson discovered the electron because he was the first to design and carry out experiments in which electrons were manipulated and measured. We might recall that, on Hacking's view, to which Falconer alludes, "if you can spray them they are real."[9] On the more sophisticated version suggested by Falconer in this passage, if you can manipulate them in such a way as to produce some measurements they are real; and if you are the first to do so, you are the discoverer. Of course, such a view needs expanding to say what counts as "manipulating" and "measuring." I will not try to do so here, but will simply take these ideas as reasonably clear. It seems obvious that Thomson manipulated electrons by means of magnetic and electric fields and that he measured their mass-to-charge ratio.

Important classification view. This view is suggested by an earlier passage in Falconer's paper. Discussing the experimental work of Wiechert, she writes:

> Wiechert, while realizing that cathode ray particles were extremely small and universal, lacked Thomson's tendency to speculation. He could not make the bold, unsubstantiated leap, to the idea that particles were constituents of atoms. Thus, while his work might have resolved the cathode ray controversy, he did not "discover the electron."[10]

This suggests that, despite the facts that both Wiechert and Thomson manipulated the electron in such a way as to obtain a mass-to-charge ratio and that both physicists claimed that cathode particles were "extremely small and universal," Thomson, and not Wiechert, is the discoverer of electrons because Thomson but not Wiechert got the idea that cathode particles are constituents

---

8. Isobel Falconer, "Corpuscles, Electrons and Cathode Rays: J. J. Thomson and the 'Discovery of the Electron,'" *British Journal for the History of Science*, 20 (1987), 241–276; quotation p. 276.
9. Ian Hacking, *Representing and Intervening* (Cambridge: Cambridge University Press, 1983), p. 23.
10. Falconer, p. 251. I would take issue with Falconer here. In a paper of January 1897 Wiechert does indeed claim that cathode particles are constituents of atoms. Emil Wiechert, "Physikalisch-Okonomischen Gesellschaft," Konigsberg, January 7, 1897, 3–16.

of atoms. Although Falconer does not say so explicitly, perhaps what she has in mind is that Thomson's identification of cathode particles as universal constituents of atoms is what is important about electrons. Generalizing from this, you are the discoverer of X when you are the first to arrive at an important (and correct) classification of X. The question remains as to what counts as an "important" classification—a major lacuna, as I will illustrate in a moment. However this is understood, it should include Thomson's classification of electrons as constituents of atoms.

Social constructivist view. Social constructivism is a very broad viewpoint pertaining to many things, including the reality of scientific objects themselves such as electrons (they are "socially constructed" and have no reality independently of this). There is, however, a much narrower social constructivist view that is meant to apply only to scientific discovery. On this view, whether some scientist(s) discovered X depends on what the relevant scientific community believes. This view is adopted by Arabatzis prior to his historical discussion of the work of Thomson and others on the electron. He writes:

> A final approach [to discovery]—and the one I favour—takes as central to the account the perspectives of the relevant historical actors and tries to remain as agnostic as possible *vis-a-vis* the realism debate. The criterion that this approach recommends is the following: since it is the scientific community (or its most eminent representatives) which adjudicates discovery claims, an entity has been discovered only when consensus has been reached with respect to its reality. The main advantage of this criterion is that it enables the reconstruction of past scientific episodes without presupposing the resolution of pressing philosophical issues. For historical purposes, one does not have to decide whether the consensus reached by the scientific community is justifiable from a philosophical point of view. Furthermore, one need not worry whether the entity that was discovered (in the above weak sense) can be identified with its present counterpart.[11]

Although in this passage Arabatzis claims that there is a discovery *only* when the community believes there is, he also says that the main advantage of his criterion is that it avoids the issue of whether the consensus reached is justified, and the issue of whether the entity that was discovered is the same as the one scientists now refer to. Accordingly, the view suggested is a rather strong one, to the effect that consensus is both necessary and sufficient for discovery. (At least, that is the social constructivist view about discovery that I will consider here.) Thomson discovered the electron if and only if he is generally regarded by physicists as having done so. The physicists who so regard him may have different reasons for doing this. But these reasons do not make him the discoverer: simply their regarding him as such does. Even if the reasons are false (in some "absolute," nonconsensual sense), he is still the discoverer, unless the physics community reaches a different consensus.

Different contributions view. According to this, there are discoveries in science, including that of the electron, that are not made by one person, or by several, or by any group, but involve various contributions by different people. We need to

---

11. "Rethinking the 'Discovery' of the Electron," p. 406.

replace the question "Who discovered the electron?" with more specific questions concerning who made what contributions to the discovery. We might note that in 1855 Geissler contributed by inventing a pump that allowed much lower gas pressures to be produced in electrical discharge tubes; that in 1859 using this pump, Plücker found by experiment that when the pressure is reduced to .001 mm of mercury, the glass near the cathode glows with a greenish phosphorescence and the position of the glow changes when a magnetic field is introduced; that in 1869 Plücker's student Hittorf found that if a solid body is placed between the cathode and the walls of the tube it casts a shadow; from which he concluded that rays are emitted from the cathode that travel in straight lines. This story could be continued with experimental and theoretical contributions by Crookes, Larmor, Lorentz, Hertz, Goldstein, Schuster, and so forth, culminating with the experiments of Thomson—or well beyond this if you like.

Now, it is not that all the people mentioned, or even several of them, or a group working together, discovered the electron. Plücker didn't discover the electron, nor was he one of several people or a group that did. Still the electron was discovered. But it was not the sort of discovery made by one individual, or several, or a group. Rather it was the sort of discovery that involved different contributions by different persons at different times. Thus, Arabatzis writes:

> Several historical actors provided the theoretical reasons and the experimental evidence which persuaded the physics community about its [the electron's] reality. However, none of these people discovered the electron. The most we can say is that one of those, say Thomson, contributed significantly to the acceptance of the belief that "electrons" denote real entities.[12]

True belief view. According to this, you have discovered something only if what you believe about it is true or substantially true. Despite Lord Kelvin's claim to know various facts about the luminiferous ether,[13] that entity was not discovered by nineteenth century wave theorists (or by anyone else), since what was believed about it, including that it exists and that it is the medium through which light is transmitted, is false. Similarly, on this view, Thomson did not discover the electron, since quite a few of his core beliefs about electrons (or what for many years he called corpuscles) were false. His corpuscles, he later thought, were entirely electrical, having no inertial mass; they were arranged in stationary positions throughout the atom; they were the only constituents of atoms; they were not waves of any sort; and they were not carriers of the smallest electric charge. So, if he discovered anything at all, it was not the electron.

---

12. "Rethinking the 'Discovery' of the Electron," p. 432. Is Arabatzis's "different contributions" view about the electron compatible with what I take to be his more general social constructivist position about discovery? I believe so. The combined view would be that in general someone is the discoverer of something when and only when there is consensus about who discovered what; in the electron case, however, there is no such consensus about any one person, only (at most) about who made what contributions toward the discovery.
13. Kelvin wrote: "We know the luminiferous ether better than we know any other kind of matter in some particulars. We know it for its elasticity; we know it in respect to the constancy of the velocity of propagation of light for different periods." *Kelvin's Baltimore Lectures and Modern Theoretical Physics,* ed. Robert Kargon and Peter Achinstein (Cambridge, MA: MIT Press, 1987), p. 14.

5. Rejection of These Views

I reject each of these five views about discovery, both in the generalized forms I have given them and as ones applicable to the case of Thomson and the electron. Although manipulation and measurement are frequently involved in a discovery, they are neither necessary nor sufficient. Galileo discovered mountains and craters on the moon without manipulating or measuring them (in any reasonable sense of these terms). Moreover, the manipulation and measurement view would too easily dethrone Thomson. Many physicists before Thomson in 1897 manipulated electrons in the sense that Thomson did; that is, they manipulated cathode rays, and did so in such a way as to produce measurements. As noted, in 1890 Schuster conducted experiments involving magnetic deflection of electrons in which he arrived at upper and lower bounds for their ratio of mass to charge. Lenard's experiments manipulated cathode rays out of the tube and measured the distance they traveled. Perhaps one can say that Thomson's manipulations yielded better and more extensive measurements. But why should that fact accord him the title "discoverer?" Manipulations and measurements after Thomson by Seitz in 1901 and by Rieger in 1905 gave even more accurate measurements of the mass-to-charge ratio. Yet none of these physicists is regarded as having discovered the electron.

The second view—"important classification"—fails to provide a sufficient condition for discovery, since you can arrive at an important classification of Xs without discovering them. You can postulate their existence on largely theoretical grounds, and describe important facts about them, without "confronting" them sufficiently directly to count as having discovered them. In the early 1930s Pauli hypothesized the existence of a neutral particle, the neutrino, in order to account for the continuous distribution of energy in beta decay. But the neutrino was not discovered until there was a series of experiments, beginning in 1938, that established its existence more directly.

Whether the important-classification view fails to provide even a necessary condition for discovery is more difficult to say because of the vagueness in the notion of "important" classification. Roentgen discovered X-rays in 1895 without knowing that they are transverse electromagnetic rays. Although he speculated that they were longitudinal vibrations in the ether, he did not claim to know this (nor could he know this), and for this reason, and to distinguish them from other rays, he called them X-rays. Did he fail to arrive at a sufficiently important classification? Or shall we say that the fact that he discovered that X-rays are rays that travel in straight lines, that have substantial penetrating power, and that cannot be deflected by an electric or magnetic field is sufficient to say that he arrived at an important classification?

Similarly, in the case of the electron isn't the fact that the constituents of cathode rays are *charged particles smaller than ordinary ions* an important classification? If so, then Crookes in 1879 deserves the title of discoverer. Is it that the classification "constituent of all atoms" is more important than "being charged particles smaller than ordinary ions," and so Thomson rather than Crookes deserves the honor? Crookes, indeed, claimed that he, not Thomson, first arrived at the classification "constituent of all atoms." Moreover, why choose this classification rather than something more specific about how these constituents are arranged in atoms? If so, then Rutherford or Bohr should be selected, not Thomson, whose plum pudding model got this dead wrong.

For me, the crucial question concerning the present view is whether you could know that X exists from observations of X without knowing an important classification of X. In the weaker sense of discovery I distinguished earlier, one could discover X without knowing very much about X, including that it is X. (Recall my discovering gold.) The stronger sense involves knowing that it is X. But what important classification one needs to know to know that something is X I'll leave to important classification theorists.

The view I am proposing also contradicts the social constructivist account of discovery, since on my view there is, or at least can be, a fact of the matter about who discovered what that is independent of whom the scientific community regards as the discoverer. This is because there is, or can be, a fact of the matter about who was the first to be in an epistemic situation necessary for discovery. Being regarded by the scientific community as the discoverer of X is neither necessary nor sufficient for being the discoverer of X. No doubt scientific discoverers wish to be recognized by the scientific community for their discovery. Perhaps for some a discovery without recognition is worthless. But this does not negate the fact of discovery itself. Nor is this to deny that a discovery that is and remains unknown except to the discoverer will have little chance of advancing science, which depends on public communication. That is one reason scientists make their discoveries public. Although publicity helps to promote the discovery and the recognition for it, neither publicity nor recognition creates the discovery. Finally, as noted earlier, one can relativize discovery claims to a group. I can be the first *in my department* to discover a certain book in the library (Columbus the first European to discover America, etc.). But this is not social constructivism, since there is a fact of the matter about discovery within a group that is independent of the beliefs of the members of the group. Either I was or I wasn't the first in my department to discover that book, no matter what views my colleagues have about my discovery.

Two of the views of discovery that contrast with mine deny the claim that Thomson discovered the electron: the "different contributions" view and the "correct belief" view. Briefly, my response to these views is this. The fact that various people made contributions to the discovery of the electron does not, on my account, necessarily preclude the fact that Thomson discovered the electron. All this means is that various people helped make it possible for Thomson to be the first to be in an epistemic situation necessary for discovery. Nor, finally, does getting into that epistemic situation regarding some X require that all or most of your beliefs about X be true. Suppose that while walking along a road I discover a person lying in the ditch beside the road. Suppose that, after observing the person, I come to believe that the person is a woman, quite tall, at least 50 years old, with blond hair, and wearing a gray jacket. Suppose, finally, that I am quite wrong about these beliefs. The person in the ditch is actually a man, 5 feet tall, 30 years of age, with dark hair, and wearing no jacket at all. I can still be said to have discovered the person in the ditch, despite the fact that what I believe about the person in the ditch is substantially false. So I reject the general rule that you have discovered X only if what you believe about X is true or substantially true. (I will return to this claim in section 7.)

6. Did Thomson Discover the Electron?

Having proposed an account of discovery, and disposed of some others, we are now in a position to take up this question. To begin with, I think my account helps us to see why we refrain from attributing this discovery to some of the other physicists mentioned. For example, claims made about the electron by Crookes, Larmor, and Lorentz, even if many were correct, were primarily theory-driven, not experimentally determined. This is not to say that Thomson had no theoretical beliefs about electrons. Falconer and Feffer[14] claim that he probably believed that they are not discrete particles with empty spaces between them, but certain configurations in an all-pervading ether. But that is not enough to put him in the same category as some of the more theoretically driven physicists. The question is whether Thomson was the first to know that electrons exist from observations of them or their direct effects.

Let me divide this question into three parts. First, in 1897 did Thomson know that electrons exist? Second, if he did, did he know this from observations of electrons or of their direct effects? Third, was he the first to know this from such observations? If the answer to all three questions is Yes, then Thomson retains the honor usually accorded to him.

In 1897 did Thomson know that electrons exist? Well, what did he claim to know in 1897? Here is a well-known passage from his October 1897 paper:

> As the cathode rays carry a charge of negative electricity, are deflected by an electrostatic force as if they were negatively electrified, and are acted on by a magnetic force in just the way in which this force would act on a negatively electrified body moving along the path of these rays, I can see no escape from the conclusion that they are charges of negative electricity carried by particles of matter.[15]

Thomson continues:

> The question next arises, What are these particles? Are they atoms, or molecules, or matter in still finer state of subdivision. To throw some light on this point, I have made a series of measurements of the ratio of the mass of these particles to the charge carried by it. (p. 384)

Thomson then proceeds to describe in some detail two independent experimental methods he employed to determine the mass-to-charge ratio. At the end of this description he concludes:

> From these determinations we see that the value of m/e is independent of the nature of the gas, and that its value 10^{-7} is very small compared with the

---

14. Stuart M. Feffer, "Arthur Schuster, J. J. Thomson and the Discovery of the Electron," *Historical Studies in the Physical and Biological Sciences*, 20 (1989), 33–51.
15. J. J. Thomson, "Cathode Rays," *Philosophical Magazine and Journal of Science* (October 1897), reprinted in Mary Jo Nye, ed., *The Question of the Atom* (Los Angeles: Tomash Publishers, 1986), 375–398; quotation p. 384.

value 10^{-4}, which is the smallest value of this quantity previously known, and which is the value for the hydrogen ion in electrolysis.

He continues:

> Thus, for the carriers of electricity in the cathode rays m/e is very small compared with its value in electrolysis. The smallness of m/e may be due to the smallness of m or the largeness of e, or to a combination of these two. That the carriers of the charges in the cathode rays are small compared with ordinary molecules is shown, I think, by Lenard's results as to the rate at which the brightness of the phosphorescence produced by these rays diminishes with the length of the path travelled by the ray. (p. 392)

After a little more discussion of Lenard's experimental results, Thomson concludes:

> The carriers, then, must be small compared with ordinary molecules.

In sum, in 1897 Thomson claimed to know these facts:

1. That cathode rays contain charged particles. (This he claimed to know from his experiments showing both the magnetic and the electrostatic deflection of the rays.)

2. That the ratio of mass to charge of the particles is approximately 10^{-7}, which is much smaller than that for a hydrogen atom. (The 10^{-7} ratio he claimed to know from experiments of two different types involving magnetic and electrical deflection.)

3. That the particles are much smaller than ordinary molecules. (This he claimed to know from his own experiments yielding a mass-to-charge ratio smaller than that for the hydrogen atom, together with Lenard's experiments on the distance cathode rays travel outside the tube, which is much greater than that for hydrogen ions.)

Did he know these facts? Well, he certainly believed them to be true. He says so explicitly. Are they true? True enough, if we don't worry about how much to pack into the notion of a particle. (Clearly Thomson had some false beliefs about his particles, in particular that they lacked wave properties.) Was he justified in his beliefs? His experimental reasons for claims 1 and 2 are quite strong, that for the smallness of the particles is perhaps slightly less so (but I think better than Heilbron alleges in his article on Thomson in the *Dictionary of Scientific Biography*, p. 367). If justified true belief is normally sufficient for knowledge, then a reasonable case can be made that Thomson knew the facts in question in 1897.

To be sure, there are *other* claims Thomson made in 1897 concerning which one might not, or could not, attribute knowledge to him. Perhaps one of the former sort is the claim that the charged particles are constituents of all atoms. Indeed, Thomson's explicit argument here seems a bit more tentative and less conclusive than those for the three claims above. It is simply an explanatory one to the effect that if atoms are composed of the particles whose existence he has already inferred, then this would enable him to explain how they are projected from the cathode, how they could give a value for m/e that is independent of the nature of the gas, and how their mean free path would depend solely on the density of the medium through which they pass. In general, explanatory reasoning does not by itself establish the claims inferred with sufficient force to yield knowledge. And, finally, there are the claims that the particles are the *only* constituents of atoms, and are arranged in accordance with a model of floating magnets suggested by Mayer. Both claims, being false, are not claims that Thomson or anyone else could know to be true.

But, like knowing that there is a person in the ditch, not every belief about that person needs to be true or known to be true. If in 1897 Thomson knew that cathode rays contain charged particles, whose ratio of mass to charge is 10^{-7} and that are much smaller than ordinary molecules, then I think it is reasonable to say that in 1897 he knew that electrons exist, at least in the weaker of the two senses discussed earlier. He knew of the existence of things that happen to be electrons. Electrons are the charged particles in question. Knowing these particular facts about them entails knowing that they exist. Whether he knew that electrons exist in the stronger sense is a question I will postpone for a moment.

The second of my three questions is whether Thomson knew what he did from observations of electrons or their direct effects. I suggest the answer is clearly yes. Those were electrons in his cathode tubes, and they did produce fluorescent effects and others that he observed in his experiments. Despite various theoretical assumptions, his conclusions about electrons are primarily experiment-driven.

The final of the three questions concerns priority. Was Thomson the first to be in the appropriate epistemic situation? Was he the first to know that electrons exist in the weaker sense of this expression? Was he the first to know of the existence of things that happen to be electrons? He was clearly not the first to know of the existence of cathode rays, which happen to be, or to be composed of, electrons. But that is not the issue here. Was he the first to know, by experimental means, of the existence of the things that happen to be the constituents of cathode rays, that is, electrons? That would be a more important question, albeit a question of discovery in the weaker sense. How do you demonstrate the existence of the constituents of cathode rays? Not simply by showing that cathode rays exist. Thomson demonstrated their existence by showing that charged particles exist comprising the rays, and he did so by means of experiments involving the direct effects of those charged particles. Was he the first to do so?

The answer I would offer is a less than decisive "maybe." Other physicists, including Schuster, Perrin, Wiechert, and Lenard, had conducted experiments on cathode rays which yielded results that gave support to the claim that the constituents of cathode rays are charged particles. Moreover, these experiments involved observing the electron's direct effects. It might be argued that although these other physicists provided such experimental support, that support was not strong enough to produce knowledge. One might claim that Thomson's refinements of Perrin's experiment, and more importantly his achievement of producing electrostatic deflection of the rays, and his determination of m/e, showed conclusively, in a way not shown before, that cathode rays contain charged particles. (This is what Thomson himself claims in his October 1897 paper.) If this is right, then one can say that, in the weaker sense of discovery, Thomson discovered the electron. Although others before him had produced experimental evidence of its existence, he was the first to produce evidence sufficient for knowledge.

This, however, is a controversial priority claim. It was vehemently denied by Lenard, who claimed that his own experiments prior to Thomson's conclusively proved the existence of electrons.[16] It was also denied, albeit less vehemently, by Zeeman, who claimed that he determined the ratio of mass to charge before Thomson.[17] Finally, Emil Wiechert makes claims about the constituents of cathode rays that are fairly similar to Thomson's, in a paper published in January 1897, before Thomson's papers of April and October of that year.[18] In this paper Wiechert explicitly asserts that cathode rays contain charged particles that are much smaller than ordinary molecules, and from experiments involving magnetic deflection of cathode rays he determines upper and lower bounds for the mass-to-charge ratio of the particles. However, unlike Thomson, Wiechert did not produce electrostatic deflection, he did not obtain two independent means for arriving at his determination of mass to charge, and he did not produce precise values. So the issue, as I have defined it, is simply this: even though others had provided some experimental evidence for the existence of charged particles as the constituents of cathode rays, were Thomson's experiments the first to *conclusively* demonstrate this? Were they the first on the basis of which knowledge of their existence could be correctly claimed? If so, he discovered the electron. If not, he didn't.

One might make another claim. Relativizing discovery to the individual, one might say that Thomson first discovered the electron (for himself) in 1897, whereas others had done so a bit earlier. One might then say that Thomson was *among* the first to discover the electron (for himself). Perhaps this is what Abraham Pais has in mind when, as reported in *The New York Times* (April 29, 1997), he claims that Thomson was a, not the, discoverer of the electron. The others Pais mentions are Wiechert and Kaufmann.

7. Strong Discovery

What about the stronger sense of discovery, the sense in which if I discover gold, then I know it is gold? To those seeking to deny the title "discoverer of the electron" to Thomson one can concede that he did not know that the constituents of cathode rays have all the properties that electrons do. If this is required, the electron has yet to be discovered, since presumably no one knows all the properties of electrons. Obviously, this is not required for knowledge in the stronger sense. I can know that I have discovered gold without knowing all the properties of gold. Indeed, I can know that I have discovered gold without knowing *any* of the properties of gold. If an expert, after examining it, tells me it is gold, then I think I know it is. Clearly, however, Thomson did not know *in this way* that the con-

---

16. Philipp Lenard, *Wissenschaftliche Abhandlungen*, vol. 3 (Leipzig: S. Hirzel, 1944), p. 1.
17. See Arabatzis, "Rethinking the 'Discovery' of the Electron," p. 423.
18. Wiechert, "Physikalisch-Okononischen Gesellschaft."

stituents of cathode rays are electrons. So what *must* one know to know that the items in question are electrons? That is a problem. (A similar problem was raised concerning the "important classification" view.)

There is a further problem here with the question of whether Thomson knew that the constituents of cathode rays are electrons. Putting the question that way presupposes some established concept of electron. And the question seems to be whether what Thomson discovered (in the weaker sense) fits that concept, and whether Thomson knew this. By analogy, to ask whether I discovered gold (in the stronger sense) is to presuppose that these objects satisfy some established concept of gold and whether I know that they do. Which concept of electron is meant in the question about Thomson? In 1897 there was no established concept. Stoney, who introduced the term "electron," used it to refer to an elementary electric charge. But Thomson was not talking about this. Nor was his claim that the constituents of cathode rays are Lorentz's electrons, which in 1895 Lorentz claimed were ions of electrolysis. (In fact, Thomson never used the term "electron" until well into the twentieth century.) Nor, of course, was Thomson claiming that the cathode ray constituents have the properties we currently attribute to electrons.

So the question "Did Thomson know that the constituents of cathode rays are electrons" is, I think, ambiguous and misleading. Instead, I suggest, it is better simply to ask what facts, if any, about the constituents of cathode rays Thomson knew, when he knew them, and when others knew them.

Very briefly, let's take four central claims that Thomson made about cathode ray constituents in his October 1897 paper: first, they are charged particles; second, their ratio of mass to charge is approximately 10^{-7}; third, they are much smaller than ordinary atoms and molecules; and, fourth, they are constituents of atoms. Earlier I said that it is reasonable to suppose that Thomson knew the first three of these facts in 1897, but not the fourth. He came to know them during that year as a result of his experiments with cathode rays. I also said that one might claim that Thomson was the first to demonstrate conclusively that the constituents of cathode rays are charged particles, though this is controversial. At least he was among the first to do so.

With regard to the second claim—that the ratio of mass to charge of these particles is approximately 10^{-7}—Wiechert had arrived at upper and lower bounds before Thomson. In defense of Thomson, one might say that his determinations were more precise and were based on two independent experimental methods.

With respect to the third claim—that the cathode particles are much smaller than atoms and molecules—perhaps Lenard is correct in claiming knowledge of this prior to Thomson. Indeed, Thomson made important use of Lenard's absorption results in his own arguments that cathode particles are smaller than atoms. And if Wiechert's arguments are sufficiently strong, he too has some claim to knowledge before Thomson.

Finally, the fourth claim—that cathode particles are constituents of atoms—is, it is probably fair to say, one that Thomson did not *know* the truth of in 1897, although he gave explanatory arguments in its favor.

8. What's so Important about Who Discovered the Electron (or Anything Else)?

This question arises especially for my account of discovery. On that account, the fact that something has been discovered by someone does not by itself imply that what is discovered, or by whom, is important or interesting, even to the discoverer. (I may have discovered yet one more paperclip on the floor.) The importance of the discovery will depend on the item discovered and on the interests of the discoverer and of the group or individual to whom the discovery is communicated. Discovering a universal particle such as the electron, which is a constituent of all atoms, is obviously more important, especially to physicists, than discovering yet one more paperclip on the floor is to them or to me.

Not only can the object discovered be of importance, but so can the method(s) employed. In his discovery of the electron (at least for himself) Thomson discovered a way to produce electrostatic deflection of the cathode rays, which had not been achieved before. Using this he devised a new independent way to obtain a fundamental measurement of mass-to-charge.

There is another point worth emphasizing about discoveries of certain entities, particularly those that are too small, or too far away, or otherwise too inaccessible to be observed directly. Scientists may have theoretical reasons for believing that such entities exist. These theoretical reasons may be based on observations and experiments with other entities. Sometimes such reasoning is sufficiently strong to justify a claim to *know* that the entity exists. Yet there is still the desire to find it, to discover it, by observing it as directly as possible. (Although the case of the electron does not illustrate this, one that does fairly closely is that of the top quark, whose existence was inferred from the "standard model" before it was detected experimentally.[19]) This need not increase the degree of confidence in its existence significantly if at all over what it was before. So why do it?

One reason may simply derive from a primal desire or curiosity to "see" or detect something by confronting it more or less directly. Another more important reason is to discover new facts about it, which is usually facilitated by observing it or its effects, and which may allow the theory that entailed its existence to be extended. It will also provide additional support for that theory without necessarily increasing the degree of probability one attaches to that theory.

Why should we care about *who*, if anyone, was the discoverer, that is, about who was the first to be in an appropriate epistemic situation for discovery with respect to some entity? It depends on who the "we" is and on what is discovered. As noted, not all discoveries and discoverers are of interest to all groups; some may be of interest to none. If what is discovered is important to some community, and if there was a discoverer, whether a person or a group, then simply giving credit where credit is due is what is appropriate and what may act as a spur to future investigations. In this regard discovery is no different from other achievements. If accomplishing something (whether flying an airplane, or climbing Mt. Everest, or discovering the electron) is valuable to a certain community,

---

19. See Staley, "Over the Top."

and some person or group was the first to do it, or if several persons independently were the first, then such persons deserve to be credited and perhaps honored and rewarded by the community, especially to the extent that the accomplishment is important and difficult. Generally speaking, more credit should be given to such persons than to those who helped make the achievement possible but did not accomplish it themselves.

Whether Thomson deserves the credit he received for being the (or a) discoverer of the electron is, of course, of interest to him and to other contemporaries such as Lenard, Zeeman, and Crookes, who thought they deserved more credit. But it should also be of interest to subsequent physicists, historians of physics, and authors of textbooks who write about the discovery. The answer to the question of who really discovered the electron, and hence who deserves the credit, is, I have been suggesting, not so simple. Part of that answer depends upon establishing who knew what, when, and how, which in the electron case is fairly complex. The other part depends on establishing some reasonably clear concept of discovery. In this chapter I have attempted to contribute to each task, particularly the latter.

Finally, credit is deserved not only for discovering the existence of an important entity, but for other accomplishments with respect to it as well. Even if Lenard has some claim to priority for the discovery that cathode ray constituents are smaller than atoms, and even if in 1897 Thomson's arguments that his corpuscles are constituents of all atoms are not conclusive, we can admire and honor Thomson, among other reasons, for the experiments leading to the conclusions he drew, for the conclusions themselves, and for proposing and defending a bold idea that revolutionized physics: that the atom is not atomic.

9. Evidence

So far I have spoken of discovery but not of evidence. Must the discoverer of X have evidence that X exists? If so, what sort? Did Thomson have such evidence in the case of electrons?

According to the account of discovery in section 2, a person P is in an epistemic situation necessary for discovering X if P knows that X exists, observations of X or its direct effects caused, or are among the things that caused, P to believe that X exists, and among P's reasons for believing X exists is that X or its direct effects have been observed. Suppose that the other conditions for discovery are satisfied, so that it is true that P discovered X (in the weak or the strong sense). Does P have evidence that X exists? If so, what sort?

My answer to the first question is: usually, but not always; and the nature of that evidence, if it exists, cannot be determined simply from the fact that P is in an epistemic situation necessary for discovering X. My answer to the second question is that if P discovered X, and if P has evidence that X exists, that evidence is of all four types distinguished in chapter 2. These claims will now be defended.

For the sake of argument, let us suppose that Thomson did discover electrons (in both the weak and strong senses) in performing his cathode ray experiments. Then among his reasons for believing that

> h: Electrons (or "corpuscles") exist

is that

> e: Electrons, or their direct effects, were observed in his experiments.

But e entails h, so that, on my view, e would not be potential or veridical evidence that h is true. (It is too close to be such evidence.) And if we may assume that Thomson knew that e entails h, then it would not be Thomson's subjective evidence that h. One's reason for believing a hypothesis is not necessarily one's evidence that it is true. Only reasons of a certain type will do. Did Thomson have such reasons?

Among Thomson's reasons for believing h that would count as (part of his) evidence that h was that

> e': "Cathode rays carry a charge of negative electricity, they are deflected by an electrostatic force as if they were negatively electrified, and are acted on by a magnetic force in just the way in which this force would act on a negatively electrified body moving along the path of these rays."

This proposition does not entail h, although (Thomson believed) its truth makes h probable.[20] Indeed, Thomson believed that e' is (veridical) evidence that electrons exist, and an important part of his reason for believing that electrons exist is that e' is true. Accordingly, e' constituted an important part of Thomson's subjective evidence that electrons exist.

The next question is whether one who believes that X exists, and does so for a good reason, must have subjective evidence that X exists, whether or not that evidence is the same as one's good reason. My answer is: not always. I believe that there is a blue spot on my tie, and my reason for believing this is a very good one, viz. I see a blue spot clearly. The fact that I see a blue spot clearly is not my evidence that it exists (it is too close), even though the fact that I *claim* to see it may be your evidence. Now, if there is a blue spot on my tie, then there may still be (potential or veridical) evidence that there is, such as the fact that the pen in my coat pocket that rubs against my tie has been leaking blue ink. But the fact that such evidence exists does not mean that I know or believe that it does, or that it is my reason for believing that a blue spot exists. My reason, and my only reason, for believing that a blue spot exists may simply be that I see one. If so, then I may have a good reason to believe something without having subjective evidence that it is so, and indeed without knowing any fact that is potential, or veridical, or ES-evidence that it is so.[21] This, in turn, means that one can discover X (for example,

---

20. An even simpler case was noted in chapter 8, section 4. Suppose I come to believe that (h) there is a bear in that tree, and that my reason for believing h is that (e) I see the bear in that tree. My subjective evidence that h is true, if there is such, is not that e is true. That is too close to be evidence. My evidence might be something like: there is a large, heavy animal in that tree, with long shaggy hair, eating berries from the tree. The latter does not entail h, but makes it probable.
21. To be sure, following some traditional epistemologists, one might retreat to the claim that in this case I do have subjective evidence pertaining to the way the tie looks or appears to me under appropriate conditions of observation. So the fact that there is something on the tie that looks exactly like a blue spot to me, and I am a normal observer under appropriate

a blue spot) without having (subjective or objective) evidence that X exists. There may well be objective evidence that X exists. But discovering X does not require that the discoverer of X know what it is or take it to be evidence that X exists.

Suppose that a person P, who discovered X, has among his reasons for believing that X exists some fact e that is also P's subjective evidence that X exists. In addition to being subjective evidence, is e also potential, veridical, or ES- evidence that X exists? My answer is that in such a case e is evidence of all four types. If P discovered X, then P knows that X exists, which, I will assume, requires that P's reasons for believing that X exists justify his belief, given his epistemic situation, and also that they be good reasons, in the strong (veridical) sense of "good reason" discussed in chapter 2.[22] Accordingly, in such a case, his subjective evidence will also be objective.

In accordance with my claim above, Thomson's subjective evidence that electrons exist was, or included, the fact that e' is true. Assuming that he was in an epistemic situation necessary for discovering electrons, and that his knowledge was based in part on e', the latter would be at least a part of evidence that is ES, potential, and veridical evidence that electrons exist. The fact that cathode ray experiments of the sort Thomson conducted yield the magnetic and electrical results he obtained constitutes at least part of the ES, potential, and veridical evidence that electrons exist. The probabilistic and other conditions for such evidence are all satisfied.

10. Discovery and Evidence

Finally, I will briefly note three important similarities between the concept of discovery I have outlined and the concepts of evidence developed in earlier chapters.

1. *Discovery and evidence can both be unknown.* There can be a discovery of some X without anyone (except the discoverer) knowing that X exists. The discovery may or may not be promulgated to some community. With objective evidence the situation can be even more unknown. Some fact e can be (ES, potential, or veridical) evidence that hypothesis h is true even though no one at all knows that e is true or that e is evidence that h. Even in the case of subjective evidence, e may be some person's evidence that h without anyone else knowing that it is. To be sure, unknown evidence, as long as it remains unknown (or known only to one person), just as in the case of a discovery known only to the discoverer, has little chance of advancing science. But that does not make evidence intrinsically a matter of public knowledge.

---

conditions of observation, and there are no countervailing indications that appear to me, is my subjective evidence for the hypothesis that a blue spot exists. I doubt that this is my reason for believing a blue spot exists, although it does justify my belief, and, if (continually) challenged to defend my claim that there is a blue spot on my tie, I might cite such facts about looks and appearances (even though I would be more inclined to reject the challenge). But even if this were my reason for believing a blue spot exists, this sort of "appearance" statement entails the hypothesis in question; it is too close to count as evidence.

22. Indeed, knowledge that X exists, if based on evidence, requires the strong sense of veridical evidence. (See chapter 8, section 4, and note 2 in that chapter.)

2. *There can be a discovery of something, as well as evidence that some hypothesis is true, that is unimportant to any community, scientific or otherwise.* I can discover yet one more paperclip on the floor, and the fact that I have may be evidence that others exist as well. What makes a discovery or evidence important to some community depends on the importance that community attaches to what is discovered and to the hypothesis for which it is evidence. Neither discovery nor evidence is intrinsically important for some community because it is a discovery or evidence. Thomson's discovery of the electron (if indeed he discovered it) was not important to the scientific community because it was a discovery, but because what was discovered was the electron. Similarly, Thomson's experimental evidence that cathode rays are subject to an electrostatic force was important not because it was evidence but because it was evidence *for that hypothesis*.

3. *Science can progress without discoveries or evidence, but eventually both are demanded.* In the mid-nineteenth century, Maxwell offered a sophisticated development of the kinetic-molecular theory of gases without being in the appropriate epistemic situation for discovering molecules. He did not know that they exist.[23] (A fortiori, he did not know this by being in an epistemic situation necessary for discovery.) Nor, indeed, did he have what even he regarded as evidence that they exist. Later scientists, such as Perrin, demanded and supplied both. As described above in chapter 12, the results of Perrin's experiments on Brownian motion constituted (veridical) evidence that molecules exist. And Brownian motion is just the sort of direct effect of molecules, the knowledge of which puts one into an epistemic situation necessary for discovery. (Whether Perrin was the first to be in such an epistemic situation with regard to molecules is a subject I will not explore here.) Similarly, in the case of the electron, there were physicists who developed theories about electrons without having reasons for believing in their existence sufficiently strong to call evidence. What Thomson attempted to do was to provide such reasons experimentally, and do so in such a way as to put himself in an epistemic situation necessary for discovery.[24]

---

23. In a letter to Stokes in 1859 when speaking about his first kinetic theory paper, Maxwell writes:

    I do not know how far such speculations may be found to agree with facts, . . . and at any rate as I found myself able and willing to deduce the laws of motion of systems of particles acting on each other only by impact, I have done so as an exercise in mechanics. Now do you think that there is any so complete a refutation of this theory of gases as would make it absurd to investigate it further so as to found arguments upon measurements of strictly "molecular" quantities before we know whether there be any molecules? (Reprinted in Elizabeth Garber, Stephen G. Brush, and C. W. F. Everitt, eds., *Maxwell on Molecules and Gases* (Cambridge, MA: MIT Press, 1986), p. 279.

24. In this chapter I am indebted to Wendy Harris for helping me express the views I want; to Robert Rynasiewicz and Kent Staley for stimulating discussions in which they tried their best to dissuade me from expressing those views; to Ed Manier who, when I presented an ancestor of this chapter at Notre Dame, raised the question that forms the title of section 8; and to Jed Buchwald and Andrew Warwick, editors of *Histories of The Electron: the Birth of Microphysics* (Cambridge, MA: MIT Press, 2001), where an earlier version of this chapter appeared, for helpful organizational suggestions and for convincing me to tone down my anti-social-constructivist sentiments.

INDEX

A priori assumption, 9–10
Arabatzis, Theodore, 266, 273–74
Arntzenius, Frank, 250
Avogadro's number, 243–248

Balmer's formula, 221–23
Barker, Stephen, 192n.
Bayesian, 6, 8, 118–20
Bayes' theorem, 134, 188
Belief
   and degrees of belief, 118–20
   confidence threshold view, 140
   degree of reasonable belief, 96–98, 104–06
   good reason to believe, 24–27, 115–18, 120–22
   Kaplan's assertion view, 140–43
   and likelihood, 128–31
Bootstrapping, 252–3
Brownian motion, 246. *See also* Perrin
Brush, Stephen, 210, 223–24, 256n., 260n.
Buchwald, Jed Z., 18, 286n.

Carnap, Rudolf, 5, 9, 19, 45, 47, 53n., 66–67, 72–73, 75–76, 84, 85, 99, 101–02, 103, 109n., 212
   a priori theory of probability, 49–52
Cartwright, Nancy, 249n., 265n.
Cathode rays, 4
   Hertz's experiments, 13–16
   Thomson's experiments, 16–17, 29–31

Chance, 59–60. *See also* Probability
Common-cause argument, 248–51
Confirmation. *See* Evidence
Conflicting evidence, 122–25
Confidence threshold view of belief, 140
Conjunction condition, 179
Consequence condition, 177
Correct explanation, 160–66
Crookes, William, 13, 267

Discovery, 268–71
   different views of, 271–74
   stronger and weaker senses, 271, 280–81
Disregarding condition, 107–08
Duhem, Pierre
   on holism, 231–34

Earman, John, 181n.
Edwards, A.W.F., 125, 127–28
Electron. *See* Cathode rays
   Thomson and discovery of electron, 277–81
Empirically complete, 26, 28, 215–16
Epistemic situation, 20–21
   necessary for discovery, 269
Error-statistical account, 132–40
ES-evidence, 19–22, 174
Evidence
   Bayesian, 6
   conclusive, 27
   counterexamples to probability definitions, 69–71

Evidence, *(continued)*
  empirical character, 38–39
  error-statistical account, 132–40
  ES-evidence, 19–22, 174
  explanatory connection condition, 155–60
  high probability definition, 46, 114–16, 145–46, 156–57
  hypothetico-deductive definition, 6, 147
  likelihood view, 125–131
  no entailment requirement, 169
  objective evidence, 21
  potential evidence, 27–28, 170
  subjective evidence, 22–24, 174
  veridical evidence, 24–27, 174
Evidential
  flaws, 25, 39–43
  holism, 231, 237–41
Explanation
  and content, 161–64
  correct, 163
  deductive-nomological (D-N), 160–61, 165
  NES requirement, 165
  vs. prediction, 210–30
Explanatory connection condition, 148–51, 155–60

Falconer, Isobel, 272–73
Feffer, Stuart M., 277n.
Foley, Richard, 119

Gert, Josh, 98n.
Gimbel, Steven, 72n.
Glymour, Clark, 5, 9, 212
  bootstrapping, 252–53
  on "old evidence," 226–29
Goldstein, Adam, 79n.
Goldstein, Eugen, 14
Good reason to believe, 24–28, 34
  degree of reasonableness of belief, 95–98, 104–06
  and evidence, 120–22
  and explanation, 152–156
  requires probability greater than one-half, 115–16
Goodman, Nelson
  grue paradox, 192–209
Gouy, Leon, 246, 254

Hacking, Ian, 125, 131, 272
Hanson, N. R., 161
Harman, Gilbert, 119n., 148n., 161
Harris, Wendy, 286n.
Hempel, Carl G., 5, 6, 9, 160–61, 165
  conditions of adequacy for evidence, 177–80
  paradox of the ravens, 185–92
Hertz, Heinrich, 4, 13–19, 224–25
High probability
  definition of evidence, 46
  necessary condition for evidence, 114–16, 156–57
Historical thesis of evidence, 210–11, 227
Hittorf, Johann Wilhelm, 13
Holism
  new-age, 236–39
  old-age, 231–36
Homogeneous reference class, 55
Hood, Thomas, 139
Howson, Colin, 101n., 102n., 104n., 227n.
Hullett, James, 194n.
Hunt, Morton, 123–24
Hunter, Daniel, 118n.
Hunter, John E., 123, 125
Hypothetico-deductive
  evidence, 6, 147–48
  reasoning to molecules, 251–52

Instrumentalist definition of evidence, 180

Jeffrey, Richard, 102n.

Kaplan, Mark, 140–43
Kelvin, Lord, 274
Keynes, John Maynard, 226
Kronz, Frederick, M., 72n.
Kyburg, Henry E., 80n.

Lelorier, Jacques, 124n.
Lenard, Philipp, 267, 280
Levi, Isaac, 76n., 102n., 119n.
Lewis, David, 59–60, 62, 67, 68n., 69n., 111, 112
Light, Richard, 123n., 124–25

Likelihood
  definition of evidence, 125–31
  law of, 125
Lipton, Peter, 148n., 149n., 161
Lottery paradox, 80, 140–41

Maher, Patrick, 73, 169n.
  positive relevance account, 86–94
  on prediction vs. explanation, 217–23
Manier, E., 286n.
Maxwell, James Clerk, 286
Mayo, Deborah, 132–40, 247n., 249n.
Mendeleyev, Dmitri, 221n., 222–23n.
Merricks, Trenton, 21n., 26n.
Mill, John Stuart, 147
Morgan, Gregory, 21
Musgrave, Alan, 211

NES requirement, 165
Newton, Isaac, 147, 235
No interference condition, 106–07
Nye, Mary Jo, 260n.

Objective epistemic probability, ch. 5
  degree of reasonableness of belief, 96–98, 104–06
  vs. propensity, 110–13
  relativizations, 106–08
Old evidence. *See* Paradoxes of evidence
Ostwald, Friedrich Wilhelm, 260

Paradoxes of evidence
  grue, 192–209
  lottery, 80, 140–41
  old evidence, 226–229
  ravens, 185–92
Peirce, Charles, 161
Perrin, Jean
  on cathode rays, 16
  on molecules, 243ff
  reconstruction of Perrin's argument, 254–58, 262–63
  and realism, 263–65
Plantinga, Alvin, 77n.
Plücker, Julius, 13, 274
Pollock, John, 26n.
Popper, Karl, 59, 210
Positive relevance definition, 45
Potential evidence, 27–28, 170

Principal Principle, 62–63, 112
Probability
  Carnap's a priori theory, 49–52, 66–67, 99
  frequency theory, 52–58, 67, 99–100
  imprecise, 102–03
  objective epistemic, 95–113
  propensity theory, 59–63, 67, 99–100, 110–13
  rules of probability, 48
  subjective theory, 63–65
Probability definitions of evidence
  counterexamples, 69–71, 114–15, 145–46, 151–52
  high probability, 46, 84–86
  increase in probability (positive relevance), 45, 83–84

Quine, W.V., 231–35

Realism and Perrin, 263–65
Reciprocity, 175–77
Redhead, Michael, 265n.
Reichenbach, Hans, 53, 54, 57
Relativization of evidence statements, 106–08, 171–73
Royall, Richard, 125–31
Rutherford, Ernest, 30
Rynasiewicz, Robert, 265n., 266, 286n.

Salmon, Wesley C., 53, 55, 58, 67, 72, 75, 248–51, 263
Sanford, David H., 194n.
Schmidt, Frank L., 123, 125
Schuster, Arthur, 13, 267
Schwartz, Robert, 194n.
Selection procedure, 40, 186, 212–14
  strongly biased, 187
Semat, Henry, 30
Smith, Paul V., 123n., 124–25
Snyder, Laura J., 25n., 212n., 265n.
Staley, Kent, 138n., 269n., 270n., 282n., 286n.
Sturgeon, Scott, 21n.
Subjective evidence, 22–24

Thomson, J.J., 4, 16–17, 29–31, 266ff, 277–281

Thomson, Judith, 198n.
Threshold concept, 7, 73–74, 77–80

Urbach, Peter, 101n., 102n., 104n

Van Fraassen, Bas, 181n. 250n.
Veridical evidence, 24–27, 174

Walley, Peter, 102n.
Warwick, Andrew, 286n.

Wave theory of light, 158–60, 225–26
Weakness assumption, 6–9
Weight of hypothesis, 53
Weiner, Steve, 5n.
Whewell, William, 148n., 161, 210
Wiechert, Emil, 272, 280
Wiedemann, Gustav Heinrich, 14
Williams, Bernard, 98n.